Lecture Notes in Geoinformation and Cartography

The Lecture Notes in Geoinformation and Cartography series provides a contemporary view of current research and development in Geoinformation and Cartography, including GIS and Geographic Information Science. Publications with associated electronic media examine areas of development and current technology. Editors from multiple continents, in association with national and international organizations and societies bring together the most comprehensive forum for Geoinformation and Cartography.

The scope of Lecture Notes in Geoinformation and Cartography spans the range of interdisciplinary topics in a variety of research and application fields. The type of material published traditionally includes:

- proceedings that are peer-reviewed and published in association with a conference;
- post-proceedings consisting of thoroughly revised final papers; and
- research monographs that may be based on individual research projects.

The Lecture Notes in Geoinformation and Cartography series also includes various other publications, including:

- tutorials or collections of lectures for advanced courses;
- contemporary surveys that offer an objective summary of a current topic of interest; and
- emerging areas of research directed at a broad community of practitioners.

More information about this series at http://www.springer.com/series/7418

Jürgen Döllner · Markus Jobst
Peter Schmitz
Editors

Service-Oriented Mapping

Changing Paradigm in Map Production and Geoinformation Management

Springer

Editors
Jürgen Döllner
Computer Graphics Systems,
Hasso Plattner Institute
University of Potsdam
Potsdam
Germany

Peter Schmitz
CSIR Built Environment
Pretoria
South Africa

Markus Jobst
Research Group Cartography,
Department of Geodesy
and Geoinformation
Vienna University of Technology
Vienna
Austria

ISSN 1863-2246 ISSN 1863-2351 (electronic)
Lecture Notes in Geoinformation and Cartography
ISBN 978-3-030-10199-2 ISBN 978-3-319-72434-8 (eBook)
https://doi.org/10.1007/978-3-319-72434-8

Printed on acid-free paper

This Springer imprint is published by the registered company Springer International Publishing AG part of Springer Nature
The registered company address is: Gewerbestrasse 11, 6330 Cham, Switzerland

Preface

Science is not only a disciple of reason but, also, one of romance and passion.

Stephen Hawking

The way how maps are currently produced and geoinformation is managed seems to change rapidly. The geospatial community, like many other IT-depending communities, face rapid technological developments; open geoinformation access and wide distribution; user participation for content creation; and semantic cross-domain networks. All these aspects have a relevant impact on the requirements of the system design, architecture principles, interface specification or even sustainable map-access provision and archiving perspectives. From this comprehensive point of view, the paradigm of map production and geoinformation management changes. Map production is not an exclusive domain for mapping experts, but is an inclusive procedure that opens up to the public: anyone can record, modify and analyse geoinformation. Geoinformation management has to deal with decentralized data sources, overlapping competences, spatial collaboration and opened up dissemination. Geoinformation and especially maps play the role of an associational framework for cross-domain topics.

This book on "Service-Oriented Mapping: Changing Paradigm in Map Production and Geoinformation Management" gathers various perspectives of modern map production and tries to verify that a new paradigm has to be accepted. This new paradigm directs the focus on "sharing and reuse", which bases on decentralized, service-oriented access as well as distributed spatial data sources. Service-Oriented Mapping is the pattern of thought which allows to access data without owning them. This pattern could be helpful for accessing big data or even establishing interrelated knowledge graphs. Colleagues from different domains, such as cartography, statistics, data engineering or archiving, report latest developments from their working field and relate the outcomes to Service-Oriented Mapping.

The book is partitioned into three parts concerning the new paradigm:

Part I: "Exploring a New Paradigm in Map Production" explores the new paradigm for map production and geoinformation management. Selected examples reaching from service-oriented processing of massive point clouds, the spatial data infrastructure INSPIRE in Europe, distributed geoprocessing chains to the depiction of multivariate flow maps, humanitarian demining and the cloud and military requirement development for geospatial information validate the existence of the new paradigm.

Part II: "Importance and Impact of the New Map Production Paradigm" of this book highlights the importance of this pattern of thought by highlighting the development of statistical geospatial frameworks, purposeful maps to achieve the sustainable development goals, geospatial data mining for real-estate applications, SDI evolution, meteorological mapping and the development of datacubes. The given examples show that the new paradigm is already in use and influences technological development, which can be interpreted as geospatial evolution.

Part III: "Requirements of the New Map Production Paradigm" focuses on selected requirements which apparently occur with the new paradigm. The chapters in this segment illustrate the importance of standards and how they work, clarify the needs for a geospatial semantic web, explicate ethical concerns and the mechanisms of disclosure control, accentuate supply chains in decentralized map production and characterize potential user groups for versioned geodata.

The editors would like to thank all contributing authors and their organizations. In addition to the scientific background of most of the authors, also the authors of public authorities and private sector have been invited to describe their perception of modern map production and geoinformation management. This inclusion completes an overall understanding of the topic of Service-Oriented Mapping and even allows to explore hidden requirements.

We thank our families and friends, who accepted our absence for editing this book additional to our regular occupation, especially in times when joint activities would have been so important. Thank you for your patience!

Potsdam, Germany — Jürgen Döllner
Vienna, Austria — Markus Jobst
Pretoria, South Africa — Peter Schmitz
February 2018

Contents

Editors and Contributors

About the Editors

Jürgen Döllner studied mathematics and computer science at the University of Siegen, Germany (1987–1992). He got his Ph.D. in computer science from the University of Münster, Germany, in 1996; there he also received his habilitation degree in 2001. In 2001, he became Full Professor for computer science at the Hasso Plattner Institute at the University of Potsdam, where he is leading the computer graphics systems department.

Markus Jobst is technically coordinating the INSPIRE directive at the Austrian Federal Office of Metrology and Surveying and providing lectures on cartographic interfaces and geospatial data infrastructures at the Vienna University of Technology in Vienna and Hasso Plattner Institute in Potsdam. He finished his Ph.D. in 2008 with the focus on semiotics in 3D maps at the Vienna University of Technology. His main focus in scientific work is the communication of spatial-related data, cartographic heritage, spatial data infrastructures (SDIs) and the management of geospatial processes in Service-Oriented Mapping.

Peter Schmitz holds a Ph.D. in Geography and a B.Sc. (Hons) in geoinformatics. He is a registered professional Geographic Information Science practitioner in South Africa. He is a principal researcher at CSIR Built Environment. He is active in geospatial analysis and forensic geography. He is currently the Commision Chair of the ICA Commission on Map Production and Geoinformation Management.

Contributors

Raffaele De Amicis is the Managing Director of Fondazione Graphitech. He holds a MEng in Mechanical Engineering and a Ph.D. on Surface Modeling in Virtual Environments from University of Bologna, Italy. His ongoing research foresees work on the visualization of big data that are made interactively available to decision makers. Within this topic, his focuses are on the area of interactive visualization and intelligent complexity reduction within virtual environment.

Sergio Anguita Startup Prophet, has been working at ICGC since 1999 leading different departments related to Information Technologies, Web Development and Innovation Management. Entrepreneurship passionate, co-organizer at Startup Weekend Barcelona (2013-2016) and Lean Startup Circle Community in Barcelona (2012–2014).

Fabian Bach European Commission, Eurostat, Luxembourg City, Luxembourg

Harald Bamberger holds a master's degree in Computer Science from the Technical University of Vienna. In 2009 he joined the IT-Departement of the Austrian National Weather Service "ZAMG Zentralanstalt für Meteorologie und Geodynamik". There he is working mostly on Web-Applications and Web-Services for internal and external purposes. Topics are among others OGC-WebServices and the Implementation of the European INSPIRE directive.

Peter Baumann Jacobs University, rasdaman GmbH, Bremen, Germany

Isaac Besora Pixel Ninja, computer engineer that has always loved maps. Started his career as a research scientist in the Moving computer graphics research group, and later focusing on web technologies and apps, he is now working at Geostart trying to mix maps, CG and web technologies into something that can be useful to everyone.

Jean Brodeur obtained a Ph.D. in geomatics science in 2004. He worked for Natural Resources Canada (1985–2015), was adjunct at Laval University (2005–2016), and is member of the Centre de recherche en Géomatique (2005–). He has contributes actively to the standardization of geographic information in ISO/TC 211 and OGC and has published wdiely on interoperability of geographic information. He founded his own consultation organization "GéoSémantic Research" in 2015.

Sanet Carow is a lecturer at the University of South Africa and has more than 15 years' experience in teaching at tertiary education institutions. She worked at Esri South Africa as the Education and Training manager for 3 years. Other research interests include the complementary use of GIS and artificial intelligence in decision making.

Vlado Cetl Ph.D. is a Scientific/Project Officer at the Joint Research Centre of the European Commission. Before joining the Commission (2016) he was an associate professor of geodesy and geoinformatics at the Faculty of Geodesy, University of Zagreb, Croatia. His current work is focused on implementation and evolution of INSPIRE. He is also working with the UN-GGIM: Europe as Observer in the regional working group on Data integration.

Arie Claassens is a Technical Advisor at iMMAP Iraq, using GIS in support of humanitarian mine action and holds an MSc (GIS) from the University of Leeds. He worked as a Senior GIS Specialist at Esri South Africa, GM of an e-commerce company and intelligence analyst. He is married to Nelmari and the father of Vivienne and Griffin.

Serena Coetzee is associate professor in the Department of Geography, Geoinformatics and Meteorology and Director of the Centre for Geoinformation Science (CGIS) at the University of Pretoria in South Africa. Her research focuses on spatial data infrastructures, address data, standards, open principles, geoinformatics education and geovisualization.

Alberto Debiasi is a technologist at the Fondazione Bruno Kessler, in the Embedded System research unit. He is working in several research projects, in the context of Cyber-Physical Systems. He holds a Ph.D. in Computer Science from the University of Trento, Italy. He worked as developer at Fondazione Graphitech, center for computer graphics. His major research interests include Geovisualization and Information Visualization.

Sören Discher studied computer science at the University of Potsdam and the Hasso Plattner Institute, Germany (2007-2013) where he is currently working as a Ph.D. student. His research interests include scalable GPU-centric processing and analysis techniques as well as real-time rendering, visualization and interaction techniques for large-scale 3D point clouds.

Jürgen Döllner studied mathematics and computer science at the University of Siegen, Germany (1987–1992). He got his Ph.D. in computer science from the University of Münster, Germany, in 1996; there he also received his habilitation degree in 2001. In 2001 he became full professor for computer science at the Hasso Plattner Institute at the University of Potsdam, where he is leading the computer graphics systems department.

Sergio Estella is Chief Design Officer and co-founded Vizzuality with the goal of making beautiful data design a true driver of change and impact on people's lives. His focused creative direction combined with a deep understanding of how humans communicate have made Vizzuality one of the most respected data visualization firms.

Georg Gartner is a Full Professor for Cartography at the Vienna University of Technology. He holds graduate qualifications in Geography and Cartography from the University of Vienna and received his Ph.D. and his Habilitation from the Vienna University of Technology. He was awarded a Fulbright grant to the USA and several research visiting fellowships. He was Dean for Academic Affairs for Geodesy and Geoinformation at Vienna University of Technology. He is Editor of the Book Series "Lecture Notes on Geoinformation and Cartography" by Springer and Editor of the "Journal on LBS" by Taylor & Francis. He serves currently as Past President of the International Cartographic Association.

Jamie Gibson is a Social Scientist working for Vizzuality. His background is in Human Geography and how humans engage with and manage the environment. Jamie investigates out how people think about the world around them, so that the tools Vizzuality builds are useful in their daily lives.

Roland Grillmayer works for the Environment Agency Austria and is leading the working group for INSPIRE data harmonisation in Austria which supports the federal provinces with the INSPIRE implementation process. He is involved in the work of ISO/TC 211 since 2010. Furthermore he is lecturer for GIS standards and GML-modelling at different Austrian universities.

Marie Haldorson holds a position as director at Statistics Sweden responsible for the Regions and Environment Department, which includes Statistics Sweden's International Cooperation Office. She has been with Statistics Sweden for more than 25 years, since 2006 in management positions. Her academic background is in Political Sciences with a Major in Human Geography. She is involved in the work of UN Global Geospatial Information Management on global and European level with a focus on integration aspects: how to increase the use of geospatial information in the statistical production. Together with Mexico, she is co-chairing a global working group on geospatial information supporting the UN Inter Agency and Expert Group on Sustainable Development Goal Indicators. The international engagement also includes coordination of European GEOSTAT projects funded by Eurostat to do with integration of statistical and geospatial information.

Lauren Hankel is a researcher in the Earth Observation Science and Information Technology (EOSIT) competency area at the Meraka Institute of the Council for Scientific and Industrial Research (CSIR) in South Africa. Her research interests include geographic information science/systems and software development.

Johannes Heymans is a GIS Project Manager registered at APM, and have MSc GIS and M.IT degrees. He has more than 16 years of experience in mine action, mainly in the domain of information management and GIS. He has a keen interest in using GIS to assist mine action to become more effective and efficient.

Jan Hjelmager works with architecture and standardization in the Danish Agency for Data Supply and Efficiency. He has been active in national and international standardization since 1993 and since 1994 in ISO/TC211 where he has participated in numerous standardization projects, serving as Head of Delegation since 2002. He has also been active in developing INSPIRE metadata implementation rules.

Bang Pham Huu Jacobs University, rasdaman GmbH, Bremen, Germany

Christine Hörfarter studied earth sciences at the University of Vienna and obtained her Master's degree in petrology. She joined the Geological Survey of Austria (GBA) in 2010 and is now concerned with the coordination and modeling of geoscientific data. She is the co-editor of the GBA-Thesaurus and a contact person in case of implementing INSPIRE.

Knut Jetlund works for the Norwegian Public Roads Administration, mainly working with the Norwegian Road Database (NVDB) and with standards for geographic information and ITS. He is also PhD student in Geomatics at the Norwegian University of Science and Technology. Knut has been active in ISO/TC_{211} since 2009, and has been co-convenor of the Harmonised Model Maintenance Group (HMMG) since 2015.

Markus Jobst is technically coordinating the INSPIRE directive at the Austrian Federal Office of Metrology and Surveying and providing lectures on cartographic interfaces and geospatial data infrastructures at the Vienna University of Technology in Vienna and Hasso Plattner Institute in Potsdam. He finished his PhD in 2008 with the focus on semiotics in 3D maps at the Vienna University of Technology. His main focus in scientific work is the communication of spatial related data, cartographic heritage, spatial data infrastructures (SDI's) and the management of geospatial processes in Service-Oriented Mapping.

Andrew Jones has been active in the fields of surveying and spatial data management since 1972. He became involved in the work of ISO/TC_{211} Geographic Information/Geomatics in 1997, serving as Australian Head of Delegation between 2000 and 2004. He became convener of the ISO/TC_{211} Terminology Maintenance Group in 2001.

Claire Knight is a graduate with a Bachelor of Science with Honours degree in Geography and Ecology from the University of the Witwatersrand, South Africa. She is currently working in the education and training field of the Geographic Information Systems industry, focusing on school to post-university level.

Alexander Kotsev Ph.D. after defending a dissertation in GIScience from the University of Sofia, was awarded a senior Fulbright scholarship at the Ohio State University on geocomputation and geovisualisation of spatio-temporal data. Since 2012 he works at the European Commission Joint Research Centre. Alexander's interests include spatial analysis, sensor web, open source technology and data interoperability.

Vanda Nunes de Lima is a Senior Expert at the Joint Research Centre. Before joining the European Commission (1988) she was an associated professor at the Agronomic Institute of the Technical University of Lisbon with MSc in earth observation and Land Evaluation. At the JRC she has been responsible for several European project regarding earth observation, statistics, geoinformation and SDI. Vanda was the coordinator for the development of INSPIRE data specifications and Regulation on interoperability of data sets and services, and a member of the EU INSPIRE Coordination team, currently addressing policy and governance issues on SDI in the digital government and digital transformation. She is working with the UN-GGIM: Europe as Observer in the Executive Committee and in the regional working group on Core reference data for Sustainable development goals.

Anita E. Locher has studied library and information science in Geneva, Switzerland. She decided for a master in digital content management in Barcelona and subsequently earned the title of Doctor of the University of Barcelona in 2017 where she has been involved in digital preservation teaching and projects.

Joan Masó is a geospatial scientific researcher at CREAF (Ecological and Forestry Applications Research Centre) in Spain. He holds a PhD in Geographic Information Science and Geography from Universitat Autònoma de Barcelona. He represents OGC and Spain in GEOSS, and is an IEEE member and Spain's representative to ISO TC_{211}.

Graeme McFerren is a senior GIS Researcher, with 20 years GIS industry experience in research, software development and professional services roles. Graeme has worked on a number of national scale web-based GIS systems. His interests include spatial databases and web services, spatial data management and geodata streaming systems.

Lance McKee is an independent writer and consultant. He was on the OGC start-up team and served OGC until 2016, when he retired as Senior Staff Writer. He is now an OGC member. He initiated and co-chairs the OGC Electromagnetic Spectrum Domain Working Group.

Vlad Merticariu Jacobs University, rasdaman GmbH, Bremen, Germany

Dimitar Misev Jacobs University, rasdaman GmbH, Bremen, Germany

Jerker Moström is a senior official and spatial analyst working in Statistics Sweden, Regions and Environment Department, since 2011. He has an academic background in Social Sciences with a Major in Human Geography. Jerker has one foot in regular production of land use and water statistics, and another foot in R&D projects and European cooperation. The European cooperation includes coordination of European projects funded by Eurostat dealing with integration of statistical and geospatial information and participation in various working groups (under Eurostat and UN GGIM Europe). He is representative of the GISCO working party and national contact point for the European Forum for Geography and Statistics (EFGS).

Gerhard Muggenhuber has obtained a diploma in surveying, a postgraduate degree on geoinformation and a doctoral degree from the Technical University of Vienna. He has some 30 years of professional experience in management of cadastre and geoinformation in Austria as well as abroad. In his present function as vice-head of international affairs at BEV he actively contributed to Eastern- and Central European initiatives and projects on Land Administration. He was member of the steering committee of UNECE-Working Party on Land Administration, the advisory body to the UNECE in Geneva and chairman of International Federation of Surveyors-Commission 3-Spatial Information Management. He published papers and articles on land administration, geoinformation, real property valuation and related subjects.

Montse Ortega In-house Tech Mastermind, programmer analyst with 11 years of experience in different programming languages (Java, javascript, node) and in different BBDD (postgreSQL, Oracle, mySQL). Started in a computer consultancy and currently working at ICGC in the Geostart group developing prototypes and maintaining the Instamaps web application.

Victor Pascual Master Maker of Things, geographer, currently working at ICGC as a member of Geostart, prototyping and investigating new technologies. Focused on web GIS programing and web services based on open standards.

Reese Plews is president of a geospatial services company in Tokyo Japan. Active in ISO/TC_{211} since 2004, Reese has participated in numerous projects and acted as editor on four different standards. From the summer of 2017 he became co-convenor of the Terminology Maintenance Group for ISO/TC_{211}.

Johannes Reischer studied geography and cartography in Vienna and has been working at the Geological Survey of Austria since 1996. His focus of work is GIS and raster data management.

Rico Richter studied computer science at the University of Potsdam and the Hasso Plattner Institute, Germany (2003-2009) where he is currently working as a research scientist at the computer graphics systems department. He has many years of experience in the development of complex processing, analysis and visualization techniques for massive point clouds. His main focus are research and industrial projects related to 3D and 4D point clouds.

Rafael Roset Mystic Chief Visioneer, has been working for the last 30 years in all major projects dealing with paper and digital map dissemination and difusion, as well as digitization and georeferencing of old maps, at the Institut Cartogràfic i Geològic de Catalunya (ICGC), the national mapping agency of Catalonia. Deeply involved with the geocommunity in Barcelona, was coorganizer of the FOSS4G 2010 congress in Barcelona and also cofounder, organizer and community manager of Geoinquiets Barcelona (@geoinquiets on Twitter), the local chapter of OSGeo.

Martin Schiegl Geological Survey of Austria, Vienna, Austria

Peter Schmitz holds a PhD in Geography and a BSc (Hons) degree in Geoinformatics. He is a registered professional Geographic Information Science practitioner in South Africa. He is a principal researcher at CSIR Built Environment. He is active in geospatial analysis and forensic geography. He is currently the commision chair of the ICA Commission on Map Production and Geoinformation Management.

Chris Schubert Head of the Data Centre at the Climate Change Centre Austria (CCCA). Beyond to hardware and software design, his activities focus on semantic interoperability for data, detailed data management and data life cycle development. He studied geoscience and informatics, started his professional career in 2004 in the field of semantic interoperability and web service technologies. Beginning of 2012, he joined the European Commission for conceptual modeling for INSPIRE, an EU directive on interoperable data exchange. His expertise continues to make him an expert for the Commission and in standardization activities.

Jessica Sena Mobile Platform Pioneer, computer engineer currently working at ICGC, as a member of Geostart, prototyping and investigating new technologies. Widely experienced working as a web GIS developer, but more recently focused on the mobile space, in charge of the development of several GIS related apps.

Bruno Simões holds degrees from Évora University (B.Sc. M.Sc. Computer Science, 2009) and Trento University (Ph.D Applied Mathematics, 2013). Bruno is a research scientist at Vicomtech-IK4 and is working in a number of multiple research projects, funded by the European Commission and industry, in the context of smart digital content and manufacturing industries. His research interests include GeoVisual Analytics, Interactive Visualization, and Data Physicalization.

Robin S. Smith Ph.D. is a Project Officer at the Joint Research Centre of the European Commission. With a background in ecological science and town and regional planning, his work addresses social, technical and policy issues of SDIs and interoperability, with a focus on e-government and stakeholder involvement.

Wladimir Szczerban Refactoring Messiah, computer engineer mainly focused on Geographical Information Systems. Working at ICGC since 2005 developing web mapping software. Really interested in Open Software, Open Data and Geotechnologies, active collaborator of OpenStreetMap and a founder member of Geoinquiets.

Friedrich Teichmann Director of the Geospatial Institute at the Austrian Armed Forces. Current military rank: Brigadier General. MSc and PhD in Geo-Sciences at the University of Maine (USA), the University of Salzburg (Austria) and the University of Rochester (USA). Post-graduate studies in GeoInformatics at the University of Salzburg and Telematics at the University of Krems. Space Study Program at the International Space University (France). Lecturer for Space Applications, GeoSciences, Information Technology and Geo-Informatics at four Austrian Universities: Universität Wien, FH Kärnten, FH Joanneum and FH Wiener Neustadt.

Robert Tomas Ph.D. in Geoinformatics. Prior joining the European Commission (2009) he worked for 13 years in the Czech Geological Survey as Deputy Director for Informatics. At DG-JRC as a member of the EU INSPIRE Coordination Team he was responsible for e.g. development of the data specifications for Earth science related themes, data quality aspects as well as coordination of the support to the INSPIRE communities – Thematic Clusters.

Marc Torres Lord Geospatial Dreamer, geographer and land surveyor. Working at ICGC within the Geostart group of innovation and prototyping of products and services in the field of geoinformation. Member of the Geoinquiets group (www.geoinquiets.cat) and the Public Innovation Network (www.xarxaip.cat)

Matthias Trapp studied computer science at the University of Potsdam and the Hasso Plattner Institute, Germany (2000-2007) where he received his Ph.D. in Computer Science (2013). During his post-doctoral studies, he was heading the junior research group on "4D-nD Geovisualization" (2012-2017). Since 2017, he is a senior researcher at the Hasso Plattner Institute. His major research areas are computer graphics, image and video processing, geovisualization, software visualization, and information visualization with a focus on GPU-based techniques.

Christina Wasström is since 2017 chair of ISO/TC 211. Since 1990 she works at Lantmäteriet, the Swedish the Swedish mapping, cadastral and land registration authority, in the field of spatial data. 2008 she become a national SDI coordinator and act as the national contact point towards the European Commission regarding the EU INSPIRE directive.

List of Figures

Part I
Exploring a New Paradigm in Map Production

Chapter 1
Changing Paradigm in Map Production and Geoinformation Management—An Introduction

Markus Jobst and Georg Gartner

Abstract Maps influence our daily life, either by being an interface that we consult, or in form of precise geospatial information in the background that influences our daily devices, such as the smartphone or car. The access to maps evolves into basic rights that e.g. is realized with volunteered geographic information. The concepts of processing geospatial data, creating geospatial information and transmitting geospatial knowledge make use of service-oriented architectures. The paradigm seems to change from a "collecting—assembling—storing" to a "reusing—assembling—sharing" methodology. This chapter explores the new paradigm in map production and geoinformation management, highlights its relevance and discusses the main important requirements that need to be considered.

Keywords Service-oriented mapping · SDI · Map production · Data integration Geoinformation management

1.1 A New Paradigm in Map Production

1.1.1 Role of the Map Today

The availability of maps and its use cases have been extensively expanded over the last years (Caquard and Cartwright 2014). Starting with the digital era, maps spread in everyday lifes and help to understand complex spatial as well as temporal situations (Peterson 2003). The artefact map is widely and publicly used for various situations spanning from navigation, topographic and environmental documentation, thematic mapping and spatial management. All these application fields are

M. Jobst (✉)
Austrian Federal Office for Metrology and Surveying, Vienna, Austria
e-mail: markus@jobstmedia.at

M. Jobst · G. Gartner
Research Group Cartography, Department of Geodesy and Geoinformation, Vienna University of Technology, Vienna, Austria

J. Döllner et al. (eds.), *Service-Oriented Mapping*, Lecture Notes in Geoinformation and Cartography, https://doi.org/10.1007/978-3-319-72434-8_1

cross-border and cross-thematic, which enhance the importance of collaboration for the geo-communication process on one hand, and within the creation of the artefact map on the other.

The wide application fields are enhanced by the fact that maps are becoming publicly well known and used ubiquitously. Owing to smart devices and increasing mobile network broadband coverages, it is convenient to access digital maps independent from an actual location almost everywhere (Gartner and Rehrl 2008). Even more, maps embed users emotions and situational context (Klettner et al. 2013).

Maps evolve as tools to access information about spatial related situations and developments on a short notice. Therefore the content complexity has been further reduced and enhanced for high pacing interaction modalities (Tidwell 2010). The reduction of map content also supports the perception on graphical restricted devices, like small smartphone displays.

The media used for the transmission of spatial information are rapidly developing in the field of "near-reality presentations" such as virtual reality glasses, augmented reality and accompanying human user interfaces (Coffman and Klinger 2015). These devices and interfaces allow for a more naive transmission of spatial content and therefore may increase the cognitive impact of the map. In the end further requirements for map production and geoinformation management are defined by these additional frameworks of naive geo-communication with 3D (Jobst and Germanchis 2007).

Spatial information shown in modern digital maps more often inherit a "short-life and information-news" character, which is in contradiction with maps as art and long-term artefacts. The main reason for this characteristic is excessive information supply. The term "big data" is increasingly used and leads to continuous data streams that may also change the information visualisation in "real-time" map applications (see Chap. 10 in this book). One drawback of these mapping portals is their lacking of long-term availability, which makes it hard to refer to after several years or even decades (Lauriault et al. 2010). The documentation character of maps, especially service-oriented maps, need to be explicitly enhanced if the map should still play its role as documentation media. This specific role of modern cartography is underpinned by missing geoinformation- or digital map legacy management and therefore stewardship for the curatorships of important spatial content (see Chap. 20 in this book). In the context of service-oriented maps we embed decentralized information sources, spatial data services and spatial infrastructures with high actuality.

The main question that arises from the viewpoint of map production asks if we still use the same paradigm in map production or do we have to rethink our production behaviour? Is it still appropriate to collect, process and harmonise data sources for specific applications? Or shall we explore new procedures on the basis of a new paradigm of sharing and reuse (Maguire and Longley 2005)? In order to explore these questions, a more detailed understanding of modern map production frameworks is needed.

1.1.2 Map Production (What Does This Mean Today?)

The activities of modern map production relate to FAIR data, decentralized data sources, data integration, service-oriented architectures, collaboration, commonly usable knowledge networks, design principles and impact on user's behaviour and knowledge. Therefore "map production" is developing to a production-process network with logistical dependencies, where monolithic sequences are a minority.

In order to evaluate existing procedures in map production and geoinformation management on the basis of a service-oriented paradigm, we need to highlight the most important map use characteristics. Map production is closely related to map use and its use cases (Cash et al. 2003; Verburg et al. 2008). As result we expect an indication that map production in service-oriented infrastructures is different. At least a difference or extension in some parts of the actual map production methodology should indicate an argument for a new map production paradigm in service-oriented infrastructures.

An important characteristic of map-use is the increasing "short-life and information news" character of increasing number of maps. For this aspect we can observe that the communication issue of maps has to be evolved more precisely than ever before. The transmission of spatial information calls for high effectiveness and expressiveness (Jobst 2006) in digital maps. Discrepancies become visible with the integration of various sources and influence user's information cognition.

Effectiveness regards aesthetic concerns as well as perceptional information acquisition, immersive interface use, optimisation processes for data simplification and visual rendering improvements. The quality of presentation and thus success of communication process is mainly depending on the understanding and acceptance of the user.

Expressiveness refers to visualisation capacity of the interface, which concerns the semiotic question of representing all important and necessary details of recorded objects in order to preserve semantic. Is it possible to present all the detailed information with the "low" resolution and "few" communication parameters the interface offers? For instance, if the resolution of the interface (e.g. screen) is lower than the number of desired detail values, the expressiveness criterion will not be met. Some detail values will then not be perceivable (Jobst and Doellner 2008).

Together with the demand for high actuality and therefore a focus on rapid mapping, the map production seems to contradict with recursions in map- and usability design. Any evaluation of a map‘s communication impact will lead to a recursion of map design (Gundelsweiler et al. 2004). Likewise, usability studies will observe room for improvement and thus lead to recursions of usability design (Coltekin et al. 2009; Slocum et al. 2001). Both cycles are associated. Their embedment require additional time in the map production procedure. For many use cases in service-oriented mapping these studies are neither included in the map production procedure nor in map application maintenance or -governance procedures. Instead, an evaluation of a map's impact and success is observed via the social web. Because maps are widely used as spatial information transmission vehicle, the trace of the map, the map's understanding or its data can be observed in social media and therefore indicate the success of geo-communication.

One may argue that design- and usability recursions are only important before the development of a map. This statement could be true with restricting maps to few communication devices. Modern map usage is characterized by an increasing variety of communication devices, reaching from smart information screens to mobile devices and individual glasses for augmentation and virtuality. These devices bring in their specific characteristics and resolutions that effect the design and usability of the map. From the map production point of view the combination of communication interface, use cases of the map and distinct user situations will introduce more detailed design (and map production) requirements that have to be considered in the preparation and production procedures (Montello 2002; Lechthaler and Stadler 2006).

The output of a successful map also requires well selected and prepared information—in addition to design and usability. Spatial data infrastructures, as the main source of data in service-oriented mapping, are characterized by providing a vast amount of data sources in different qualities, structures and coverages. The term "big data" is widely used. According to (Hashem et al. 2015) this term is comprehensively used and requires definitions in more detail, which may help to assign the different views correctly. The content spans from reliable authoritative data with constant life-cycles to single data recordings without quality control, from free and open data without any restrictions to data regulating licenses and even to closed data registers for exclusive usage in authorities. At present most of the content in spatial data infrastructures is a mixture of primary and secondary data, which complicates the identification and extraction of important data sources for a specific map product. This situation becomes even more complicated when data are not consistently covering their field of activity, but miss regions. One can highlight this topic by comparing the homogeneous base data collection "Natural Earth" prepared for map production in scales up to 1:10 million (Kelso and Patterson 2009) against the collection of authoritative data in the INSPIRE geoportal (Maguire and Longley 2005; EC 2018a). In terms of using the above mentioned spatial data, the general requirement for FAIR data is then valid (Wilkinson et al. 2016).

In this decentralized spatial data network, the cartographer acts once more as the person who semi-automatically integrates available sources for communication success and thus produces the map. This activity of map compilation/production atop spatial data infrastructures requires a comprehensive and mature understanding of cross-thematic data sources, their qualities, processing abilities, requested communication aim and transmitting device characteristics. The activities of data identification and information preparation/wrangling may be supported and automatised by Natural Language Processing (Collobert et al. 2011), but will not relieve the "map maker" from design based and integration dependent assessment decisions.

1.1.3 Importance of (Spatial Data Engineering) Geoinformation, Data Integration and Transmission

Different thematic communities make use of maps, which highlights the importance of involved geoinformation. The spatial dimension is mainly used to structure information by space, to show spatial dimensions or to make it comparable in time, respectively for multiple dimensions, like in datacubes (see Chap. 14 in this book). This enrichment of thematic information by its spatial dimension could be called "geocoding" or "georeferencing" (see Chap. 9 in this book), which enables additional integration and comparison of data by its location.

From a more generic viewpoint of geolocated information in a service-oriented information network, meaning decentralised and loosely coupled data sources via spatial data services, we have to ask for the categorisation of geoinformation in order to support a more efficient map production. Is geoinformation delivering simple features that act as anchor to space or is there the need that thematic information is incorporated in geometric features? In fact both views are valuable, but make a difference in their usability. Whereas thematic enriched geoinformation has been modified for a specific use case such as thematic recording, analysis or information transmission, fundamental spatial information tries to provide as many anchors for consistent georeferencing as possible. The fundamental spatial informations' geometric feature quality may vary according to specified scales, for example reduced density of points to enhance processing speed, which also restricts a general usability, but supports a scale dependent data integration (Lee et al. 2000; Cohen and Richman 2002). For example a very detailed large scale spatial dataset with all feature points may not be used for a quick geographical search, because performance will be slowed down. Map production for fundamental spatial information aims at general requirements for georeferencing, consistent and persistent identifiers as well as performance optimized data structures. The creation of the spatial information is reduced to its own data structure and the variety of adaptors that can be used by various thematic domains. In contrast, thematic spatial information comes along with an aggregation of spatial features with thematic values. Generally the thematic dimension influences data structures, depending on its use for analysis or visualisation. Persistent identifiers are not that important within the dataset, because the thematic aggregation has been done or is closely coupled. For example a "join" operation is done on the data level with specific knowledge of the keys and not via an abstracted service interface (Hacigumucs et al. 2002).

Different scientific and thematic communities make use of spatial information. Therefore different spatial infrastructures start to grow. In addition the focus on fundamental spatial information for the geospatial domain triggers a cross-domain movement, which enhances spatial competence. As result of the creation of the spatial competence and domain-specific competences accompanied with well defined connections (such as adaptors, keys, interfaces, …) a shift to cross-domain networks can be observed. One outstanding example is the Group of Earth Observations in which different thematic groups and the geospatial domain create a scientific observations and analysis network (GEOSS 2018). The next arising requirement of this scientific networks is the mutual understanding of all embedded information. This call for easy accessible and automatic processable knowledge networks are the pragmatic start of a spatially related semantic web (Egenhofer 2002), which increases availability and accessibility for cross-domain information.

1.1.4 Data Integration and Its Need for Consistent and Time-Appropriate Geolocation

To a large extend, the main aim of "(spatial) information networks" is data integration, which creates new information mixtures and -structures that are used for geo-communication, especially maps. Data integration needs consistent and time-appropriate geolocation, beside persistent identifiers and well-formed adaptors/keys, in order to establish valid relations to the thematic content. It is not only the valid geolocation, but mainly the time reference that counts for geocoding. For example the names and extension of authorities may change during time. Several times a year these changes could affect 5–20% of existing authority structures as this is the case in Austria. These changes lead to an improper adaptor/key with the effect that thematic information cannot be georeferenced (joined). Only a time consistent adjustment of geolocation and thematic content, a valid space-time relation, will allow for a correct matching and data integration.

Valid space-time relations are the foundations for spatial analytics and the spatial data engineering working areas like simulation, prediction and spatial optimisation (Kitchin 2014). Spatial analytics covers data mining, pattern recognition, data visualisation and visual analytics, which is a specific discipline in cartography that makes sense of space and time (ICACI 2018). Simulations on the basis of valid space-time relations aim at determining a systems' functionality and how its behaviour changes for different scenarios. While doing so, the performance of analytical methodology is observed for further improvement of its efficiency and effectiveness (Robinson 2004), which may lead to changes in procedure. Predictive models could be improved and produce less errors and more accurate predictions with valid space-time relations. Basically the quality of prediction is a variation of the problem type and data (Seni and Elder 2010). Optimisation deals with improving performance or identifying an optimal course with the help of predictive models, simulations, other kinds of algorithms or statistical tests, which make use of valid space-time relations.

Simulation, prediction and spatial optimisation deliver new results that need to be transmitted to the users knowledge and therefore expand map usage. Most of predictions on the basis of valid space-time relations and its functioning data integration will result in new maps. Other applications may just make internal use of the simulation- or prediction results and trigger functionalities of the map application.

Both perspectives influence map production, namely on one hand widening map usage with visual analytics, on the other hand establishing intelligent functionalities, which can be embedded as spatial data service in future map applications.

1.1.5 Service-Oriented Mapping

The map production with distributed sources, real-time data collection and in a highly automated way can be described as Service-Oriented Mapping. Service-Oriented Mapping offers the creation of maps and complex geoinformation analysis by use of the Service-Oriented Architecture paradigm, which covers specific information technology characteristics (Bieberstein et al. 2008) in a decentralized network. Decentralized networks, which may also contain centralized components, help to access more diverse data sources, support flexible production flows and enhance storage needs as well as processing power if appropriate standards, interfaces and connections are established.

Due to growing data collections and diversity of data sources, automation of geoinformation processing is needed. It follows aspects of data integration, data interpretation (Sester 2010), data "semantification" (such as adding multilingual and—contextual meaning and importance to data) and data transmission preparation in terms of efficient communication for information handlers. Service-Oriented Architecture (SOA) is the main paradigm to connect and explore everything (Hendriks et al. 2012).

Service-Oriented Mapping, which makes use of aforementioned spatial data engineering, spatial analytics and data integration, depends on decentralized and loosely coupled data- and service sources (Lucchi et al. 2008). This characteristic results in a variety of data competences at the data provider side. These competences are combined in a technological thematic network, a network of spatial data infrastructures (Chan et al. 2001; Vandenbroucke et al. 2009). Data competences describe the relevance of a data producer, who is responsible for a dataset and its quality. Likewise the service competence describes the importance of a functional components' providing site, which is responsible that SDI functionality is available whenever it is needed. Functionality includes data access, transformations, processing steps and so forth. All competences are autonomous within the conceptual network (Jin and Cordy 2005), which means that they can be modified, expanded or exchanged without destabilizing the network.

The different components and characteristics of a Service-Oriented map production line points the challenge of governing this system out. Governance in this context means that the maintenance of system components is done in a planned and controlled manner. From an overall system perspective this maintenance could be supported by balanced scorecards (Kaplan and Norton 1995) for decisions that rule technological and organisational changes as well as developments. For example the development of new standards of an existing system component could have enormous impact on the decentralized map production line and therefore needs a precise assessment of all impacting factors: Which other components are affected and need to be replaced, changed or enhanced? Is the change useful from a technological point of view? Is the change and all the impact feasible?… and many other questions have to be answered. When we enhance once more the decentralized and loosely coupled character of a Service-Oriented map production network, we can observe that it is not possible to force autonomous nodes (such as participants, data and service providers) to do a change or follow technological recommendations.

A tier approach, which splits the Service-Oriented production network into different levels may support change management, but definitely is not a single rule solution.

Spatial data infrastructure follows a tier approach, which means that data, including metadata and registries, are hosted in the data tier, services and functionalities in general are hosted by the logic tier and applications such as map clients are hosted by the presentation tier (Evans and Szpoton 2015). This multi-tier architecture design creates independence between data, functionality and the presentation, which means that changes in one tier can be made without affecting the other tiers. All three tiers are involved in map production. For that reason this IT architecture concept is most relevant for the design of map production and geoinformation management (Fig. 1.1).

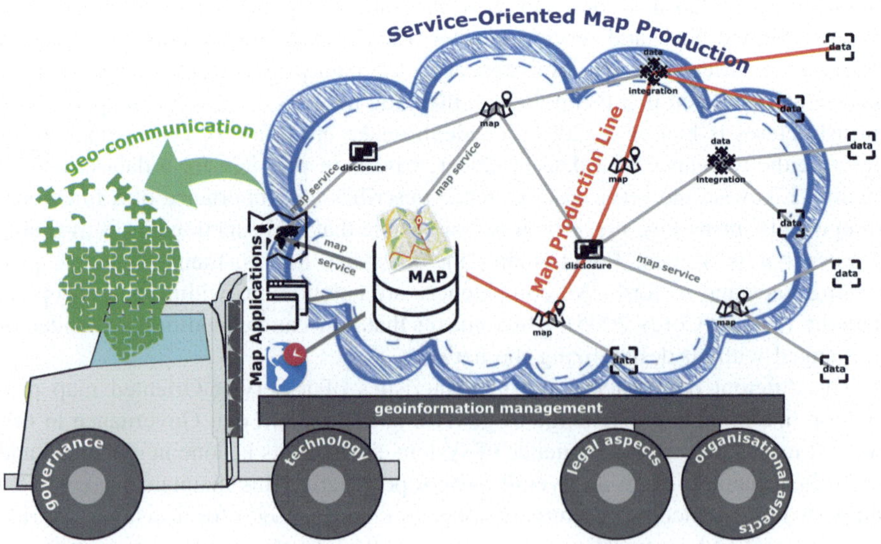

Fig. 1.1 The service-oriented map production truck. It shows dependencies and the main terms of the service-oriented map production

1.1.6 Concluding the Introduction Section

Map production and geoinformation management have to massively deal with the principles that are built upon a transparent and well understood spatial data infrastructure. Additionally the aim of geo-communication, which still is the design of graphics and coding of information for a better transmission, has to be considered. This aim could be achieved in a more effective, expressive and manifold way if we can make use of the new map production paradigm by exploiting the multifaceted spatial environment.

In this contribution we refer to different terms of cartography, which are widely and often synonymously used. Therefore these terms are described as they are used in this contribution. The most important terms are map, geo-communication, map production, map production line, service-oriented map production, map service, map application and geoinformation management.

Map. A map is an artefact that carries a set of geospatial information in a way that it can be perceived, distinguished and that its crosslinks can be observed. The design and graphical language used in the map should support geo-communication.

Geo-communication. The process of transmitting geospatial data to the individual knowledge of the user can be defined as geo-communication. This process covers, among others, activities of cartographic generalisation, geo-media evaluation and communication requirement collection. The transmitting process makes use of a map. An impact assessment could deliver a statement for the success of geo-communication.

Map production. All requirements, processes, actions, functionalities, interfaces, information and data that are used to produce a successful map are components that are covered by map production. It is the general procedure how a map is compiled.

Service-oriented map production. Map production that makes use of the service-oriented methodology for all, or at least for most, of its components can be called service-oriented map production. It is characterized by a high automation level and distributed storage as well as processing nodes.

Map production line is a defined set of sequences that start at data collection and result in the map.

Map service describes a spatial data service that delivers a stored map to an application. It is the interface between the maps' data tier and the map application. It may contain styling instructions, e.g. with CSS technologies.

Map application is the presentation tier of a map. It embeds interaction, multimedia and other perception promoting mechanisms that shall support geo-communication.

Geoinformation management. The overall fundament that covers the legal-, organisational-, technical- and governing aspects of service-oriented mapping can be called geoinformation management. It will need leadership, governance and maintenance for all required aspects in order to ensure the existence of service-oriented map production.

In the following section we explore the question if there is something like a new paradigm for map production and geoinformation management. With the assumption that there is something like a new paradigm, we will investigate its relevance and try to define the most important requirements in order to make use of this paradigm in map production and geoinformation management.

1.2 Exploring a New Paradigm in Map Production and Geoinformation Management

After this general introduction on map production and geoinformation management we could assume that this working area of Service-Oriented Mapping becomes more demanding in terms of technology use, collaboration, semantic knowledge and geo-communication support. The main question that arises for this chapter (and book) concerns a new paradigm: Is there a new paradigm to be followed for modern map production and geoinformation management? A new paradigm may have impact on alternative procedures and bring in specific requirements. In this section we will observe indicators to answer this question by

- a selection of use cases and map applications,
- embedding cross-domain information in a near real-time manner,
- exploring the demand for enhanced information quality,
- observing user participation mechanisms in map recording and design,
- highlighting datacubes and their perspectives and
- touching on the data revolution and its consequences.

As result of this section we expect to have some observations or reasons that support the idea of having a new paradigm in map production.

1.2.1 Use Cases and Map Applications

Map applications and their use cases from the map provider's perspective should allow to observe a new situation in map production. This perspective goes beyond the situation of geoinformation providers, which focuses on the harmonisation of data, data integration and data exchange. It is focusing on the communication with a digital map product in terms of basemaps or online map portals. A glance on IT architecture could explain a big technology change and the challenge modern map production is confronted with. More and more geoinformation systems make use of the Internet, which concerns the access to functionality and the access to basemaps.

The aspect of functionality aims at bringing Geo Information Systems (GIS) to the WWW and make those systems available on a subscription basis. Depending on the subscription the variety of GIS functionality changes or the amount of data to be processed on the WWW may be restricted (ArcGIS 2018). The main advantage of this business model is its independence from the operating system, because the software runs on the GIS provider's server. Any update of the software is done by the GIS provider and not by the user. This portfolio of online-GIS may be related to cloud storage solutions, in which data are managed. Again this geoinformation management is independent from a local machine and maintained by the storage provider (CloudStorage 2018). The chapter by Schmitz on humanitarian demining explores this option (see Chap. 7 in this book). From an IT-architectural point of view these solutions follow a "centralized" approach, in which the functionalities and storage are centralized in the warehouses of a single provider. Although the processing is split to several servers, the maintenance, provision and access control is done by a single provider or provider consortium. Therefore the provider is capable of the data model in the databases and the application schemas of the service interfaces.

Similarly the access to basemaps aims at the centralized provision of maps (map services) via the Internet. The product provided has been harmonised and delivers a consistent presentation. For example Open European Location Services of Eurogeographics will deliver web map services of a European base map that is periodically created by the NMCA's of Europe (Jakobsson 2012). A central storage and access point allows for a homogeneous access whereas instead decentralized webservices will make inhomogeneous service providing visible. This means that the transmission time of the map product depends on the processing of the provider's servers and bandwidth. Having one single product delivered by several providers will show different delivery times and some kind of "flickering" for the image tiles of the maps (Cronin and Hornbaker 2003). This technical shortcoming may influence the geocommunication process and its impact (Brodersen 2007).

From a map designer's perspective a decentralized information access must not disturb the map transmission process. Therefore appropriate IT architectural structures have to be put in place, which are for example centralized architectures for functionality or storage. Other mechanisms explore "caching" and stewardship contracts.

Caching mechanisms will create a centralized storage buffer at the main access point which ensures permanent delivery even with high load. The main questions for pragmatic cache creation performance deal with the time for creation, regional coverage of the cache and its covered zoom level (Campos Bordons 2013).

Stewardship contracts establish an agreement between the main access point and content providers (data and services), which ensures data and service availability, performance and capacity (Baranski et al. 2011).

A successful map transmission process and its failures particularly stand out in use cases of everyday life, when a wide audience makes use of the map for their daily decisions. Therefore the most relevant use case is navigation. Navigation reduces a map to its most needed content and spatial links for a specific situation, interest and preferences (Park et al. 2012). The needs for spatial information, data integration and decentralized data access are high, because outdated data, fake information and situation irrelevant coding will increase frustration and lower acceptance of the map.

Highest available actuality and reliability for geoinformation is a key for the functioning of self-driving vehicles. For many issues these vehicles perform their own measurements with different sensors. But when it comes to swarm intelligence any exchange of information in the regional network, but also in a wider, sometimes global, network is needed. One most pre-eminent example is the management of traffic flows and traffic jam warnings. Just the position information of a swarm of sensors (vehicles, smartphones,…) in a network, like the Internet, allow for the precise input in navigation and map applications (Tan 2013). In order to make this information relevant, it uses spatial links between the processed/aggregated information and the location and situation of the sensor. These spatial links are highly formalized for self-driving vehicles, e.g. street infrastructures, kilometrages and so forth.

The formalization of spatial links may not work for the human interaction, which bases upon the individual knowledge of the map user (Andrews and Delahaye 2000). From a navigational point of view, the "where to find" or "where to go" communication requires spatial links, mostly landmarks, to describe the spatial answer. Spatial links highlight decision relevant objects, which could also enhance relevance of the overall map context and its importance for the given situation. Actual and precise geoinformation as well as context-aware representation types are needed in these map applications. These requirements can be established by decentralized data sources, service-oriented mapping, swarm based data integration and user preference tracking.

Further more sophisticated aspects of modern mapping are thematic maps with their visualisation of complex information themes. In terms of service-oriented mapping the production steps need to be automated in order to be available as functional building blocks, often called spatial data services. The production steps reach from data acquisition (by requesting standardized services), data wrangling (mining, sorting, manipulating and preparing of data) to data integration (combining and enriching data) and coding for communication, which includes generalisation to support perception and considers geomedia characteristics. In the following three selected prominent examples, the mapping of environmental topics, demography, health and distribution, should help to explain the paradigm of service-oriented mapping.

1.2.1.1 Mapping Environmental Topics

The mapping of environmental topics is often used to support environmental policy. Therefore environmental developments and their forecast are needed. In a uniform manner short term environmental developments, for example the spreading of a radioactive cloud, is of public interest and needs to be communicated. Most of the data are delivered in near real-time by different sensors or sensor networks. These spatial data undergo different modelling steps before any data analysis and information preparation can be done. Because those spatial data are time relevant, all involved processes are automated and delivered/combined by standardized services. In many cases processing power is shared by various data centres in order to be fast enough in the modelling step. There is the need to represent the most actual data model available: for example we are interested in the actual weather situation for a specific location and its future development, but we will generally not ask for the weather situation of the week before. The chapter of ZAMG (see Chap. 13 in this book) introduces to the topic of meteorological mapping, which requires performant modelling procedures, different data sources as well as standardized interfaces for map production.

Environmental map production in a mid term or long term perspective has to create indicators and stable geometries that allow for comparison of an observed and communicated status. Geographical grid systems help to correspond an observation to a regional extend in a uniform manner. This is how regions become directly comparable in space and time.

The European Environment Agency (EEA) provides independent information in form of maps, diagrams and tables. This information is created by the environmental agencies of European member states, integrated at EEA and result in indicators that are designed to answer key policy questions. The continuous monitoring and evaluation checks the indicators against settled targets (Gabrielsen and Bosch 2003).

In addition to the very clear demarcation of administrative units and geographical grid systems, vague geographical areas are used to create the georeference. For example the description of marine location uses vague geographical area like "Celtic Sea", "Ionian Sea", "Mediterranean Sea" and so forth (EEA 2018). Vague demarcation causes more challenges for a formalized service-oriented map production.

Due to the fact that decentralized European environmental agencies provide the data via spatial data services, this example describes on use case of service-oriented mapping. Although the central data warehouse at EuroSTAT prepares all the information for the map, the main principles of service-oriented mapping are established for parts of the map production line.

1.2.1.2 Mapping Demography

Similar to environmental topics, a service-oriented thematic mapping of demographic topics, which is a use case of statistics, struggles with the decentralized and inhomogeneous structure of spatial recording. Although the national statistical offices harmonise data for the main demographic indicators, which are delivered to the central data warehouse at the European Statistical Office during their reporting activities, in some cases the data may contain semantic inconsistencies if originated by thousands of municipalities. For example a point for an address could be placed as a centroid of the parcel or as a centroid of a house, near the entrance of the house or the entrance of the parcel. This interpretation is up to the registering person, if not further defined.

Any observation of demographic development is the result of population change within a regional unit and defined time period. Two main components can be observed, namely a natural population change and net migration. Natural population change results from the number of live births minus the number of deaths. Net migration is the difference of the number of immigrants minus the number of emigrants (including statistical adjustment). The recording of numbers for births, deaths, and so on generally happens within municipalities. Whenever the regions of municipalities change, the reference region changes and may distort the values. Therefore the values are time and reference region dependent, which leads to comparison complexities.

Statistics define indicators with the aim to be persistent and clearly defined. Hence map production has to consider persistent definitions (including the time relation) and value definitions. Global activities create the framework for superior definitions for the purpose of data, service and semantic interoperability (Troltzsch et al. 2009). For instance the initiative "gapminder" collects statistical data "as is" and creates global development maps to explore the given indicators (Gapminder 2018; EuroSTAT 2018a, b).

Again this example shows that data flows for thematic map production are massively decentralized. In addition to environmental topics, the georeferences for demographics may change during time and therefore destroy comparability. Statistical grid systems try to overcome this effect. Another issue that is introduced in statistics are privacy issues: as soon as we are able to identify individuals, the processing of information has to intervene for ethical reasons (see Chap. 18 in this book).

1.2.1.3 Mapping Health and Distribution

From the ethical as well as security point of view the handling of spatial health and distribution in service-oriented mapping is a delicate issue. These data and outcomes will for most cases have some impact on privacy and may be relevant for public security. Therefore data sources for the health status are still rare.

Maps for health and its development as well as distribution will make use of a variety of data sources: transport networks, hydrography, meteorology and even real-time data of humans as sensors. Topographic information like transport networks in combination with population allow for detailed distribution models (Bell et al. 2006). Smartphones and health trackers deliver individual information which furthermore could be one component of human swarm intelligence.

This selected topic on health shows that beyond an ethical dimension, cross domain relations, real-time health status tracking and data integration for human swarm intelligence could characterize service-oriented map production. The main challenges from an IT point of view are disclosure control (see Chap. 18 in this book), the variety of sensor protocols, inhomogeneous data and almost countless data streams (see Chap. 13 in this book).

The selected examples of modern use cases and map applications extensively show that real-time processing, cross domain relations, high quality data as well as processing and user participation are the foundation for service-oriented mapping. Resulting from the use cases, these characteristics seem to be a dominant part of the new map production paradigm.

1.2.2 Embedding of Real-Time and X-Domain Information

In the previous section we have observed main characteristics of the service-oriented map production paradigm by examples. In this section we would like to have a closer look on these characteristics by ongoing discussions and developments from a global perspective. The main focus is on cross-domain information usage and real-time embedding, which can be understood as "on-the-fly" data integration.

1.2.2.1 Data Integration

Whenever requirements for a smooth data integration are discussed, we need to understand the different classes of geospatial related data. A smooth data integration concerns persistent and consistent identifiers for a distinct correlation of diverse (geospatial) datasets in the data integration procedure. The class of the dataset defines an "identifier requirement" at the different levels of georeferencing. For example thematic data will relate to addresses, administrative units or regional raster. This relation is performed with identifiers. Therefore the moulding of the thematic dataset requires specific identifier information in order to be georeferenced and spatially processed.

The expert group of the UN Global Geoinformation Management Regional Committee Europe (UN-GGIM: Europe, http://un-ggim-europe.org/) discusses data combination with a quite technological mature view of Europe and creates recommendations for policies. In this discussion "data combination" is understood as the merging, processing and creation of datasets (UNGGIM 2018), which do not restrict sources, characteristics, qualities, extensions or specifications. From this wide-open point of view, data combination can be done from various sources. Even data aggregation of the same dataset with itself at different quality levels will deliver a new combined dataset. It incorporates lists, tables, geometric simple features and complex thematic objects and therefore goes beyond simple data table joining. For instance the geocoding process may transform any administrative, business or thematic dataset into geoinformation by integrating core geographies with their unambiguous geospatial references. Those references represent name or code conventions, like ISO abbreviations for country names, and need to be persistent during time and semantic (Scharl and Tochtermann 2009).

Because data sources for combination are not restricted, it makes sense to establish a core classification, which allows for a better side-effect management, requirement evaluation and quality estimation. Side-effects concern an "a priori" classification of side effects. Their direct and indirect impact and occurring relations describe a side-effect severity.

A data classification could differ between three different formings: core reference geographies, core thematic geographies and spatial related data (UNGGIM 2018).

Core reference geographies are used as common spatial reference that are used by different stakeholders, in various location strategies and use cases. For example geographical grids or statistical units are polygons with persistent identifiers that are used as georeferencing links.

Core thematic geographies embed also theme specific features. They consist of georeferencing geometries that include theme specific attributes. For example hierarchy levels, codes, geographical naming and so forth are included in administrative units to enhance existing statistical datasets or to elaborate production of new statistical content (e.g. land use and value indicators).

Spatially related data cover documents, tables and other features that can be related to space or core geographies in order to enrich a combined dataset.

This observed classification of UN-GGIM delivers a perspective with regards to content. For the map production we have to add the production processing perspective, which adds attributes for the automated cartographic modelling (Mackaness et al. 2011). These processing attributes are derived either from existing thematic attributes by formal classification, aggregation and similar or from semantic modelling that introduces its own dimension. For example the classification of street vectors may use rank attributes to support cartographic visualisation (Kelso and Patterson 2009).

1.2.2.2 Geoinformation Management

Data integration in a real-time- and cross-domain manner, as it is needed in service-oriented mapping, requires geoinformation management. Based on the UN point of view, different states of geoinformation management can be observed on a global, continental and regional level.

On a global level recommendations are formulated for common structures and standardisation. Common structures explain how data are accessed, postulate FAIR data (Wilkinson et al. 2016), define a tier approach for loosely coupled interaction (Papazoglou 2003) and establish cross domain frameworks for a better interoperability.

On a continental level, the global recommendations shall be considered. Inhomogeneous realisations with a similar direction of development can be observed on this level. In many cases the developments are triggered by legal or organisational frameworks. For example the environmental policy in Europe is driven by the spatial data infrastructure INSPIRE as legal framework (see Chap. 3 in this book).

On a regional level geoinformation management is driven by pragmatic implementations and follow tangible use cases. Therefore these specific realisations can often not be embedded in other (infra)structures. For instance very specific spatial data services that do not follow the service-oriented paradigm (Michlmayr et al. 2007) are useful in their framework but cannot be accessed from other domains.

A dominant solution for geoinformation management is the creation of collaborative frameworks. As example the statistical and geospatial domain develop a high level, generic framework that follows five principles in order to cover all phases of statistical production, from data collection to dissemination (see Chap. 9 in this book). The framework is brought to the regional level with more applicable subprojects of the European project GEOSTAT, where a comprehensive and systematic view at the statistical geospatial framework is implemented.

1.2.2.3 Data Integration for the Common Good

Data integration and modern geoinformation management is in use for various specific topics. Especially advising citizens and social actions are main topics of capturing, sharing, analysing and integrating spatial related data. In general analysing and using data for the development of new products and services of well understood customers are core to this world's fastest growing businesses (Baeck 2015).

There is the intention to provide insights on the needs of citizens, their behaviour and infrastructures usage. One advanced technique that gives insights from textual data, such as twitter, facebook or similar, is natural language processing (Collobert et al. 2011).

Following up the exploration of textual data, a visualisation through (spatial) dashboards brings impenetrable text data to life. It makes these integrated data searchable, understandable and spatially related for the first time and is a big source of inspiration for resulting applications or knowledge products.

Knowledge products may incorporate spatial dynamics of social actions and therefore form key components for supporting public, social and environmental innovation. This innovation counts on spatial communication and knowledge transmission as interdisciplinary approach that helps to understand what data needs to be communicated, how and with whom. It encompasses psychology, design, data science and software development (see Chap. 10 in this book).

The processing, searching, distribution and integration of data bring up additional challenges when it comes to modelling and using of voluminous data streams. This characteristic of a constant data flow prevents from copying and storing, but calls for powerful processing performance and algorithms. The spatial theme of meteorology shows up corresponding use cases where these challenges have to be resolved (see Chap. 13 in this book).

The given views on data integration and geoinformation management with their real-time and cross-domain character unambiguously confirm a new paradigm in this area. Especially the intention to gain new insights out of impenetrable information collections as well as continuous processing of voluminous data streams demonstrate the paradigm-move from copy, store and process to continuous sharing and processing.

1.2.3 Demand for Enhanced Data and Information Quality

The extensive processing for geospatial data integration demands enhanced data and information quality. Quality disturbances and errors at different levels can occur and can be grouped to obvious discrepancies, missing value integrations and hidden occurrences.

Obvious discrepancies cover data quality mismatches of the different sources. For example a scale mismatch will result in inconsistent structures. For example mixing a primary data model and cartographic information model does generally not fit. Therefore it will negatively influence visualisation and analysis (Jakobsson and Tsoulos 2007).

Missing value integrations embed missing or wrong values within data processing. For example data that were modified with disclosure mechanisms (see Chap. 18 in this book) are integrated in a dataset. Its "wrong" values cannot be identified anymore and will cause faulty results in subsequent data processing activities.

Hidden occurrences may concern errors in the semantic-, temporal- or geometric dimension of the used data and cannot be directly observed. Discrepancies may occur in subsequent processing steps of combination and analysis.

Metadata in spatial data infrastructures host attributes for "scale of use" and "lineage" of a described dataset. But in general it does not offer or maintain an extensive assessment of selected data sources for automated production. This geoinformation quality assessment for a map production line will have to consider allowed data combinations and deal with the question if there is an acceptable flexibility in handling inhomogeneous spatial data quality. At this starting point we will need to discuss the mechanism from lineage description to usage prohibition and if this decision should be enabled for a system automatically or for non-experts.

The role of "non-experts", in terms of being not an expert in geoinformation processing and cartography, is of increasing importance. User participation enhances quality due to local knowledge and at least indicates correction activities of the crowd. Even within authoritative data the crowd might increase actuality and correctness of geoinformation if appropriate feedback mechanisms exist in the service-oriented mapping framework (Mooney et al. 2010; Barron et al. 2014).

The demands for enhanced data and information quality have to be amended by infrastructural quality assurance. Particularly because any functionality and data will be accessed by spatial data services, the quality of service needs to be defined. The most important quality of service parameters are interoperability, availability, reliability, performance, capacity and security (Baranski et al. 2011). These parameters ensure that a service could be found, understood and used as described in these parameters and therefore build the framework of a service-oriented infrastructure for map production. Any definition, register or code list are accessed by services and embedded in applications and workflows which presume the existence of these contents.

The urgent request for quality documentation for service as well as for data is another indicator for the change of paradigm in map production and geoinformation management. Although quality documentation existed before, it is now the main enabler of automated processing and thus a service-oriented production pipeline.

1.2.4 User Participation in Map Recording and Design

User participation itself is not only a quality enhancement, but also a main requirements developer in spatial data recording and map design.

The most prominent example of crowd-based, spatial data recording is Open Street Map (Haklay and Weber 2008), a spatial database populated and maintained by volunteering participants. No in-depth knowledge is needed to fill the database. Just simple, well documented, rules to classify new information and little location knowledge are needed to participate this global initiative.

OSM is an active participator in spatial data recording. Using e.g. smartphones to collect location behaviour of their users is a passive method to create spatial information. This is done by using humans, especially their mobile phones' embedded sensors, as sensors (Goodchild 2007).

New technologies expand cartography and art for map design issues. Users are able to influence map design as needed via cartographic stylesheet definitions. These style definitions are directly embedded with the shown vectors of a map. This flexible technology, which is generally called vector tiles, can be observed/tried with applications like mapbox and observed with design creations of the studio Stamen (Murray 2014). In the end this open technology leads to an unclear border between cartography and art. Both are influenced by inputs of a wide public when creating new graphics and communication issues (Brodersen 2007).

The technical possibility for a wide public/crowd to directly participate data recording as well as map production via service based architectures is another characteristic indicator for the change of map production paradigm. It highlights the importance of public participation and -integration.

1.2.5 Geospatial Datacubes and Their Perspectives

The creation of data networks by using data integration, minimum quality levels and geographical crowd participation aims at knowledge frameworks that even allow for predictions. For example weather forecast models using meteorological networks (see Chap. 13 in this book) are already precise enough to predict the weather for the next few days. For other domains, knowledge networks are being built up. The main requirement for predictions is comparability. Comparability is achieved with identical geographical references through time such as a persistent raster. A persistent raster during time and other dynamical dimensions can define a datacube that allows for direct comparison, analysis and prediction (see Chap. 14 in this book). Generally a datacube represents a multi-dimensional object that can be sliced, diced, aggregated and analysed by spatial data services and therefore unfold its value in comparison to classical archives with masses of inhomogeneous structured files, but which are also described by metadata (see Chap. 14 in this book).

The geometry of the defined raster is the connecting element for data comparison. In order to enable comparison at different scales, the raster cell is a multiple of the other scales and its identifier can be deviated automatically (Steinnocher et al. 2006). These rasters are consistent throughout different scales and their defined region.

The perspective of large-scale time series analysis with distributed datacubes is an actual scientific working field. New storage and processing paradigms are required in order to make queries equally fast along all dimensions of the datacube. Additionally the distributed query processing relies on acceptable quality of services and network connections (see Chaps. 14 and 15 in this book).

For the field of geospatial datacubes we can deviate that new storage and processing paradigms are needed for enhanced querying and processing algorithms. Thus a foundation for the change of the map production paradigm is given, because extracting and analysing datacubes delivers information for geo-communication.

1.2.6 The Data Revolution and Its Consequences

Modern map production builds on data processing, information preparation and aims at knowledge creation. According to Weinberger (2011) the various kinds of data are united by a knowledge pyramid: data precedes information, which precedes knowledge, which precedes understanding and wisdom (Weinberger 2011). Therefore the bedrock of map production is involved in the data revolution and its consequences (Fig. 1.2).

The data revolution is founded by the development of information and communication technologies (ICTs), in which enhancements and challenges of data production, -management, -analysis or -archiving have been faced. Data are not limited in their availability or accessibility any more, in contrast a torrent of timely, diverse, exact and related data are increasingly open and wait for their processing. The dimension of exploding data presence is illustrated by Hal Varian (Smolan and Erwitt 2012): "*between the dawn of civilisation and 2003, we only created five exabytes of information, now we are creating that amount every two days*".

The increasing availability of spatial data does have some impact on societal knowledge building (Wiig 2007), on imageries of facts and on citizen's behaviour. Societal knowledge is built by linking-up data and registers. For example the link between population register and the register of health facilities allows for analysis of the regional coverage, which may be important especially in rural regions. Another example of register integration and data mining in the domain of real estate and land administration is given in Chap. 11 (see Chap. 11 in this book).

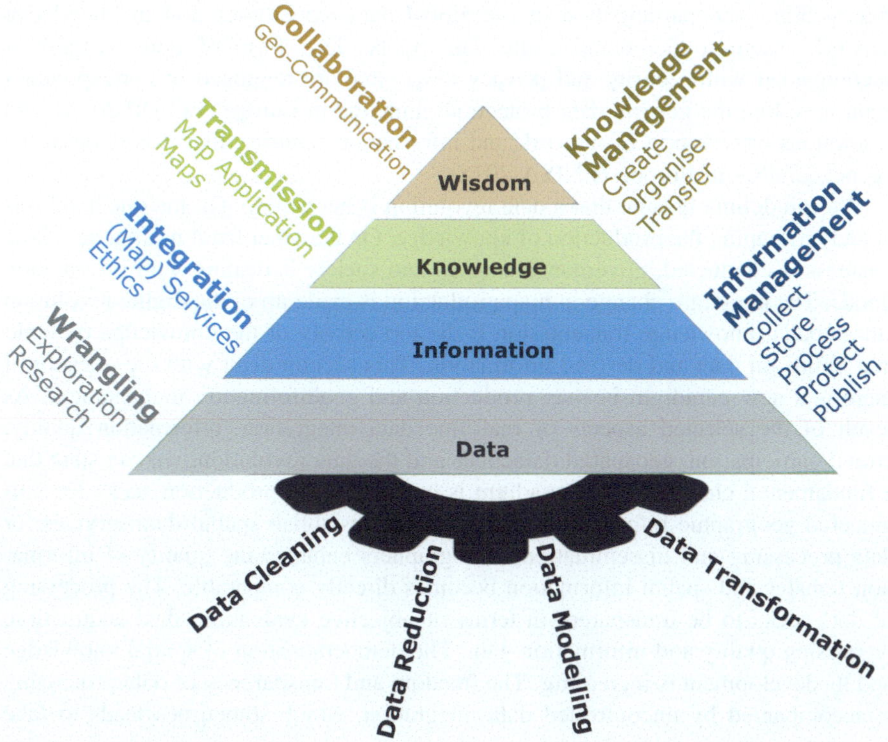

Fig. 1.2 Knowledge Pyramid adopted from Weinberger (2011). It shows the different levels of knowledge creation, which makes use of map services, maps and geo-communication

Imageries of facts are influenced by cartographic information generalisation, which aggregates, highlights or selects content. This kind of filtering by an editor now moves to the individual user (Dodge et al. 2011). Social media and open data streams can be accessed without restriction and possibly require some thematic knowledge before creating one's thematic imagination. For example combining weather and climate phenomenons to generate a temperature thematic image without a proper knowledge with regards to temperature, weather and climate may lead to the unintended conclusion that climate change does not exist, but in fact this conclusion is false and mixes definitions.

Citizen's behaviour can be influenced by delivering specific data integration results. For example showing traffic jams on digital maps will influence road users to take an alternative route.

Generally policy makers are convinced that data revolution is for the common good and therefore support open data policy. A positive influence for the societal development is expected by sharing and reuse as well as a single digital market, which are now the main strategies of the European Commission (EC-Strat 2018).

Accessibility and participation in the global data repositories and in knowledge networks seem to democratize the data pools. This form of data freedom is accompanied with security and privacy issues that are regulated in corresponding legal acts, like the general data protection guideline in Europe (EC-DP 2018), and continuous observation of network and information security, like it is done in the agency ENISA in Europe (ENISA 2018).

We can definitely state that a data revolution is happening. On the one hand it is already reshaping the production of knowledge. On the other hand it influences how business is conducted, governance enacted and society is dealing with information floods. The paradigm change in map production is made up of the digital revolution since spatial knowledge transmission is the top activity of the knowledge pyramid that builds on data and derived information. This section dealt with the question if there is a new paradigm in map production and geoinformation management. As result of the selected aspects of real-time data integration, information quality, crowd participation, geospatial datacubes and the data revolution, we can state that a fundamental change of the paradigm is ongoing. Map production seeks for volunteered geographic information, data engineers establish spatial data services for data processing and dissemination, cartographers enhance the quality of information transfer and spatial information becomes directly comparable. The processing of data seeks to be transparent in terms of objective evaluation, data acquisition, processing quality and information gain. The democratisation of spatial knowledge and its development is increasing. The freedom and transparency of data processing is accompanied by uncontrolled data integration, which sometimes leads to fake information or misinterpretations.

When we take the new map production paradigm for granted, we have to investigate the importance of the new map production paradigm and explore its most important requirements.

1.3 Relevance of the Map Production and Geoinformation Management Paradigm

In this section the importance of the new paradigm shall be highlighted. Some of the aspects have been used as arguments for the existence of the new paradigm in map production, but others express the importance and perspectives that are supported by this new way of fragmented collaboration.

The relevance of the map production and geoinformation management paradigm is predominantly expressed by

- a change of processing chains and its related supply chain management,
- extensive widening of the scope in geoinformation management, which has to cover legal, organisational and technological aspects in addition to the management of information processing,

- the creation of the geospatial semantic web, which established a knowledge network across domains and
- the challenge to support sustainable development goals with geospatial knowledge networks.

1.3.1 *Observing Monolithic Versus Networked Process Chains*

The paradigm of decentralized-, service-oriented map production in terms of “sharing and reuse” does not only concern the frontend with the resulting map or webmap application. It can also be adopted for the backend and production lines for creating the base material for a map. The production line spans from data pre-processing, -wrangling, -normalisation, -aggregation and -reduction to data integration and communication/transmission preparation following perceptive parameters of transmission media (Kandel et al. 2011).

Usually a map production line follows a monolithic procedure, which spans from selecting of data sources, copying, harmonisation, information modelling and map compilation. Most of the steps are in sequential order (Yoshizawa and Sakurai 1995). The time requested to complete the whole procedure is the sum of the single production steps.

Otherwise, a service-oriented decentralization allows for parallelisation of single production steps and networked process chains. This architecture may reduce production time. It will enhance flexibility of the production chain and autonomy of single steps, which helps to maintain a complex production environment. For example when a production step needs to be renewed, it does not impact on the overall procedure. The result of the missing step could be split to other units or it is merged into the product after its renewal. For example the processing of an orthophoto product, which is split to different regional processing steps, will be merged for a homogeneous dissemination in the end (Dean and Ghemawat 2008).

The hard- and software architecture as well as the process chain needs to be adopted for the decentralized architecture accordingly. Beside loosely coupled system components, smart caching mechanisms support a continuous flow of production. An error handling for each single component and production step establishes risk analysis, which delivers the basis for decisions if a specific result will be omitted for a given product version. Mainly, the risk analysis defines the production-cycle time spans, in which all the results should be accomplished. From the overall production point of view, which will take all sources to the final product, the dependencies and deliveries of the single steps should be managed as supply chain. This will help to control the production system by identifying caching failures and varying dependency times. It helps to estimate the overall production time and manages the production system.

Some key dimensions of supply chain management have to be elaborated before this methodology can be used. It reaches from technical process management (Benner and Tushman 2003) to the conceptual and organisational structures of the supply chain SCOR model (see Chap. 19 in this book).

The new paradigm has enormous impact on existing map production procedures, especially when the time span to market needs to be reduced, resources in the production management require improvement or a balanced governance of IT architecture shall be established.

1.3.2 Potential for Data Integration, Supported Information Compilation and Power of Naive Communication

The new paradigm is an enormous challenge for modern geoinformation management. As we highlighted data integration in Sect. 1.2, on the one hand the flexibility and access to various data-sources with different qualities leads to integrated products with an apparent objectiveness. Without knowing the data sources and their lineage, a resulting product may pretend a higher quality than effectively exists. Geoinformation management needs to consider this characteristic of decentralized data sources.

On the other hand, service-oriented access to geoinformation supports different modern geomedia techniques. Some of them enable naive information transmission (Baik et al. 2004) with high perceptive impact factors by using actual data, data streams and real-time adoption for semi-realistic renderings (Jasso and Triesch 2005). In order to support naive information transmission and its requirement of high rendering cycles, geoinformation management has to provide planning, implementation and controlling methodologies for an appropriate usage of distributed spatial data services.

A first step to confront geoinformation management with the requirements of service-oriented map production is the creation of seamless services and data flows. In Europe the research and development programme ISA2 (EC-ISA 2018) focuses on these interoperability solutions for public administrations, businesses and citizens. The main aim is to promote the benefit from interoperable cross-border and cross-sector public services.

The digital government of a near future makes use of geospatial and location intelligence. Barriers for citizens and authorities are reduced by access to and automated use of knowledge networks. The European Location Interoperability Solutions for e-Government (ELISE) Action is a compilation of policy, organizational, semantic and technical interoperability solutions to facilitate more efficient and effective digital cross-border interaction as well as data reuse (ELISE 2018).

In a near future, geoinformation management in the service-oriented environment will have to serve all aspects of spatial data processing, location intelligence and the transmission into knowledge networks.

1.3.3 A Geospatial Semantic Web

Knowledge networks have been developed in the domain of artificial intelligence, where the description of knowledge in a machine processable way is investigated. These efforts of extending the Web with semantic processing shall also extend spatial data infrastructures in order to find, extract and link spatial content more efficiently. The W3C defines the semantic web as "*The Semantic Web provides a common framework that allows data to be shared and reused across application, enterprise, and community boundaries*" (W3C 2018). Most parts of this definition are identical to the main intention of spatial data infrastructures for service-oriented mapping, namely sharing and reusing spatial data across application, community and business restrictions. If we agree with this equality then the research questions will be similar. The most important research questions for the geospatial semantic web are (Janowicz and Hitzler 2015):

- What kinds of geospatial classes should be distinguished?
- How to refer to geospatial phenomena?
- How to perform geo-reasoning and querying over the semantic web?
- How to discover events and how to account for geographic change?
- How to handle places and moving object trajectories?
- How to compare, align, and translate geospatial classes?
- How to process, publish and retrieve geodata?

With the geospatial semantic web a common language for representing spatial data can be understood. This language should be used by software agents, ontologies (sets of statements and relations) and frameworks in a way that disparate data sources can be translated into common terms, rules and knowledge concepts. Any object, such as text, images, video, maps and communications, described in this language can be understood, linked and used for reasoning across the World Wide Web (WWW). Therefore the semantic web is a global framework that uses service-oriented technologies and covers various digital objects, also spatial data and artefacts (Feigenbaum et al. 2007).

The main important issue in the semantic web is related to more persistent ways of storing and cataloguing digital content (Cathro 2009; Koch and Weibel 2006). Initiatives from the digital library community have studied the persistent linking, dissemination and development of knowledge structures on the top of the current WWW infrastructure. The represented content is processable by both humans and machines (Shadbolt et al. 2006).

The persistent storage and identifiers as well as the common language of the semantic web enable a clear and precise understanding of modelled objects. This clear understanding—or better: clearly defined semantic for objects—is a critical aspect beyond data harmonisation that supports correct relationships across domains. One example in the field of geology and the challenges of global- versus regional semantics is highlighted in Chap. 17 (see Chap. 17 in this book).

1.3.4 Supporting the Global Sustainable Development

The geospatial semantic web as common knowledge framework and especially the new paradigm in map production as representation framework are needed to support a global sustainable development in all its dimensions: data management, information derivation and knowledge communication.

In 2015 most of the United Nations countries adopted a set of goals to end poverty, protect the planet and to ensure prosperity for all in a newly defined sustainable development agenda (SDG 2018). Each of the defined goals has specific targets which have to be achieved over the next 15 years. The targets are monitored by indicators, which are computed from different (spatial) data sources. The need for its global comparability requires a standardised definition of these indicators and their calculation. In addition a common understanding, as in the geospatial semantic web, for all indicators and their sources is needed. The finding, accessibility, interoperability and reusability (FAIR) of data (Wilkinson et al. 2016) is an indispensable precondition for the global creation of the SDG indicators and their temporal datacubes.

From the map production point of view the information and communication models can count on comparable SDG indicators and directly use them in mapping. This mapping intention is a very pragmatic activity within the statistical and geospatial community, which has been established the conceptual structure of a statistical geospatial framework (SGF) (see Chap. 9 in this book). The basic recommendations of the SGF should be implemented in order to use this framework on a global level. From the time the SGF is productive and usable for map production, the collected indicators should be stable in time and therefore provide a constant basis for map production. Then the new challenges for map production will be more than ever before the communication of spatial comparison, -relation and -development (Olson and Bialystok 2014).

The service-oriented paradigm for map production and geoinformation management is already in use for actual research and development. The superior aim to establish a common language for digital objects with a (geospatial) semantic web already makes use of distributed resources and expects persistent identifiers and storage locations. The basic organisational and technological methodologies of spatial data infrastructures and their service-oriented approach are supporting requirements for the semantic web and its feasibility.

The fragmented character of service-oriented map production urgently needs supply chain methodologies incorporated with process-, requirement- and change management. These dimensions are part of geoinformation management and will support a precise documentation of processes, a comprehensive assessment of direct and indirect requirements or a balanced evaluation of dependencies in the change process.

1.4 The Main Important and Obvious Requirements Within the New Paradigm

A changed way of thinking for map production and geoinformation management is on its way. Selected examples in this contribution highlighted, what kind of frameworks are needed in a near future. In addition the big picture of the semantic web and supporting the SDG delivers a vision and impulsive motivation for the implementation of the requested frameworks. The geospatial domain is an important part in these frameworks because most topics are referenced to and structured with space.

The new geoinformation management as well as map production's paradigm of "decentralized geosource networks" comes along with specific requirements (Nebert 2004). Some of them have been considered for implementation on one hand or identified for further research on the other hand. These requirements are relevant at various organisational levels—global, regional or local ones.

One of the best known efforts for a decentralized spatial information network for the environmental domain is INSPIRE (see Chap. 3 in this book). It uses a top-down approach and pushes regulations to its public stakeholders in order to fulfil the most important requirements. Those requirements concern interoperability, basic network services, metadata, data specifications and registries (EC 2018b). The existence of those components in combination with the legal acts does lastly not guarantee success for the INSPIRE framework. Success for this framework means that broad use as well as sustainable availability can be established. Success could be assured if the INSPIRE framework takes the steps towards a main access point for environmental map production. The community has observed missing priorities and focuses on "INSPIRE in practice" (EC 2018c) for the second half of its implementation.

Other requirements have not been identified yet. Their absence may prevent from the success of a spatial framework, like INSPIRE. For this reason the most important requirements for a service-oriented map production architecture shall be accentuated.

1.4.1 Legal and Organisational Challenges

Legal, organisational and even technological challenges are driven by a common consensus. Whereas technical aspects are covered in standardisation initiatives, legal and organisational aspects are driven by local policies.

Technical standardisation for service interfaces are in development at the International Standardisation Organisation (see Chap. 16 in this book) or the Open Geospatial Consortium (see Chap. 15 in this book). Working groups take identified challenges and try to develop solutions on a consensual basis. These solutions should be functional and a documentation is delivered with the standard.

For example the development of datacube services result in a standard at OGC (see Chap. 14 in this book). As result a standard way of solving a problem is for most cases considered in software products, software clients or server applications. Derivations of a given standard may help to overcome specific local challenges or use cases. In the best case these use cases will be brought back to a global standardisation procedure and come up with a revision of the standard if it is important enough.

The technical standardisation of data structures, such as metadata or thematic data in a specific encoding (XML, GML, GeoJSON, RDF, …), is also a consensus on a global level. Especially thematic data may bring in local knowledge and therefore request specific data structures. The minimum global consent will support interoperability, but may be not enough to describe a phenomenon. Local extensions are valuable to expand the global consent and to describe a given phenomenon, but they have to be matched against a global—possibly aggregated—vocabulary in order to link to a global knowledge. The development of geological vocabularies as discussed in Chap. 17 is one thematic example thereof (see Chap. 17 in this book).

Policy making is driven by society and national systems. Even local administrations in federated nations will incorporate their needs into the legislation and organisational implementation. For example if the responsibility for transport network is at the federal state level, not all federal states may provide the same access, licensing and usage conditions. A nationwide product will be suppressed. At least national agreements will be needed for a homogeneous transport network product. This well understood but complex situations underpin the importance of common agreements for e-government and global policy recommendations, like the ones the UN-GGIM releases (UN 2018).

A common understanding of an agreement within spatial frameworks cannot be implemented by global/regional recommendations itself. There is a strong requirement for harmonisation and management boards at the local levels in order to build a common understanding. These boards are the best places of departure for interrelated steering committees in continuous change processes of a technical framework. All upcoming ambiguities can be discussed and solved in a collaborative manner.

We can observe similar structures in the core development of open source software components, where a steering committee, made up of the most active developing members, decides on the implementation of new code if its importance is high enough for the community. If there is no community interest, the software product is developed upon request of the developers or single customers.

1.4.2 Privacy Issues and Disclosure Mechanisms

Data integration in a decentralized production network uses data from autonomous and independent sources. In these sources privacy issues have been considered and disclosure mechanisms were adopted. This does not prevent from disaggregation in the process of data combination. Privacy issues may occur again. Humans have a right for privacy. Therefore data integration, especially with open data licenses, requires to include the ethical dimension and to add existing disclosure mechanisms from the statistical domain in the map production workflows.

Currently the statistical domain includes disclosure in their data processing and map production (see Chap. 18 in this book). All statistical products will include ethical considerations. This awareness does not exist for all domains that make use of map production. With an increasing usage of decentralized map production networks, disclosure and privacy become an important factor in other domains as well. An indicator for the danger of privacy violation in data integration is required to overcome restrictions in data access and -quality. The indicator could regulate if a dataset has to be distorted for a planned data combination in order to preserve privacy.

1.4.3 The Role of Administrative Geoinformation

Volunteered geographic information and open data often lead to the question: What is the role of administrative data? Decentralised registered data sources from often undocumented data providers make it difficult to evaluate data quality in terms of actuality, precision or correctness. Although it is very important to have this extensive data offering, the meaning of a normative role of administrative data is rising. Administrative geoinformation that is produced under constant conditions with persistent high qualities offers comparison for assessment procedures. Examples show that reliability of Open Street Map can be assessed by comparison with authoritative geoinformation (Mooney et al. 2010).

For this reason the requirement for administrative geoinformation to define itself as a normative parameter and quality control exists especially for service-oriented map production.

1.4.4 Side Effects Induced by Data Integration

From the elaborations in the previous sections, we can assume that side effects may be mainly induced by data combination. But side-effects do not only occur in map production, but also for the dissemination and stage of usage (UNGGIM-Europe 2018).

Side-effects generally occur by accident. If a side-effect re-occurs and becomes used in map production, it is not a side-effect any more. Instead it has transformed to a production component. Repetition is a key aspect. Side-effects can result in beneficial outcomes that need to be reproduced and exploited. On the opposite also drawbacks can be retraced and avoided during future iterations. Benefits as well as drawbacks of the side-effects should be considered during production management. Very often the documentation of side-effects leads to change proposals for the production environment (Anderson and Anderson 2010).

A big challenge for the documentation of side-effects is their weighting. Not all observations can be identified as a side-effect, because their impact on availability, accessibility, interoperability and acceptance changes. A comprehensive requirement engineering (Kotonya and Sommerville 1998) would classify side-effects by their occurrence level, namely primary-, secondary- and third level occurrence.

A side-effect as a primary-level occurrence has direct impact on the technical or semantic dimension of the data combination. It can be directly assigned to the used data and services and thus in the resulting product.

A side-effect as a secondary-level occurrence concerns organisational or legal dimensions that are not directly related to the resulting product of the data combination. For instance legal actions, stewardship programs and other organisational activities may be derived from observing secondary-level side-effects.

Side-effects as a third-level occurrence are related to observations in societal areas. Ethical restrictions, definitions of privacy or educational maturity levels are based on observations of third-level occurrences. For example the minimum level of education that is required to work with interoperability frameworks is issued by observing educational gaps.

Positive as well as negative side-effects occur during data combination. The aim of geoinformation management is to maximise the positive effects and to minimize the negative ones. For example semantic interoperability is supported by UML models that describe the understanding and structure of thesauri, codelists, registers, data structures and semantic descriptions. A positive effect of UML is that the validation of conformance can be automated since it is machine readable. A negative effect is that humans require knowledge to read UML models in order to understand its content and data requirements. Thus additional activities may be needed to create a mature level for semantic understanding.

1.4.5 Archiving, Licensing Topics and Sustainable Access

A geospatial semantic web depends on persistent locations and identifiers. The reason is the sustainable access to spatial objects and data over time, which is needed when those objects are embedded in knowledge networks and linked by others. Missing objects will result in dead links. For this reason archiving aspects for digital geoinformation need to be standardised. This standardisation affects best practice procedures of geoinformation archiving, archiving formats for spatial information, application requirements, long-term service interface specification, storage methodologies and many more (Jobst 2010).

Once all technical requirements for archiving have been solved, the rating for important content and requirements for use have to be defined. This geospatial legacy management calls for a very specific curatorship as decision point for useful geoinformation, for its access regulation and appropriate license definition (see Chap. 20 in this book).

1.4.6 The Term Governance for Interrelated Production Infrastructures

Service-oriented map production frameworks are complex technical networks. Although the nodes are loosely coupled by standardized interfaces, a high dependency exists for content deliveries within the map production process. Supply chain estimations quantify these dependencies for content delivery and will indicate the needs for improvement.

Technological advance for information and communication technologies is extremely speedy. Once changes are implemented on some production framework components, other changes will be needed for others. Inter-component dependencies create an inscrutable situation. On this account the maintenance of infrastructure components call for governance and well balanced planning. One promising tool to support governance in this networked environment are balanced scorecards, which summarize indicators in four areas: position and quality, actual processes, potential of change and importance of a component. "Position and quality" concentrates indicators for the covered area of a product, quality constraints or market impact. "Actual processes" merges indicators for dependencies, resources or complexity. "Potential of change" unites indicators for organisational- and financial effort, feasibility or additional demands. "Importance of a component" combines indicators for usage, external market situation, relevance or external image of a product. According to the strategic vision of governance indicators of all four areas are weighted and result in a scorecard, which ranks future changes accordingly (Fig. 1.3).

1.4.7 Cybersecurity Frameworks and Information Security

Sustainable knowledge frameworks on the basis of service-oriented architectures make use of loosely coupled interfaces and distributed information sources. This open network architecture could be "attacked" in order to eliminate connections (to services, metadata, data, registries), to extract privacy data or to introduce misinformation. The more our knowledge relies on information technology networks, all the more data-, network- and information security guidelines are requested. Security guidelines recommend specific architectures, cybersecurity frameworks as well as prevention activities (ENISA 2018). A service-oriented map production framework needs to trust its sources of information, which are the ingredients of the product.

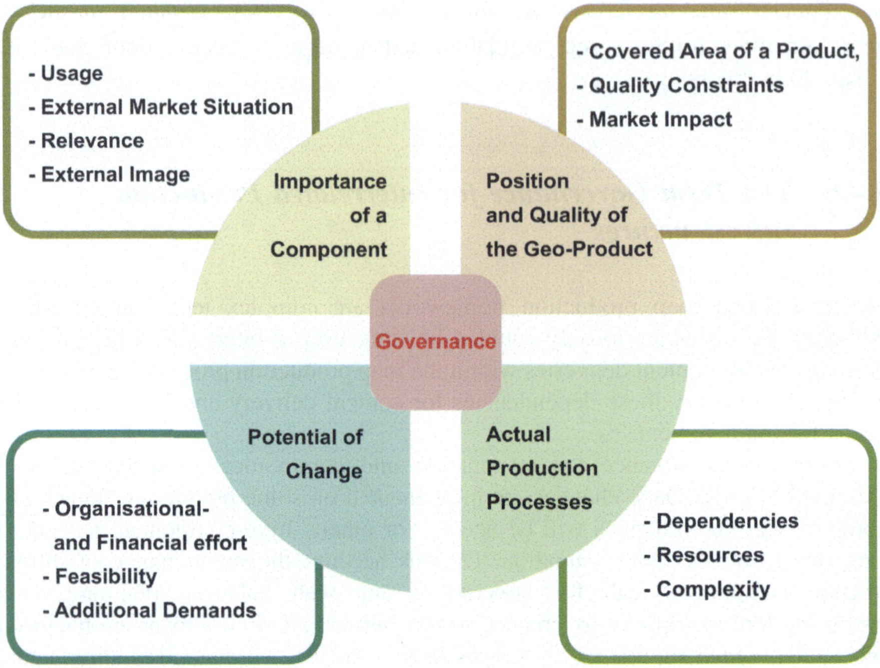

Fig. 1.3 Service-oriented map product governance with the help of a balanced scorecard. The assessment indicators are grouped into four main areas of the production environment

1.4.8 Geospatial Maturity, Education and Their Measurement

A service-oriented map production framework is a complex system of different technologies, methodologies, components, thematic data and the one central paradigm of decentralisation as well as sharing and reuse. Generally a user does not know about registries, UML or requests-to-service endpoints. The client application will have embedded most of the infrastructure functionality. As a developer or purchaser for specific additional components more understanding of the overall framework is useful. If parts of a service-oriented map production framework are operated, the main standards, interfaces, technologies and their actual developments are required.

The disseminating of spatial data via spatial data infrastructures (SDI) requires specific knowledge. Also making use of spatial data in SDI without relying on predefined clients will require mature knowledge for finding, binding and exploiting the data source.

Operating a service-oriented infrastructure in a long-term perspective causes continuous effort for adopting technology and education. Particular the decentralized characteristic with many stakeholders will not allow to stick the software on an old version. Continuous updates are the result. Therefore continuous learning and expanding knowledge is part and parcel of the system and financial input.

It makes sense to prove geospatial- and service-oriented maturity. A comparable measurement discovers educational gaps which will be useful for the support (Babinski 2009). For example using GIS tools for geospatial analysis is well established. But how does a user know if the knowledge is sufficient for answering geospatial questions with GIS? Which would be the right courses to fill educational gaps? The maturity assessment model of the Geoinformation Management Institute (GMI) is a step in this direction and helps users to orientate themselves in geoinformation management (GMI 2013).

This model needs to be expanded for spatial data infrastructures and geospatial knowledge frameworks.

1.5 Concluding and Arguing the Usefulness of Extending the Way of Thinking in Modern Map Production

There are changes in the way we have to deal with modern map production and geoinformation management. New possibilities with regards to data access, -integration, real-time analysis or data mining extend traditional production lines and create challenges for a useful geoinformation management. The basic map production paradigm moves from "collecting—assembling—storing" to "reusing—assembling—sharing". This movement is driven by a continuous flow of data (big data) that cannot be copied and stored easily. But the data torrent can be accessed, structured and distributed by services. Therefore a network of service-oriented architectures increasingly performs as a foundation for modern map production.

We have shown by several examples that this change or paradigm exists. Additionally we could demonstrate by several application fields that the new paradigm is already in use. Map production and the whole geospatial domain get roped into these procedures of decentralized data mining and semantic knowledge networks. Maps are part of a geospatial semantic web, which delivers global access and the foundation of a cross-domain digital knowledge.

Various thematic domains are touched by modern map production and geoinformation management. The role of spatial information for knowledge structuring, -referencing, -combining and -communication is central and requires a mature understanding of geoinformation and cartography. The following points summarize the discussed change of paradigm from their perspective. This may help to comprehend the dimension of modern map production, geoinformation management, its requirements and increasing importance.

1.5.1 From the Geovisualisation Point of View

From a geovisualisation point of view service-oriented geoinformation frameworks provide a very selectable access to distributed information. Assuming that detailed metadata and a common semantic language exists, proper sources may be found for complex geovisualisation activities.

Geospatial knowledge networks serve as a foundation for dynamic geovisualisation. Their persistent objects and relations enable time-based and comparing visualisations. Even the outcome of prediction models and calculated future states can be embedded in these opinion-forming map applications.

In certain cases the discussion of flexibility versus consistency arises. Especially when fast-paced data refreshing cycles do not match with other data themes (and sources), discrepancies in the visualisation may occur. Then a decision is needed if the cycle times have to be decelerated for the geovisualisation in order to establish consistency in the data integration. In general the cycle times for map production will be determined by the cycle times of the main components.

1.5.2 From the Data Quality Viewpoint

From a data quality viewpoint any referencing across domains and data integration will bring out quality constraints. Hence any combination of data makes data providers aware of their product and service quality. At least the consumer will claim for better quality if discrepancies occur. For example if the building footprint in the ortho-imagery does not fit with the cadaster, clients will flag the product as of inferior quality.

Once a valuable quality has been achieved, consumers get used to it and will not accept any degradation. At this point stewardship for the published data is initiated. The responsible party is the data provider, who needs to construct a sustainable character for the products.

1.5.3 From the Geo-Communication Point of View

Geo-communication dilates geovisualisation in geospatial knowledge frameworks. Whenever a common language and mutual understanding (across disciplines) exists, the communication with maps could be personalised. This means that individual relevant information enhances involvement and geospatial imagery at the recipient. Personalised information could be extracted by machine learning according to preference groups out of the geospatial knowledge framework, which is a network of semantically enriched spatial data infrastructures.

A numerousness of data sources integrated in an appropriate network could establish geospatial swarm intelligence. Then it is not individual information which is of interest, but an aggregated group of individual information which has been analysed for a specific topic, which could be of individual interest. For example the composite position of mobile phones in use in a traffic jam allows individuals to find an alternative route.

1.5.4 From a Sustainable Point of View

Service-oriented architectures do not innately provide time-persistent information. Instead the most actual data status will be published. Nowadays, more and more historical data become available via spatial data services, but their accessibility varies: some publish historical data as spatial data services, others embed historical data in the spatial data service of the actual ones. This inhomogeneous procedure makes it impossible to have a common access procedure for historical data. Furthermore, historical data of different data sources may refer to different points in time and consequently prevent from data combination.

From another viewpoint a periodic service-oriented map production could freeze one frame in time and make the resulting spatial data service archivable. This means that the storage of the map product is cached in an archivable way. Raster- as well as vector tile -based technologies support this concept. An identifier system for the archived map product, the frozen time frame, still needs to be defined. Archetypes exist for webpages as the Archive (Archive 2018) shows us.

1.5.5 From the Process and Collaboration Point of View

Service-oriented map production frameworks provoke collaboration and corporate action. The development of standards for service interfaces, storage mechanisms, spatial encodings or data structures is based on collaborative maintenance. The maintenance activities consider local or regional requirements and establish at least a common consensus that is published as standard.

Other dimensions of collaboration in service-oriented frameworks are driven by process-, change- and supply chain management. Collaboration in planning, controlling, development and adoption rests upon a common understood documentation, accepted indicators and mutual standard agreements for future changes. Adopted balanced scorecards help to receive an objective decision.

We have seen that a lot of positive arguments exist for service-oriented map production frameworks. At the same time visions for (geospatial) semantic knowledge frameworks (the Semantic Web) exist in which service-oriented mapping could play an essential role.

On the opposite also resentments exist in terms of security, privacy and false data. But also in these aspects some solutions can already be observed. For example could block-chain technology provide a method for secure transactions? Or could service-oriented mechanisms for disclosure control automatically incorporate the ethical dimension from statistics in order to establish privacy control?

The big picture of the geospatial semantic web as georeferenced knowledge network of humanity still delivers a lot of challenges. It is on us and our flexibility in expanding our minds to take the next steps of geospatial democratisation and thereby reach the sustainable development goals of the UN and beyond.

References

Anderson D, Anderson LA (2010) Beyond change management: how to achieve breakthrough results through conscious change leadership. Wiley, New Jersey

Andrews KM, Delahaye BL (2000) Influences on knowledge processes in organizational learning: the psychosocial filter. J Manage Stud 37(6):797–810

ArcGIS (2018) ArcGISonline in ESRI. https://doc.arcgis.com/de/arcgis-online/reference/what-is-agol.htm, visited Jan 2018

Archive (2018) Archive of the web. https://www.archive.org/, visited Jan 2018

Babinski G (2009) A proposed municipal GIS capability maturity model. In: Proceedings of Washington GIS conference, Center GKCG

Baeck P (2015) Data for good: how big and open data can be used for the common good. NESTA, Londres

Baik S, Bala J, Hadjarian A, Pachowicz P (2004) A naive geography analyst system with cognitive support of imagery exploitation. In: Mexican international conference on artificial intelligence, pp 40–48

Baranski B, Foerster T, Schдffer B, Lange K (2011) Matching INSPIRE quality of service requirements with hybrid clouds. Trans GIS 15(s1):125–142

Barron C, Neis P, Zipf A (2014) A comprehensive framework for intrinsic OpenStreetMap quality analysis. Trans GIS 18(6):877–895

Bell BS, Hoskins RE, Pickle LW, Wartenberg D (2006) Current practices in spatial analysis of cancer data: mapping health statistics to inform policymakers and the public. Int J Health Geographics 5(1):49

Benner MJ, Tushman ML (2003) Exploitation, exploration, and process management: the productivity dilemma revisited. Acad Manag Rev 28(2):238–256

Bieberstein N, Laird R, Jones K, Mitra T (2008) Executing SOA: a practical guide for the service-oriented architect. Addison-Wesley Professional, Boston

Brodersen L (2007) Geokommunikation. Livstegn, Haase, pp 91–93

Campos Bordons C (2013) Web map tile services tiny tile server. Ph.D. thesis, HSR Hochschule fur Technik Rapperswil

Caquard S, Cartwright W (2014) Narrative cartography: from mapping stories to the narrative of maps and mapping. Cartographic J 51(2):101–106

Cash DW, Clark WC, Alcock F, Dickson NM, Eckley N, Guston DH, Jдger J, Mitchell RB (2003) Knowledge systems for sustainable development. Proc Natl Acad Sci 100(14):8086–8091

Cathro W (2009) Metadata: an overview. National Library of Australia Staff Papers

Chan TO, Feeney M, Rajabifard A, Williamson I (2001) The dynamic nature of spatial data infrastructures: a method of descriptive classification. Geomatica 55(1):65–73

CloudStorage (2018) Cloud storage solutions. CloudNewsDaily. http://cloudnewsdaily.com/cloud-storage/, visited Jan 2018

Coffman T, Klinger MB (2015) Google glass: using wearable technologies to enhance teaching and learning. In: Society for information technology & teacher education international conference, pp 1777–1780

Cohen WW, Richman J (2002) Learning to match and cluster large high-dimensional data sets for data integration. In: Proceedings of the eighth ACM SIGKDD international conference on Knowledge discovery and data mining, pp 475–480

Collobert R, Weston J, Bottou L, Karlen M, Kavukcuoglu K, Kuksa P (2011) Natural language processing (almost) from scratch. J Mach Learn Res 12(Aug):2493–2537

Coltekin A, Heil B, Garlandini S, Fabrikant SI (2009) Evaluating the effectiveness of interactive map interface designs: a case study integrating usability metrics with eye-movement analysis. Cartography Geogr Inf Sci 36(1):5–17

Cronin JC, Hornbaker CV (2003) Network image view server using efficient client-server, tiling and caching architecture. Google Patents, US Patent 6,510,459. https://www.google.com/patents/US6510459

Dean J, Ghemawat S (2008) MapReduce: simplified data processing on large clusters. Commun ACM 51(1):107–113

Dodge M, Kitchin R, Perkins C (2011) Rethinking maps: new frontiers in cartographic theory. Routledge, UK

EC-DP, European Commission (2018) Data protection guideline of the European Commission. http://ec.europa.eu/justice/data-protection/reform/index_en.htm, visited Jan 2018

EC-ISA, European Commission (2018) Research and development programme ISA2. https://ec.europa.eu/isa2/, visited Jan 2018

EC-Strat, European Commission (2018) Main strategies of the European Commission. https://ec.europa.eu/info/strategy_en, visited Jan 2018

EEA (2018) Vague Geographic Regions by EEA. https://www.eea.europa.eu/data-and-maps/indicators/sea-level-rise-5/assessment/#generic-metadata, visited Jan 2018

Egenhofer MJ (2002) Toward the semantic geospatial web. In: Proceedings of the 10th ACM international symposium on advances in geographic information systems, pp 1–4

ELISE, European Commission (2018) European location interoperability solutions for e-Government. https://ec.europa.eu/isa2/actions/elise_en, visited Jan 2018

ENISA, European Commission (2018) Network and information security in Europe. https://www.enisa.europa.eu/, visited Jan 2018

European Commission (2018a) Geoportal for the spatial data infrastructure INSPIRE. http://inspire-geoportal.ec.europa.eu/, visited Jan 2018

European Commission (2018b) Regulations of INSPIRE. https://inspire.ec.europa.eu/inspire-legislation/26, visited Jan 2018

European Commission (2018c) INSPIRE in practice. https://inspire-reference.jrc.ec.europa.eu/, visited Jan 2018

EuroSTAT (2018a) Dedicated section on population (demography, migration and projections). http://ec.europa.eu/eurostat/web/population-demography-migration-projections/overview, visited Jan 2018

EuroSTAT (2018b) REF EuroSTAT: Atlas. http://ec.europa.eu/eurostat/statistical-atlas/gis/viewer/?year = &chapter = 02&mids = BKGCNT,C02M02&o = 1,1&ch = POP, C01&center = 48.01006,12.05668,5&lcis = C02M02&, visited Jan 2018

Evans E, Szpoton R (2015) Domain-driven design. Helion

Feigenbaum L, Herman I, Hongsermeier T, Neumann E, Stephens S (2007) The semantic web in action. Sci Am-Am Ed 297(6):64

Gabrielsen P, Bosch P (2003) Environmental indicators: typology and use in reporting. EEA, Copenhagen

Gapminder (2018) Gapminder statistical data. http://www.gapminder.org/, visited Jan 2018

Gartner G, Rehrl K (2008) Location based services and telecartography: from sensor fusion to context models. Springer Science & Business Media, Berlin

GEOSS (2018) Group of earth observations. http://www.geoportal.org/, visited Jan 2018

GMI, Geoinformation Management Institute (2013) GIS capability maturity model. http://citeseerx.ist.psu.edu/viewdoc/download?doi=10.1.1.687.1725&rep=rep1&type=pdf, visited Jan 2018
Goodchild MF (2007) Citizens as sensors: the world of volunteered geography. GeoJournal 69(4):211–221
Gundelsweiler F, Memmel T, Reiterer H (2004) Agile usability engineering. In: Mensch & computer, pp 33–42
Hacigumucs H, Iyer B, Li C, Mehrotra S (2002) Executing SQL over encrypted data in the database-service-provider model. In: Proceedings of the 2002 ACM SIGMOD international conference on management of data, pp 216–227
Haklay M, Weber P (2008) Openstreetmap: user-generated street maps. IEEE Pervasive Comput 7(4):12–18
Hashem IAT, Yaqoob I, Anuar NB, Mokhtar S, Gani A, Khan SU (2015) The rise of big data on cloud computing: review and open research issues. Inf Syst 47:98–115
Hendriks PH, Dessers E, Van Hootegem G (2012) Reconsidering the definition of a spatial data infrastructure. Int J Geogr Inf Sci 26(8):1479–1494
ICACI, Visual Analytics (2018) ICA commission on visual analytics. https://www.researchgate.net/topic/Geovisualization/, visited Jan 2018
Jakobsson A (2012) Introducing a new paradigm for provision of european reference geo-information-case study of the european location framework concept. Serv Oriented Mapp 2012:51
Jakobsson A, Tsoulos L (2007) The role of quality in spatial data infrastructures. In: Proceedings of the 23rd international cartographic conference, Moscow
Janowicz K, Hitzler P (2015) Geospatial semantic web. The international encyclopedia of geography: people, the earth, environment, and technology. SAGE Publications, Thousand Oaks
Jasso H, Triesch J (2005) A virtual reality platform for modeling cognitive development. In: Biomimetic neural learning for intelligent robots, pp 98–98
Jin D, Cordy JR (2005) Ontology-based software analysis and re-engineering tool integration: the OASIS service-sharing methodology. In: ICSM'05. Proceedings of the 21st IEEE International Conference on Software maintenance, 2005, pp 613–616
Jobst M (2006) Semiotic considerations for 3D cartography; workshop on geographical visualisation, ISPRS commission II, WG II/5, Vienna, 8 July 2006
Jobst M (2010) Preservation in digital cartography: archiving aspects. Springer Science & Business Media, Berlin. www.springer.com/de/book/9783642127328
Jobst M, Doellner J (2008) 3D city model visualization with cartography-oriented design, CORP 2008. https://hpi.de/fileadmin/hpi/FG_Doellner/publications/2008/JD08b/2008-03-15_corp08_jobst_preprint.pdf
Jobst M, Germanchis T (2007) The employment of 3D in cartography—an overview. In: Multimedia Cartography, Springer, Berlin, pp 217–228
Kandel S, Heer J, Plaisant C, Kennedy J, van Ham F, Riche NH, Weaver C, Lee B, Brodbeck D, Buono P (2011) Research directions in data wrangling: visualizations and transformations for usable and credible data. Inf Vis 10(4):271–288
Kaplan RS, Norton DP (1995) Putting the balanced scorecard to work. In: Performance measurement, management, and appraisal sourcebook, vol 66, p 17511
Kelso NV, Patterson T (2009) Natural earth vector. Cartographic Perspect 64:45–50
Kitchin R (2014) The data revolution: big data, open data, data infrastructures and their consequences. Sage Publications, Thousand Oaks
Klettner S, Huang H, Schmidt M, Gartner G (2013) Crowdsourcing affective responses to space. Kartographische Nachr 2(3):66–72
Koch T, Weibel S (2006) Introduction to a special issue on metadata: selected papers from the Dublin core 2001 conference. J Digital Inf 2(2)
Kotonya G, Sommerville I (1998) Requirements engineering: processes and techniques. Wiley, New Jersey

Lauriault TP, Pulsifer PL, Taylor DF (2010) The preservation and archiving of geospatial digital data: challenges and opportunities for cartographers. In: Preservation in digital cartography, pp 25–55

Lechthaler M, Stadler MA (2006) Cross-media publishing over scales. https://www.researchgate.net/publication/255583041_CROSS-MEDIA_PUBLISHING_OVER_SCALES

Lee SH, Malallah A, Datta-Gupta A, Higdon D (2000) Multiscale data integration using markov random fields. In: SPE annual technical conference and exhibition

Lucchi R, Millot M, Elfers C (2008) Resource oriented architecture and REST In: Assessment of impact and advantages on INSPIRE. European Communities, Ispra

Mackaness WA, Ruas A, Sarjakoski LT (2011) Generalisation of geographic information: cartographic modelling and applications. Elsevier, The Netherlands

Maguire DJ, Longley PA (2005) The emergence of geoportals and their role in spatial data infrastructures. Comput Environ Urban Syst 29(1):3–14

Michlmayr A, Rosenberg F, Platzer C, Treiber M, Dustdar S (2007) Towards recovering the broken SOA triangle: a software engineering perspective. In: 2nd international workshop on Service oriented software engineering: in conjunction with the 6th ESEC/FSE joint meeting, pp 22–28

Montello DR (2002) Cognitive map-design research in the twentieth century: theoretical and empirical approaches. Cartography Geogr Inf Sci 29(3):283–304

Mooney P, Corcoran P, Winstanley AC (2010) Towards quality metrics for OpenStreetMap. In: Proceedings of the 18th SIGSPATIAL international conference on advances in geographic information systems, pp 514–517

Murray S (2014) Changing minds to changing the world: mapping the spectrum of intent in data visualization and data arts. https://repository.usfca.edu/art-architecture/1/

Nebert DD (2004) SDI cookbook: infrastructures, developing spatial data. GSDI

Olson DR, Bialystok E (2014) Spatial cognition: the structure and development of mental representations of spatial relations. Psychology Press, UK

Papazoglou MP (2003) Service-oriented computing: concepts, characteristics and directions. In: Proceedings of the fourth international conference on web information systems engineering, 2003 (WISE 2003), pp 3–12

Park E, Kim KJ, Jin D, del Pobil AP (2012) Towards a successful mobile map service: an empirical examination of technology acceptance model. In: International conference on networked digital technologies, pp 420–428

Peterson M (2003) Maps and the internet. Elsevier Science, The Netherlands

Robinson S (2004) Simulation: the practice of model development and use. Wiley, Chichester

Scharl A, Tochtermann K (2009) The geospatial web: how geobrowsers, social software and the Web 2.0 are shaping the network society. Springer Science & Business Media, Berlin

SDG, UN (2018) Sustainable development agenda. http://www.un.org/sustainabledevelopment/sustainable-development-goals/, visited Jan 2018

Seni G, Elder JF (2010) Ensemble methods in data mining: improving accuracy through combining predictions. Synth Lect Data Min Knowl Discov 2(1):1–126

Sester M (2010) Geoinformatics. In: Handbook of geomathematics, Springer, Berlin, pp 1313–1329

Shadbolt N, Berners-Lee T, Hall W (2006) The semantic web revisited. IEEE Intell Syst 21(3): 96–101

Slocum TA, Blok C, Jiang B, Koussoulakou A, Montello DR, Fuhrmann S, Hedley NR (2001) Cognitive and usability issues in geovisualization. Cartography Geogr Inf Sci 28(1):61–75

Smolan R, Erwitt J (2012) The human face of big data. In: Against all odds production. O'Reilly, USA

Steinnocher K, Banko G, Kuistl M, Petrini-Monteferri F (2006) European spatial indicators–temporal development and quality aspects. In: Schrenk M (ed) Proceedings of CORP 2006, pp 201–208

Tan Y (2013) Swarm robotics: collective behavior inspired by nature. J Comput Sci Syst Biol 6: e106

Tidwell J (2010) Designing interfaces: patterns for effective interaction design. O'Reilly Media, Inc., USA
Troltzsch K, Van Brusselen J, Schuck A (2009) Spatial occurrence of major tree species groups in Europe derived from multiple data sources. For Ecol Manage 257(1):294–302
UN, United Nations (2018) Regional entities of UNGGIM. http://ggim.un.org/regional-entities/, visited Jan 2018
UNGGIM-Europe (2018) UNGGIM Europe WG B3: report on side-effects induced by data combination. http://un-ggim-europe.org/sites/default/files/20160909_ReportonManagingSideEffects_UNGGIM-Europe_WGB3_v1.0.pdf#overlay-context=content/wg-b-data-integration, visited Jan 2018
Vandenbroucke D, Crompvoets J, Vancauwenberghe G, Dessers E, Van Orshoven J (2009) A network perspective on spatial data infrastructures: application to the sub-national SDI of flanders (Belgium). Trans GIS 13(s1):105–122
Verburg PH, Eickhout B, van Meijl H (2008) A multi-scale, multi-model approach for analyzing the future dynamics of European land use. Ann Reg Sci 42(1):57–77
W3C (2018) Semantic eeb definition by W3C. https://www.w3.org/2001/sw/, visited Jan 2018
Weinberger D (2011) Too big to know: rethinking knowledge now that the facts aren't the facts, experts are everywhere, and the smartest person in the room is the room. Basic Books, New York. ISBN-13: 978–0465085965
Wiig KM (2007) Effective societal knowledge management. J Knowl Manage 11(5):141–156
Wilkinson MD, Dumontier M, Aalbersberg IJ, Appleton G, Axton M, Baak A, Blomberg N, Boiten J, da Silva Santos LB, Bourne PE (2016) The FAIR guiding principles for scientific data management and stewardship. Sci Data 3:160018
Yoshizawa M, Sakurai T (1995) Production management system and its application method. Google Patents, US Patent 5,442,561

Chapter 2
Service-Oriented Processing and Analysis of Massive Point Clouds in Geoinformation Management

Sören Discher, Rico Richter, Matthias Trapp and Jürgen Döllner

Abstract Today, landscapes, cities, and infrastructure networks are commonly captured at regular intervals using LiDAR or image-based remote sensing technologies. The resulting point clouds, representing digital snapshots of the reality, are used for a growing number of applications, such as urban development, environmental monitoring, and disaster management. Multi-temporal point clouds, i.e., *4D point clouds*, result from scanning the same site at different points in time and open up new ways to automate common geoinformation management workflows, e.g., updating and maintaining existing geodata such as models of terrain, infrastructure, building, and vegetation. However, existing GIS are often limited by processing strategies and storage capabilities that generally do not scale for massive point clouds containing several terabytes of data. We demonstrate and discuss techniques to manage, process, analyze, and provide large-scale, distributed 4D point clouds. All techniques have been implemented in a system that follows service-oriented design principles, thus, maximizing its interoperability and allowing for a seamless integration into existing workflows and systems. A modular service-oriented processing pipeline is presented that uses out-of-core and GPU-based processing approaches to efficiently handle massive 4D point clouds and to reduce processing times significantly. With respect to the provision of analysis results, we present web-based visualization techniques that apply real-time rendering algorithms and suitable interaction metaphors. Hence, users can explore, inspect, and analyze arbitrary large and dense point clouds. The approach is evaluated based on several real-world applications and datasets featuring different densities and characteristics. Results show that it enables the management, processing, analysis, and distribution of massive 4D point clouds as required by a growing number of applications and systems.

Keywords 4D point clouds · Analytics · Visualization · Service-oriented infrastructure

S. Discher (✉) · R. Richter · M. Trapp · J. Döllner
Faculty of Digital Engineering, Hasso Plattner Institute, University of Potsdam, Potsdam, Germany
e-mail: soeren.discher@hpi.uni-potsdam.de

J. Döllner et al. (eds.), *Service-Oriented Mapping*, Lecture Notes in Geoinformation and Cartography, https://doi.org/10.1007/978-3-319-72434-8_2

2.1 Introduction and Problem Statement

Over the last decades, 3D point clouds, i.e., discrete, digital representations of 3D objects and environments based on unstructured collections of 3D geometric points, have become an essential data category for geospatial and non-geospatial applications in diverse areas such as building information modeling (Pătrăucean et al. 2015; Stojanovic et al. 2017), urban planning and development (Musialski et al. 2013), or the digital preservation of cultural and natural heritage (Rüther et al. 2012; Hämmerle et al. 2014). A major cause for that popularity surge have been technological advances in remote and in situ sensing technology, allowing us to capture assets, buildings, cities, or complete countries with unprecedented speed and precision. As an example, state-of-the-art laser scanners may capture millions of points per second, generating highly detailed point clouds of individual sites (e.g., several points per square centimeter) within few hours (Rüther et al. 2012). Larger areas can be covered by attaching the scanning device to a moving vehicle, such as cars or unmanned aircraft systems (UAS), resulting in massive point clouds that may contain several billion points and terabytes of raw data (Leberl et al. 2010; Martinez-Rubi et al. 2015). Due to its increased affordability and effectiveness, communities worldwide have started to intensify the use of in situ and remote sensing technology by conducting scans more regularly (e.g., once a year) and by combining different capturing methods, such as aerial laser scanning and mobile mapping (Nebiker et al. 2010; Puente et al. 2013). The resulting dense and multi-temporal datasets, commonly referred to as *4D point clouds*, may be visualized and used as interactive models that document how a given site or land-scape has changed over time (Discher et al. 2017). Furthermore, they allow us to automate and speed up common geoinformation management workflows: in urban planning and development for example, existing geodata such as official tree cadasters or terrain models, can be efficiently updated based on 4D point clouds (Oehlke et al. 2015). In the wake of natural disasters, remote sensing can be applied to quickly gather up-to-date information about affected areas and to conduct an automated damage assessment based on a comparison to previous scans (Richter et al. 2013b).

However, existing geoinformation systems (GIS) are often limited by processing and visualization techniques that do not scale for such massive datasets. To handle limited processing and memory capabilities (i.e., main and GPU memory), data quality is often reduced, either by thinning the respective point clouds (Peters and Ledoux 2016) or by converting them into generalized 3D meshes (Berger et al. 2014). To overcome these limitations and to make use of the full potential and resolution of 4D point clouds, external memory algorithms are required, which dynamically fetch and unload subsets of the data based on their relevance to the task at hand, limiting the memory usage to a predefined budget. An efficient access to the relevant subsets can be ensured by organizing point clouds using appropriate spatial data structures and level-of-detail (LoD) concepts (Elseberg et al. 2013; Goswami et al. 2013). With data sources becoming more numerous and diverse, an ever-increasing number of stakeholders requires access to the data. Hence, there is a growing demand for a seamless integration of point-based processing and rendering functions into existing workflows and systems. As a result, the efficient and scalable management of 4D point clouds becomes an increasingly relevant aspect (van Oosterom et al. 2015; Cura et al. 2017). Available research solutions use service-oriented design principles to implement scalable systems. As an example, Martinez-Rubi et al. (2015) describe a system that integrates massive point cloud data and makes it publicly accessible via a web interface, allowing users to interactively explore or download arbitrary subsets of the data. Richter and Döllner (2014) propose a system for the management of multi-temporal point clouds, addressing integration, processing, and rendering functionality. They discuss a service-oriented architecture that focuses on a seamless integration of system components into existing workflows: each component can be accessed individually via standardized web services.

We extend the architecture proposed in Richter and Döllner (2014) by improving processing performance and interoperability. We introduce a *modular processing pipeline* that implements parallel processing strategies to speed up individual tasks and facilitate the distribution of those tasks alongside corresponding datasets within a distributed infrastructure for data storage. The use of external memory algorithms allows us to handle arbitrary large 4D point clouds. In addition, existing web services for geodata can be seamlessly integrated into the processing pipeline to improve processing results. The article is structured as follows: Sect. 2.2 discusses the acquisition and management of massive, heterogeneous 4D point clouds, focusing on the characteristics of different data sources and consequential system requirements. The pipeline architecture is presented in Sect. 2.3. Section 2.4 discusses web-based rendering techniques and interaction metaphors that facilitate the exploration and inspection of resulting datasets. In Sect. 2.5, we present case studies based on real-world applications and datasets. Section 2.6 gives conclusions and an outlook on future challenges.

2.2 Data Acquisition and System Requirements

Systems for the acquisition of point clouds are manifold and allow us to capture real-world surfaces at all scales, ranging from small objects over individual buildings to complete cities or even countries. In this section, we describe the characteristics of common, established acquisition systems and identify requirements for systems aiming to enable the efficient management, processing, and distribution of the resulting data on a massive scale as vital for a variety of applications. Based on these requirements, a corresponding system architecture is presented.

2.2.1 Data Sources and Characteristics

Today, real-world surfaces are captured using active or passive sensors. Active sensors (e.g., LiDAR, radar, stripe projection systems) emit electromagnetic radiation to directly measure the distance between surface and sensor, generating so-called range data (Eitel et al. 2016). Passive sensors (e.g., aerial cameras, digital cameras, hyperspectral radiometers) on the other hand (Remondino et al. 2013b) solely rely on natural radiation, most notably sunlight, and generate series of images used as input for dense image matching algorithms to derive 3D information (Rothermel et al. 2012). Both active and passive sensors can capture individual objects with point densities of up to a few micrometers. The resolution depends on the distance between sensor and captured surface. On a larger scale, point clouds of entire rooms, buildings, or facilities can be efficiently generated by placing sensors at key positions within the site in question (Remondino et al. 2013a). By attaching the sensors to moving vehicles such as cars, trains, UAS, or satellites, data for even larger areas such as infrastructure networks, cities, or countries can be efficiently collected, although at reduced point densities (Ostrowski et al. 2014).

Each acquisition system comes with specific advantages and disadvantages, affecting its suitability for different use cases. Passive sensors tend to be more affordable, portable, and easier to use than their active counterparts, as evidenced by their frequent integration into state-of-the-art consumer electronics (Kersten et al. 2016) and UAS. Furthermore, image-based methods allow for potentially higher point densities, e.g., up to 100 points/m^2 for aerial photographs as in contrast to typically 1–25 points/m^2 for aerial laser scans (Remondino et al. 2013b). However, the quality of the resulting point clouds is significantly influenced by surface materials (e.g., shininess, texturedness) and image properties (e.g., shadows, color variety, depth of field). This can be especially noticeable when capturing glassy surfaces, where range-based approaches tend to generate much cleaner point clouds (Remondino et al. 2013b). With respect to performance, passive sensors collect data faster; however, a computation-intense post-processing of the generated images is required to compute 3D points, whereas active sensors provide those directly (Remondino et al. 2013a). In practice, both sensor categories are frequently used in parallel. Advanced driver assistance systems for example, combine varying active and passive sensors to observe a car's immediate surroundings (Langner et al. 2016). Table 2.1 provides an overview of several common acquisition systems and their specific characteristics.

2.2.2 Challenges and System Requirements

Traditionally, point clouds are captured, processed, analyzed, and visualized in the scope of only one specific application. However, each dataset might also contain relevant information for completely different use cases. For example, point clouds

Table 2.1 Commonly used acquisition systems for point clouds and their characteristics

Acquisition system	Typical scale	Typical density (pts/m^2)	Costs
Airborne laserscanning	Infrastructure networks, urban + rural areas	1–25	Very high
Aerial photography	Infrastructure networks, urban + rural areas	25–100	High
Mobile mapping (rails, roads)	Infrastructure networks, urban areas	200–1400	Medium
UAS	Buildings, facilities, infrastructure networks	500–6000	Medium
Static terrestrial laserscanning	Indoor scenes, buildings, facilities	4000–20,000	Medium
Smartphone cameras/ DSLRs	Individual objects, indoor scenes, buildings	4000–40,000	Low
Depth cameras	Individual objects, indoor scenes	4000–20,000	Low
Stripe-projection systems	Individual objects, indoor scenes	100,000–400,000	Low

generated by advanced driver assistance systems, can be of immense value for urban planning and development, as they provide up-to-date information about the infrastructure of a city on a frequent basis and from an often supplementary perspective (i.e., from a pedestrian's perspective instead of a bird's-eye view). Similarly, a frequent scanning of a site raises valuable insights about its history, which has the potential to greatly benefit common geoinformation management workflows such as predictive maintenance and the automated updating of official cadaster data (Sect. 2.5). Nonetheless, traditional GIS still operate primarily on 2D or 3D meshes (Musliman et al. 2008; van Oosterom et al. 2008) derived from point clouds in a time-consuming process (Berger et al. 2014). A direct processing of point clouds would be more efficient because point clouds have no limitations regarding model, geometry, structure, or topology and can be updated automatically in contrast to meshes (Paredes et al. 2012). Hence, the use of point-based instead of mesh-based models within geoinformation management workflows has the potential to speed up common tasks and applications notably (Richter and Döllner 2014).

Establishing 4D point clouds as a basic data type to provide geographic content requires systems that facilitate the efficient management of such datasets. In particular, this refers to the integration of heterogeneous point clouds from various data sources, the updating, processing, and analysis of massive datasets, as well as the provision and visualization of arbitrary subsets based on varying per-point attributes (e.g., spatial, temporal, or semantic information). To ensure an efficient access to the data, suitable spatial data structures and Level of Detail (LoD) concepts are required. Parallel processing strategies need to be implemented and combined with a distributed storage of the point data to significantly improve the performance of a system. To facilitate the exploration of analysis results, suitable visualization techniques and interaction metaphors must be applied that enhance the recognition of objects, semantics, and temporal changes within point cloud depictions.

To ease the integration of such systems into existing workflows and process chains, their interoperability must be guaranteed by making their functionality available via standardized web services (van Oosterom et al. 2008), likewise the integration of existing web services for geodata into the system should be supported. In summary, the following requirements need to be addressed:

R1. Integration of point clouds from heterogeneous data sources into a homogeneous spatial data model,
R2. distributed storage of point clouds,
R3. processing services for the distributed, scalable, adaptive, and selective updating, processing, and analysis of massive 4D point clouds,
R4. support to integrate existing web services for geodata into the system to further increase interoperability,
R5. visualization services to provide 4D point clouds for heterogeneous clients (e.g., desktop computers, mobile devices),
R6. visualization and interaction techniques to enable a task-specific and application-specific visualization of massive 4D point clouds.

The next section presents a system architecture that fulfills these requirements.

2.2.3 System Architecture

We propose a service-oriented architecture that builds upon the system proposed by Richter and Döllner (2014), but further improves its scalability and interoperability (Fig. 2.1). Generally, managing 4D point clouds comprises three stages:

(1) **Data Integration**. Components that implement the integration of point clouds from heterogeneous data sources into a homogeneous spatial data model (R1). This includes georeferencing of acquired data as well as preparing or updating LoD data structures to ensure an efficient data access in subsequent stages. Another aspect is the filtering and quality control of the data (e.g., thinning, noise reduction, and outlier filtering).
(2) **Data Analytics**. Components for common and domain-specific analyses and simulations (R3). Those typically require additional per-point attributes (e.g., normal, topological, or surface category information) that are either computed by specific preprocessing components or provided by external web services, which can be seamlessly integrated into the system (R4). Depending on the use case, analysis results can be stored as additional per-point attributes or exported as 2D or 3D meshes (e.g., shapes).
(3) **Data Provision**. Components providing access to data and analysis results for external processing and visualization tools. The data is either provided in standardized geodata formats via Web Feature Services (Rautenbach et al. 2013) or

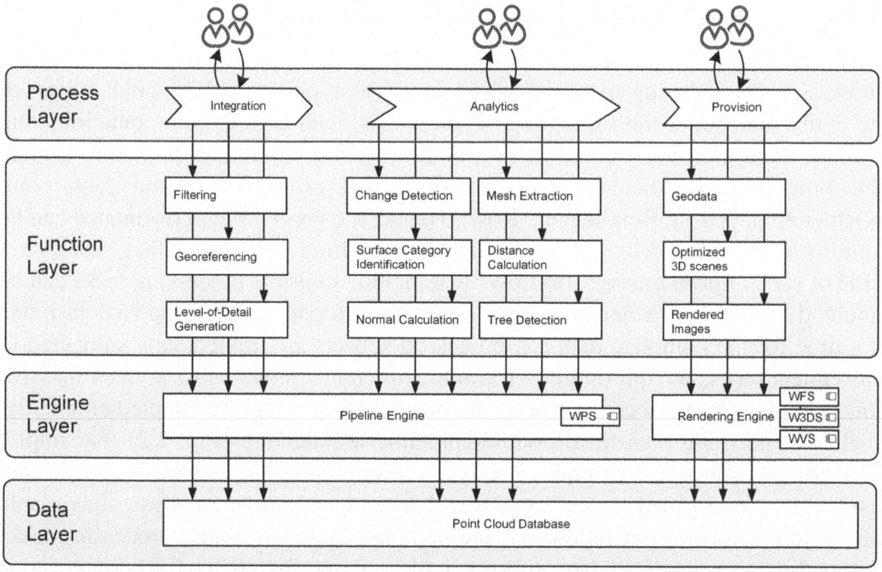

Fig. 2.1 Service-oriented architecture of our system with arrows indicating usage relations between components

optimized for a web-based visualization using Web 3D Services or Web View Services (Klimke et al. 2013). This allows for the interactive exploration of the data on heterogeneous client systems (R5), ranging from high-end workstations to low-end mobile devices with very limited computing capabilities, network bandwidth, or memory resources.

The individual components are chained together in a modular processing pipeline that implements parallel and distributed computing concepts to al-low for the efficient and scalable execution of updating, processing, and analysis tasks (R3). The processing pipeline can be accessed and reconfigured via a Web Processing Service (Mueller and Pross 2015) to dynamically adapt the applied analyses and simulations to the field of application and current use cases (Sect. 2.3). Our system stores point clouds in a distributed way: If required, additional computing and storage resources may be added at runtime, maximizing its scalability and availability (R2). A standardized interface to integrate, update, and access arbitrary subsets of the data is provided by a central system component, which we refer to as *point cloud database*. Spatial data structures, that hierarchically subdivide the spatial area, and LoD concepts, that provide subset representations, are used by that database to reduce the amount of data adaptively for data queries and processing tasks. The suitability of different *LoD data structures* (Richter and Döllner 2010; Elseberg et al. 2013; Goswami et al. 2013) varies based on the application scenario (e.g., real-time visualization or analysis) and spatial distribution of the data (e.g., terrestrial or airborne). Evaluating, generating, and maintaining them is also in the responsibility of the point cloud database.

2.3 Service-Oriented Point Cloud Analytics

Processing or analyzing 4D point clouds requires typically only a small subset of the entire data set at the same time. Typical tasks and complex computations that must be computed for each point in a dataset such as duplicate detection, change detection, and classification operate on a small proximity around each point (Richter et al. 2013a; Belgiu et al. 2014). Hence, the processing performance can be significantly increased by applying parallel computing concepts, either based on a CPU or GPU. Furthermore, workflows that include multiple processing tasks can be efficiently chained together by interleaving them. Instead of executing each task one at a time for the complete data set, processed subsets are immediately subjected to subsequent tasks. By splitting the data into sufficiently small subsets, even massive point clouds, which exceed available memory capacities, can be handled efficiently. In this section, we present a modular pipeline architecture (Fig. 2.2) that implements those concepts: Complex analyses, comprising several basic processing tasks, can be performed on arbitrary large data sets, making optimal use of available hardware resources by parallelizing, interleaving, and distributing processing tasks alongside corresponding data subsets within networks (R3). Each analysis is described by a processing pipeline that defines involved processing tasks and can be reconfigured and replaced dynamically. Also, the pipeline architecture can be easily expanded to integrate existing web services for geodata, thus, maximizing its interoperability (R4).

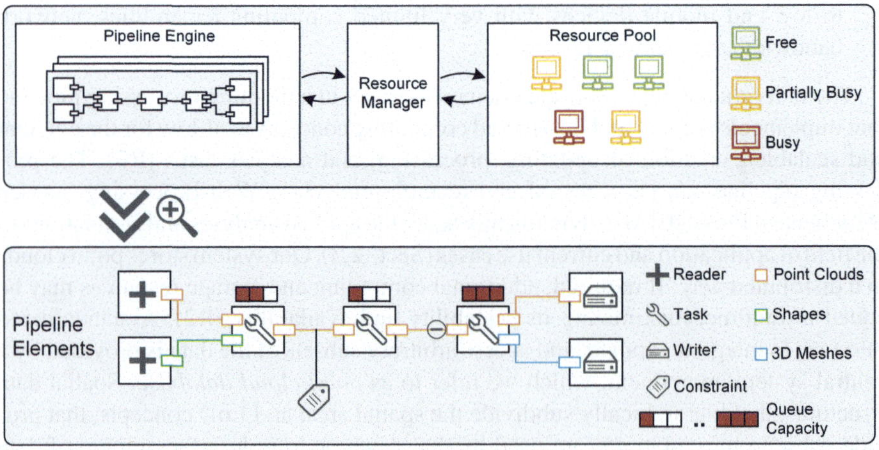

Fig. 2.2 Overview of the pipeline architecture and pipeline elements

2.3.1 Pipeline Architecture

The proposed architecture comprises two major components: first, a re-source manager, monitoring the memory and processing capacity of a system and distributing them among currently executed processing tasks (Sect. 2.3.2); second, a pipeline engine to configure and execute various processing pipelines. Each pipeline defines a specific combination of basic input, processing and output tasks. We define the elements that compose a pipeline as follows:

- **Importers**, i.e., pipeline nodes that import data from any source such as files, the *point cloud database* or other, external sources (e.g., web-services). Each importer prepares *data packages*. If the input data exceeds the maximum data package size, the importer prepares subsets by splitting the data.
- **Exporters**, i.e., pipeline nodes that export processing and analysis results into standardized formats for point clouds (e.g., LAS/LAZ), the point cloud database or other geodata (e.g., shape files, CityGML, GeoTIFFs). The latter functionality makes exporters essential in facilitating the integration of the proposed system into existing workflows.
- **Tasks**, i.e., pipeline nodes that implement a specific processing or analysis algorithm. Some algorithms operate on multiple data sets simultaneously, e.g., to compare or to merge them. Similarly, algorithms may split incoming data sets or yield multiple results, e.g., additional per-point attributes and corresponding shapes. Hence, multiple incoming and outgoing connections may be defined per task.
- **Connections**, i.e., links between two pipeline nodes for the transfer of *data packages*. They define the order of execution. A given connection transfers only packages of a specific type, e.g., point clouds or shapes. Depending on the pipeline nodes being connected, various constraints may be defined, such as defined per-point attributes that are required.
- **Data Packages**, i.e., data subsets that are transferred between pipeline nodes via connections. Similar to *connections*, a given data package may only contain a specific type of geodata. Also, the size of the corresponding data subset may not exceed a specific maximum defined by the resource manager.

External services for geodata can be seamlessly integrated by implementing tasks and importers as interfaces. The pipeline engine allows to execute multiple pipeline plans in parallel, each of which can be started, paused, and stopped dynamically. At runtime, each active pipeline node gets assigned its own set of resources by the resource manager. Processed data packages are immediately transferred to subsequent pipeline nodes. For each incoming connection, a pipeline node manages a queue of incoming data packages. The query size is limited to a defined number of data packages. If a queue reaches its maximum capacity, no additional data packages are accepted and preceding nodes are not executed. To improve their runtime performance, the most time-consuming pipeline nodes are executed in parallel by adaptively assigning additional resources (e.g., CPU or GPU cores).

2.3.2 *Memory and Resource Management*

The resources of a system may be distributed across several network nodes, each featuring different memory capacities (i.e., size of secondary storage, main memory, and GPU memory) and computing capabilities (e.g., number and clock speed of CPU and GPU cores, memory transfer rates). Network nodes and their resources are added to a global *resource pool* that is monitored by the resource manager of the system. Whenever a pipeline node needs to be executed, the resource manager assigns resources based on available system capabilities. After the execution is finished, all assigned resources are released to the resource pool and become available for other nodes (Fig. 2.2). Distributing resources requires the resource manager to make a trade-off between several, often contradicting optimization goals:

- **Exclusivity**. Exclusive access to a resource (e.g., storage or GPU) significantly improves the runtime performance of a pipeline node, e.g., by minimizing cache misses and seek times.
- **Transfer Costs**. Frequently transferring data packages via connections may notably reduce the performance if subsequent pipeline nodes operate on different network nodes. This can be avoided by executing them on the same network nodes.
- **Parallelization**. Executing pipeline nodes in parallel or interleaved is an essential mechanism to improve the overall performance of the system. Thus, available resources and network nodes should be shared among as many pipeline nodes as possible.

The runtime of nodes may vary significantly depending on the task. An adaptive resource scheduling allows to prevent bottlenecks in the processing. The execution time is tracked for each node and the number of assigned resources is adjusted dynamically.

2.4 Point Cloud Visualization

The interactive exploration of massive 4D point clouds is an essential functionality of our system, facilitating the visual recognition and interpretation of objects, semantics, and analysis results within corresponding data sets. In this section, we discuss web-based rendering approaches allowing to render arbitrary large data sets on a multitude of heterogeneous clients (R5, Sect. 2.4.1). Furthermore, we describe how visual filtering and highlighting within point cloud depictions can be facilitated by combining different visualization techniques and interaction metaphors (R6, Sect. 2.4.2).

2.4.1 Web-Based Rendering

Since point clouds with billions of points exceed typically available main and GPU memory capacities by an order of magnitude, out-of-core rendering techniques are required to decouple rendering efforts from the amount of data. Exemplary systems (Goswami et al. 2013; Richter et al. 2015) use LoD data structures to organize the data into chunks that are suitable for fast processing and rendering. The LoD node selection is performed on a per-frame basis and depends on the view position and user interaction, as well as computing and storage capabilities of the underlying rendering system. Different caching approaches are used to minimize data swapping latencies for memory transfers between secondary storage, main memory, and GPU memory.

While being applicable to arbitrary large data sets, out-of-core rendering algorithms require a direct access to the data, which generally restricts their application to systems with massive storage capacities. For other systems such as mobile devices, these algorithms must be combined with existing web-based approaches that limit workload and data traffic on client-side by using a central server infrastructure to maintain and distribute the data (R5, Fig. 2.3). Rendering directly on the server and only transferring the rendered images to the client is commonly referred to as a thin client approach (Döllner et al. 2012; Klimke et al. 2014). As an optimization, many of such approaches render and transfer cube-maps instead of individual images. Thus, new data requests are only required when the view position changes significantly. Alternatively, thick client approaches can be used that delegate the rendering to the client side. Here, the server is only responsible for selecting and transferring the data to the client (Krämer and Gutbell 2015; Martinez-Rubi et al. 2015; Limberger et al. 2016b; Schoedon et al. 2016). While a thin client approach notably reduces the minimal hardware requirements on client side, a thick client approach is usually more feasible to serve a large number of clients due to lower workload generated on server side. Furthermore, a thick client approach tends to be more resilient to unstable network connections since the visualization can still be adjusted to user interactions when the connection to the server has been lost temporarily, albeit some relevant data might be missing.

2.4.2 Semantic-Based Visualization

The visualization of point clouds can be adapted to various objectives (e.g., highlighting structural changes within multi-temporal datasets) by switching between different point-based rendering techniques and color schemes (Gross and Pfister 2011). Many rendering techniques focus on a photorealistic visualization: points are represented as splats, i.e., oriented disks, spheres, or paraboloids. An adequate size and orientation enables to render a closed surface (Preiner et al. 2012; Schütz and Wimmer 2015). However, a photorealistic representation often hinders the visual identification and categorization of structures and topological properties. In airborne datasets for

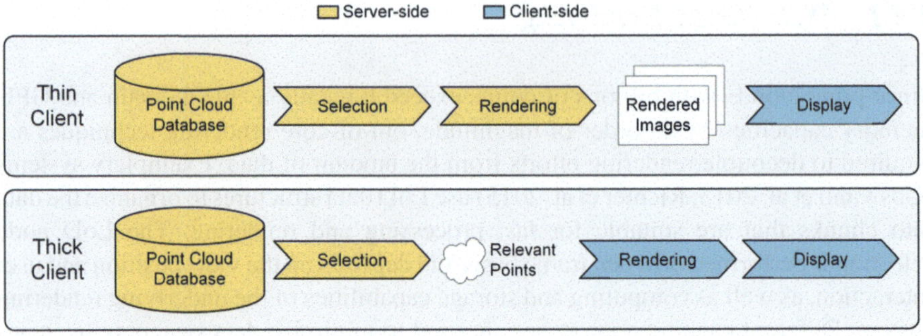

Fig. 2.3 Service-oriented rendering approaches: thin clients versus thick clients

example, vegetation can be difficult to distinguish from ground points and small structures might not be visible due to an insufficient contrast to the environment (Richter et al. 2015). Meanwhile, non-photorealistic rendering techniques (Boucheny 2009) efficiently highlight edges and structures within point cloud depictions.

In general, task-specific explorations of virtual 3D scenes can be facilitated by combining different rendering techniques, color schemes, and post-processing effects (Döllner et al. 2005; Semmo et al. 2015; Würfel et al. 2015; Limberger et al. 2016a). For point cloud depictions, this is exemplified by Richter et al. (2015) who use per-point attributes (e.g., surface category, topologic information) to adapt the appearance of a point, i.e., its color, size, orientation, or shape (Fig. 2.4). They apply multi-pass rendering utilizing G-Buffers for image-based compositing (Fig. 2.5). Points are grouped into different subsets (e.g., based on their surface category), each of which is rendered separately. A compositing pass merges the

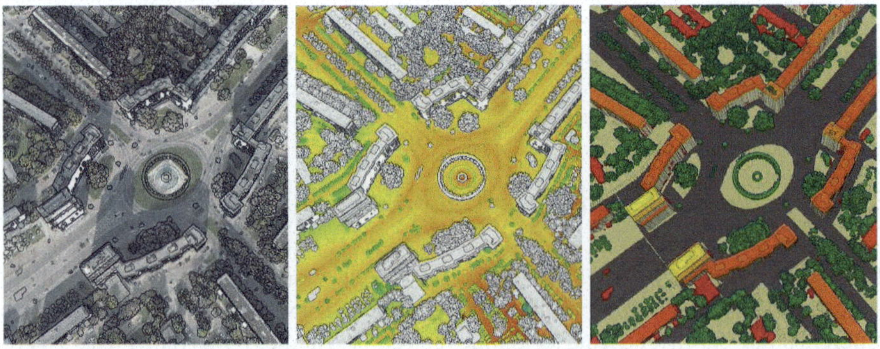

Fig. 2.4 Visualization of an airborne laser scan of an urban area, showcasing different rendering setups using per-point surface category and topological information: (left) Uniform rendering for all categories using rgb colors. (middle) Height gradient applied to ground points, uniform color and edge highlighting applied to other categories. (right) Height gradient and closed surface rendering applied to building points, category-specific colors used for ground, street and vegetation points

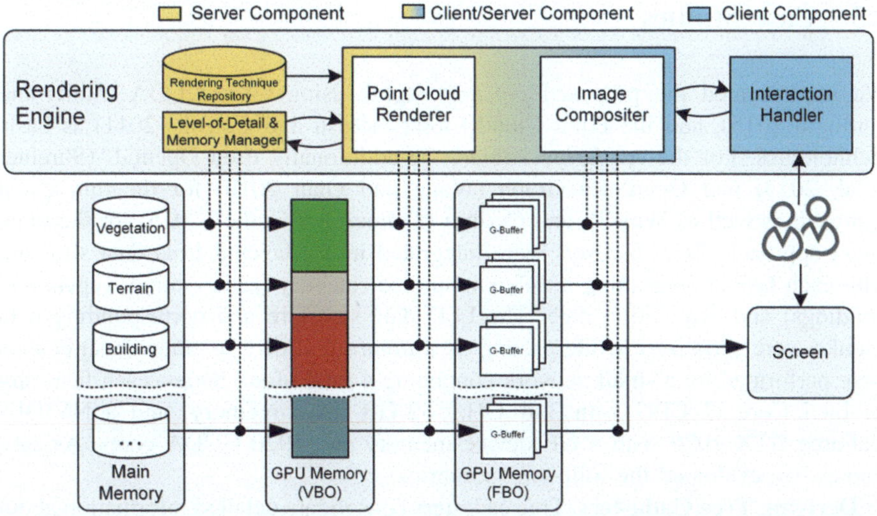

Fig. 2.5 Overview of the proposed rendering engine. Prior to rendering, points are separated based on per-point attributes (e.g., surface categories)

rendering results, enabling sophisticated focus + context visualization and interaction techniques (Elmqvist and Tsigas 2008; Vaaraniemi et al. 2013; Semmo and Döllner 2014) such as visibility masks (Sigg et al. 2012) or interactive lenses (Trapp et al. 2008; Pasewaldt et al. 2012). Thus, task-relevant structures can be easily identified even if they are fully or partly occluded (R6, Fig. 2.6). All rendering and interaction techniques can be selected, configured, and combined at run-time.

Fig. 2.6 Focus + context visualization for massive point clouds. Left: Using an interactive see-through lens, the occluded interior of a building can be inspected in the context of the overall scan. Right: A visibility mask is used to highlight otherwise hidden street points

2.5 Case Studies

We implemented the proposed *pipeline engine* using C++, CUDA (Storti and Yurtoglu 2015), and the Point Cloud Library (Rusu and Cousins 2011) as basic technologies. For the *rendering engine*, we additionally used OpenGL (Shreiner et al. 2013) and OpenSceneGraph (Wang and Qian 2010) for the thin client approach as well as WebGL and Cesium (Krämer and Gutbell 2015) for the thick client approach. Test data sets were integrated into a layered LoD data structure with each layer representing a different surface category (e.g., ground, vegetation, buildings) and acquisition date. The LoD data structure and corresponding data chunks were serialized to files acting as a *database*. All tests and measurements were performed by a small network containing a total of six nodes, each featuring an Intel Core i7 CPU with 3.40 GHz, 32 GB main memory, and a NVIDIA GeForce GTX 1070 with 8 GB device memory and 1920 CUDA cores. As case studies, we evaluated the following scenarios:

Deriving Tree Cadasters. Tree cadasters consolidate detailed information about biomass in an area and are essential for a growing number of applications in urban planning, landscape architecture, and forest management. Point cloud analytics allows for the automatic, area-wide derivation and continuation of such tree cadasters and corresponding metrics, e.g., height and crown diameter (Oehlke et al. 2015). As depicted in Fig. 2.7a, the analysis comprises three major tasks: identification of points representing vegetation, delineation of individual trees within those vegetation points, and calculation of per-tree metrics. To identify vegetation points, an iterative, segment based classification (Richter et al. 2013) was applied that distinguishes between ground, building, and vegetation by analyzing for each point the topology of its local neighborhood. To efficiently delineate individual trees, a point-based approach by Li et al (2012) was adapted, requiring only a single iteration over the point cloud. For fast nearest-neighbor queries a GPU-based implementation was used.

Monitoring Infrastructure Networks. Infrastructure networks (e.g., roads, canalizations, or power lines) are constantly subjected to environmental loads (e.g., wind, temperature changes), causing them to deteriorate over time. Regularly capturing such infrastructure networks provides efficient means for their automatic, accurate, and predictive maintenance. The corresponding analysis (Fig. 2.7b) is commonly referred to as a change detection (Kang and Lu 2011; Eitel et al. 2016). For each input point, the distance to a *reference geometry* (e.g., another point cloud or 3D model) is estimated as a metric for the degree of change within the covered area and stored as a per-point attribute. This approach allows to identify changes efficiently and establish update and maintaining workflows.

Continuing 3D City Models. Many applications in urban planning and development or building information modeling require official, state-issued cadaster data or even full-fledged 3D city models. Keeping those models up-to-date constitutes a major challenge for municipalities that can be facilitated by using point cloud analytics. As an example, a change detection as described above can be combined

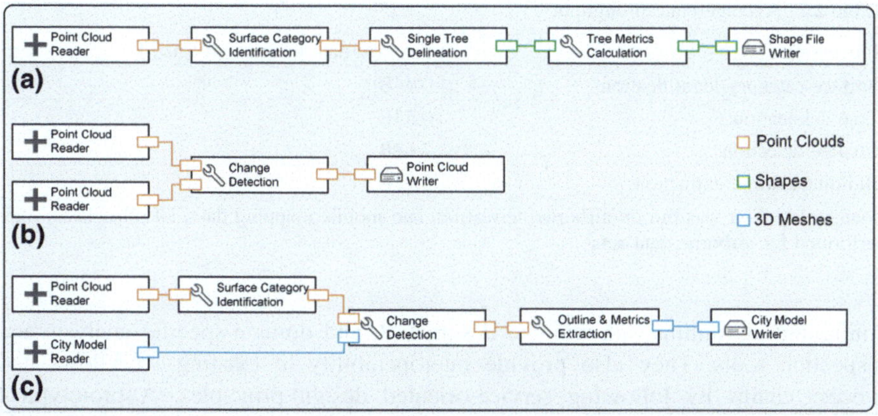

Fig. 2.7 Exemplary processing pipelines as they have been used for the case studies

with per-point surface category information to automatically identify all building points with a certain distance to a given virtual 3D city model (Fig. 2.7c), indicating newly constructed, removed, or otherwise modified buildings. The resulting points can be segmented into subsets of neighboring points, each representing a separate building. Based on these subsets, 2D building outlines and additional metrics (e.g., average distance, projected roof area) can be derived that facilitate the assessment of necessary—typically manual—modifications to the 3D city model.

Test data sets for those scenarios comprised airborne, terrestrial, and mobile scans covering different parts of three metropolitan regions with the largest point cloud featuring 100 points/m^2 for an area of 800 km^2. In Table 2.2, the data throughput for the most relevant processing tasks is specified. Data throughput for the integration and rendering of the data is defined by the overall network and memory bandwidth and was at around 80 MB/s.

2.6 Conclusions and Future Work

By applying out-of-core, parallel, and distributed processing strategies, 4D point clouds of any size can be updated and analyzed at significantly reduced processing times. Our modular pipeline architecture implements such concepts, making efficient use of available hardware resources. A database component is used for the distributed storage and efficient integration of point clouds from heterogeneous data sources, ensuring efficient access to arbitrary subsets of the data. A rendering component allows for analysis results to be visualized and inspected on heterogeneous clients with differing computing capabilities by implementing web-based and out-of-core rendering algorithms. To facilitate the inspection, specialized visualization and interaction techniques are provided. All system components can be

Table 2.2 Average data throughput

Processing task	Average data throughput (pts/h)
Surface category identification	0.44B
Tree delineation	0.53B
Change detection	7.68B
Building outline extraction	8.27B

Change detection was run on airborne, terrestrial, and mobile mapping data, all other tasks were performed for airborne data sets

configured at runtime, allowing to design task and domain-specific analysis and inspection tools. They also provide interoperability to existing workflows and process chains by following service-oriented design principles. A prototypical implementation of our system has been successfully tested for several real-world scenarios. Novel devices and sensors have the capability to generate point clouds in real-time, resulting in massively redundant point cloud streams. Hence, ad hoc data management, analysis, and visualization tools are fundamental requirements. Our future work will focus on challenges introduced by such point cloud streams to use them even for applications which are not in the core area of GIS to expedite the digital transformation.

Acknowledgements This work was funded by the Federal Ministry of Education and Research (BMBF), Germany within the InnoProfile Transfer research group "4DnD-Vis" and the Research School on "Service-Oriented Systems Engineering" of the Hasso-Plattner-Institut. We would like to thank virtualcity SYSTEMS for providing datasets.

References

Belgiu M, Tomljenovic I, Lampoltshammer TJ, Blaschke T, Höfle B (2014) Ontology-based classification of building types detected from airborne laser scanning data. Remote Sens 6(2):1347–1366

Berger M, Tagliasacchi A, Seversky L, Alliez P, Levine J, Sharf A, Silva C (2014) State of the art in surface reconstruction from point clouds. Eurographics Stars 1(1):161–185

Boucheny C (2009) Interactive scientific visualization of large datasets: towards a perceptive-based approach. Ph.D. thesis, Université Joseph Fourier, Grenoble, France

Cura R, Perret J, Paparoditis N (2017) A scalable and multi-purpose point cloud server (PCS) for easier and faster point cloud data management and processing. ISPRS J Photogrammetry Remote Sens 127:39–56

Discher S, Richter R, Döllner J (2017) Interactive and view-dependent see-through lenses for massive 3D point clouds. Adv in 3D Geoinf, pp 49–62

Döllner J, Buchholz H, Nienhaus M, Kirsch F (2005) Illustrative visualization of 3D city models. Electron Imaging 2005:42–51

Döllner J, Hagedorn B, Klimke J (2012) Server-based rendering of large 3D scenes for mobile devices using G-buffer cube maps. In: 17th international conference on 3D web technology, pp 97–100

Eitel J, Höfle B, Vierling L, Abellán A, Asner G, Deems J, Glennie C, Joerg P, LeWinter A, Magney T, Mandlburger G (2016) Beyond 3-D: the new spectrum of lidar applications for earth and ecological sciences. Remote Sens Environ 186:372–392

Elmqvist N, Tsigas P (2008) A taxonomy of 3D occlusion management for visualization. IEEE Trans Visual Comput Graphics 14(5):1095–1109

Elseberg J, Borrmann D, Nüchter A (2013) One billion points in the cloud–an octree for efficient processing of 3D laser scans. ISPRS J Photogrammetry Remote Sens 76:76–88

Goswami P, Erol F, Mukhi R, Pajarola R, Gobbetti E (2013) An efficient multi-resolution framework for high quality interactive rendering of massive point clouds using multi-way kD-trees. The Vis Comput 29(1):69–83

Gross M, Pfister H (eds) (2011) Point-based graphics. Morgan Kaufmann, USA

Hämmerle M, Höfle B, Fuchs J, Schröder-Ritzrau A, Vollweiler N, Frank N (2014) Comparison of kinect and terrestrial lidar capturing natural karst cave 3-D objects. IEEE Geosci Remote Sens Lett 11(11):1896–1900

Kang Z, Lu Z (2011) The change detection of building models using epochs of terrestrial point clouds. In: International workshop on multi-platform/multi-sensor remote sensing and mapping, pp 1–6

Kersten TP, Przybilla HJ, Lindstaedt M, Tschirschwitz F, Misgaiski-Hass M (2016) Comparative geometrical investigations of hand-held scanning systems. ISPRS Archives, XLI-B5, pp 507–514

Klimke J, Hagedorn B, Döllner J (2013) A service-based concept for camera control in 3D geovirtual environments. In: Progress and new trends in 3D geoinformation sciences, pp 101–118

Klimke J, Hagedorn B, Döllner J (2014) Scalable multi-platform distribution of spatial 3D contents. Int J 3-D Inf Model 3(3):35–49

Krämer M, Gutbell R (2015) A case study on 3D geospatial applications in the web using state-of-the-art WebGL frameworks. In: 20th international conference on 3D web technology, pp 189–197

Langner T, Seifert D, Fischer B, Goehring D, Ganjineh T, Rojas R (2016) Traffic awareness driver assistance based on stereovision, eye-tracking, and head-up display. In: IEEE international conference on robotics and automation, pp 3167–3173

Leberl F, Irschara A, Pock T, Meixner P, Gruber M, Scholz S, Wiechert A (2010) Point clouds: lidar versus 3D vision. Photogram Eng Remote Sens 76(10):1123–1134

Li W, Guo Q, Jakubowski M, Kelly M (2012) A new method for segmenting individual trees from the lidar point cloud. Photogram Eng Remote Sens 78(1):75–84

Limberger D, Fiedler C, Hahn S, Trapp M, Döllner J (2016a) Evaluation of sketchiness as a visual variable for 2.5D treemaps. In: 20th international conference information visualisation, pp 183–189

Limberger D, Scheibel W, Lemme S, Döllner J (2016b) Dynamic 2.5D treemaps using declarative 3D on the web. In: 21st international conference on web 3D technology, pp 33–36

Martinez-Rubi O, Verhoeven S, van Meersbergen M, Schütz M, van Oosterom P, Gonçalves R, Tijssen T (2015) Taming the beast: free and open-source massive point cloud web visualization. In: Capturing Reality Forum 2015

Mueller M, Pross B (2015) OGC WPS 2.0 interface standard. Open Geospatial Consortium, USA

Musialski P, Wonka P, Aliaga DG, Wimmer M, Gool LV, Purgathofer W (2013) A survey of urban reconstruction. Comput Graph Forum 32(6):146–177

Musliman IA, Rahman AA, Coors V (2008) Implementing 3D network analysis in 3D GIS. Int Arch ISPRS 37(part B):913–918

Nebiker S, Bleisch S, Christen M (2010) Rich point clouds in virtual globes–a new paradigm in city modeling? Comput Environ Urban Syst 34(6):508–517

Oehlke C, Richter R, Döllner J (2015) Automatic detection and large-scale visualization of trees for digital landscapes and city models based on 3D point clouds. In: 16th conference on digital landscape architecture, pp 151–160

Ostrowski S, Jozkow G, Toth C, Vander Jagt B (2014) Analysis of point cloud generation from UAS images. ISPRS Ann Photogram Remote Sens Spat Inf Sci 2(1):45–51
Pasewaldt S, Semmo A, Trapp M, Döllner J (2012) Towards comprehensible digital 3D maps. In: International symposium on service-oriented mapping, pp 261–276
Paredes EG, Bóo M, Amor M, Bruguera JD, Döllner J (2012) Extended hybrid meshing algorithm for multiresolution terrain models. Int J Geogr Inf Sci 26(5):771–793
Pătrăucean V, Armeni I, Nahangi M, Yeung J, Brilakis I, Haas C (2015) State of research in automatic as-built modelling. Adv Eng Inform 29(2):162–171
Peters R, Ledoux H (2016) Robust approximation of the medial axis transform of LiDAR point clouds as a tool for visualization. Comput Geosci 90(A):123–133
Preiner R, Jeschke S, Wimmer M (2012) Auto splats: dynamic point cloud visualization on the GPU. In: Eurographics symposium on parallel graphics and visualization, pp 139–148
Puente I, González-Jorge H, Martínez-Sánchez J, Arias P (2013) Review of mobile mapping and surveying technologies. Measurement 46(7):2127–2145
Rautenbach V, Coetzee S, Iwaniak A (2013) Orchestrating OGC web services to produce thematic maps in a spatial information infrastructure. Comput Environ Urban Syst 37:107–120
Remondino F, Menna F, Koutsoudis A, Chamzas C, El-Hakim S (2013a) Design and implement a reality-based 3D digitisation and modelling project. In: Digital heritage international congress 2013, pp 137–147
Remondino F, Spera MG, Nocerino E, Menna F, Nex F, Gonizzi-Barsanti S (2013b) Dense image matching: comparisons and analyses. In: Digital Heritage International Congress, pp 47–54
Richter R, Döllner J (2010) Out-of-core real-time visualization of massive 3D point clouds. In: 7th international conference on virtual reality, computer graphics, visualization and interaction in Africa, pp 121–128
Richter R, Döllner J (2014) Concepts and techniques for integration, analysis and visualization of massive 3D point clouds. Comput Environ Urban Syst 45:114–124
Richter R, Behrens M, Döllner J (2013a) Object class segmentation of massive 3D point clouds of urban areas using point cloud topology. Int J Remote Sens 34(23):8408–8424
Richter R, Kyprianidis JE, Döllner J (2013b) Out-of-core GPU-based change detection in massive 3D point clouds. Trans GIS 17(5):724–741
Richter R, Discher S, Döllner J (2015) Out-of-core visualization of classified 3D point clouds. In: 3D Geoinformation Science, pp 227–242
Rothermel M, Wenzel K, Fritsch D, Haala N (2012) Sure: photogrammetric surface reconstruction from imagery. In: LC3D Workshop, Berlin, vol 8
Rusu RB, Cousins S (2011) 3D is Here: point cloud library (PCL). In: IEEE international conference on robotics and automation, pp 1–4
Rüther H, Held C, Bhurtha R, Schroeder R, Wessels S (2012) From point cloud to textured model, the Zamani laser scanning pipeline in heritage documentation. South Afr J Geomatics 1(1):44–59
Schoedon A, Trapp M, Hollburg H, Döllner J (2016) Interactive web-based visualization for accessibility mapping of transportation networks. EuroVis 2016—Short papers
Semmo A, Döllner J (2014) An interaction framework for level-of-abstraction visualization of 3D geovirtual environments. In: 2nd ACM SIGSPATIAL international workshop on interacting with maps, pp 43–49
Semmo A, Trapp M, Jobst M, Döllner J (2015) Cartography-oriented design of 3D geospatial information visualization—overview and techniques. Cartographic J 52(2):95–106
Shreiner D, Sellers G, Kessenich JM, Licea-Kane BM (2013) OpenGL programming guide: the official guide to learning OpenGL, Version 4.3. Addison-Wesley Professional, USA
Schütz M, Wimmer M (2015) High-quality point-based rendering using fast single-pass interpolation. In: Digital heritage, pp 369–372
Sigg S, Fuchs R, Carnecky R, Peikert R (2012) Intelligent cutaway illustrations. In: IEEE Pacific visualization symposium, pp 185–192
Stojanovic V, Richter R, Trapp M, Döllner J (2017) Comparative visualization of BIM geometry and corresponding point clouds. Int J Sustain Dev Plann 13(1):12–23

Storti D, Yurtoglu M (2015) CUDA for engineers: an introduction to high-performance parallel computing. Addison-Wesley Professional, USA

Trapp M, Glander T, Buchholz H, Döllner J (2008) 3D generalization lenses for interactive focus + context visualization of virtual city models. In: 12th international conference information visualisation, pp 356–361

Vaaraniemi M, Freidank M, Westermann R (2013) Enhancing the visibility of labels in 3D navigation maps. In: Progress and new trends in 3D geoinformation sciences, pp 23–40

van Oosterom P, Martinez-Rubi O, Ivanova M, Horhammer M, Geringer D, Ravada S, Tijssen T, Kodde M, Gonçalves R (2015) Massive point cloud data management: design, implementation and execution of a point cloud benchmark. Comput Graph 49:92–125

van Oosterom P, Zlatanova S, Penninga F, Fendel E (eds) (2008) Advances in 3D geoinformation systems. Springer, Berlin

Wang R, Qian X (2010) OpenSceneGraph 3.0: beginner's guide. Packt Publishing, UK

Würfel H, Trapp M, Limberger D, Döllner J (2015) Natural phenomena as metaphors for visualization of trend data in interactive software maps. Comput Graph Vis Comput 2015: 69–76

Chapter 3
Establishing Common Ground Through INSPIRE: The Legally-Driven European Spatial Data Infrastructure

Vlado Cetl, Robert Tomas, Alexander Kotsev, Vanda Nunes de Lima, Robin S. Smith and Markus Jobst

Abstract Back in the 1990s, there were several barriers for accessing and using the spatial data and information necessary for environmental management and policy making in Europe. These included different data policies, encodings, formats and semantics, to name a few. Data was collected for, and applied to, domain specific use cases and comprehensive standards did not exist, all impacting on the re-usability of such public sector data. To release the potential of spatial data held by public authorities and improve evidence-based environmental policy making, action was needed at all levels (Local, Regional, National, European) to introduce more effective data and information management and to make data available for citizens' interest. The INSPIRE Directive, the Infrastructure for Spatial Information in Europe, directly addresses this set of problems. The Directive came into force on 15 May 2007, with full implementation in every EU Member State required by 2021. It combines both, a legal and a technical framework for the EU Member States, to make relevant spatial data accessible and reused. Specifically, this has meant making data discoverable and interoperable through a common set of standards, data models and Internet services. The Directive's data scope covers 34 themes of cross-sector relevance as a decentralised infrastructure where data remains at the place it can be best maintained. A great deal of experience has been gained by public administrations through its implementation. Due to its complexity and wide scope, this is taking place in a stepwise manner, with benefits already emerging as important deadlines approached. Efficient and effective coordination are following the participatory approach established in its design. It is timely to reflect on 10 years of progress of the "cultural change" which the European Spatial Data Infrastructure represents. We therefore, consider the lessons INSPIRE is

The views expressed are purely those of the authors and may not in any circumstances be regarded as stating an official position of the European Commission.

V. Cetl (✉) · R. Tomas · A. Kotsev · V. N. de Lima · R. S. Smith
European Commission, Joint Research Centre, Ispra, Italy
e-mail: vlado.cetl@ec.europa.eu

M. Jobst
Research Group Cartography, Vienna University of Technology, Vienna, Austria

J. Döllner et al. (eds.), *Service-Oriented Mapping*, Lecture Notes in Geoinformation and Cartography, https://doi.org/10.1007/978-3-319-72434-8_3

offering for those interested in joined-up and federated approaches to geospatial data-sharing and semantic interoperability across borders and sectors. The approach itself is evolving through this experience.

Keywords European spatial data infrastructure · INSPIRE · Data interoperability Service-oriented architecture

3.1 Introduction: The Need for a European Spatial Data Infrastructure

Climate change, natural disasters such as floods, air pollution or any environmental phenomenon do not stop at political borders and exercise complex, inter-related effects on society (Masser 2005). To take effective preventative measures, the mitigation of impacts occurring from local to global levels or to support sustainable economic, social and environmental development, information and georeferenced data must be shared across organisations and borders (Rajabifard et al. 2002). Such activities involve stakeholders from different contexts, including public administrations, businesses, research bodies and citizens and their combined efforts lead to the results that we see in practice.

Looking back twenty years to the 1990s, the accessibility and online sharing of public sector data were minimal. Finding content was very difficult. Documentation was poor or missing and data were kept in incompatible formats. It was difficult and time consuming to combine datasets from different sources (Craglia 2010). Data-sharing was also hampered by cultural and institutional barriers, including closed or non-existent data policies. At the same time, European Union (EU) policies needed data throughout the whole policy cycle (for formulation, assessment and monitoring). This situation needed to take into account the cultural, technical and political diversity of situations across Europe's regions.

In order to overcome these challenges, strong coordination was needed between stakeholders at European and national levels. The most appealing solution for all was a pan-European Spatial Data Infrastructure (SDI), leveraging on existing national and regional data infrastructures. In particular, this was addressed on a political level through the establishment of a European framework directive, Infrastructure for Spatial Information in the European Community—INSPIRE (Directive 2007/2/EC, in the following referred to as the "Directive").

Within this chapter, exactly ten years after its adoption, we take the opportunity to reflect on the development and implementation of the Directive. We also highlight the benefits it is bringing through the establishment of both, its legal and technical framework. Structurally, the chapter is organised in four sections. An introductory section, defining the context (Sect. 3.1) is followed by an extensive overview of the design and implementation of the Directive (Sect. 3.2). It continues with a discussion on the selected benefits for society (Sect. 3.3). Section 3.4

concludes with a discussion on the recognised challenges and an outlook for the future evolution of the infrastructure.

3.2 INSPIRE: An Overview

As with most SDIs, technical framework needs to be accompanied by a set of organisational rules and agreements. In the case of INSPIRE the technical framework was strengthened by political framework through European law. The setting-up of both was itself a multi-disciplinary, inclusive and transparent process, which unique experience is part of best practices. Those best practices have to be considered when such transversal innovation is in the act. This section gives an overview of how the European SDI has been developed and implemented.

3.2.1 Developing the Framework—Legal and Technological Setting

The legal framework has been set by the Directive (2007/2/EC) and interdependent legal acts, which are called implementing rules, in the form of Commission regulations and decisions. By design, the infrastructure itself is built upon the SDIs established and operated by European Union member states that are then made compliant with the implementing rules, covering its core components: metadata, network services, interoperability of spatial datasets and services, data-sharing and monitoring and reporting (see Sect. 3.2.2), together with the obligation to establish a national coordination body.

As a Directive, the legal obligations did not come into force immediately but had to be transposed into national law. Its scope applies to spatial data held by, or on the behalf of, public administrations in performance of public tasks, with a focus on environmental policies or policies which have an impact on the environment. The Directive does not require a collection of new data, instead existing data should be transformed to fit agreed data models. The SDI is also developed and implemented in a decentralised and distributed manner, mainly following fundamental principles (European Commission 2007a, b):

- Data should be collected only once and kept where it can be maintained most effectively.
- It should be possible to combine seamless spatial information from different sources across Europe and share it with many users and applications.
- It should be possible for information collected at one level/scale to be shared with all other levels/scales; i.e. detailed for thorough investigations and general for strategic purposes.

- Geographic information needed for good governance at all levels should be readily and transparently available.
- Easy to find what geographic information is available, how it can be used to meet particular needs, and under which conditions it can be acquired and used.

More specifically, its thematic scope involves 34 data themes (Fig. 3.1), divided in three annexes of the Directive that reflect two main types of data: spatial reference data presented in annex I and partly in annex II in order to define a location reference that the remaining themes can then refer to.

Such a broad, cross-sector data infrastructure was not only intended to support European policies but also national policies and requirements at all levels of government. Benefits are likely to reach actors beyond the immediate scope, including businesses and citizens.

To guide implementation, these legal acts were supplemented with a detailed technical framework. This consisted of a set of technical guidelines and tools developed in a collaborative manner, and based on international standards (see Sect. 3.2.2).

The development of implementing rules and technical guidelines for all infrastructure components (metadata, services, data interoperability, monitoring and reporting) was an interactive process (Fig. 3.2) that included many stakeholders (Fig. 3.6).

3.2.2 *Infrastructure Components*

The implementing rules for metadata, the interoperability of data theme, the network services (that help to share the infrastructure's content online) and data

Annex I

1. Coordinate reference systems
2. Geographical grid systems
3. Geographical names
4. Administrative units
5. Addresses
6. Cadastral parcels
7. Transport networks
8. Hydrography
9. Protected sites

Annex II

1. Elevation
2. Land cover
3. Ortho-imagery
4. Geology

Annex III

1. Statistical units
2. Buildings
3. Soil
4. Land use
5. Human health and safety
6. Utility and governmental services
7. Environmental monitoring facilities
8. Production and industrial facilities
9. Agricultural and aquaculture facilities
10. Population distribution – demography
11. Area management/ restriction/regulation zones & reporting units
12. Natural risk zones
13. Atmospheric conditions
14. Meteorological geographical features
15. Oceanographic geographical features
16. Sea regions
17. Bio-geographical regions
18. Habitats and biotopes
19. Species distribution
20. Energy Resources
21. Mineral resources

Fig. 3.1 INSPIRE themes, organised in three Annexes. *Source* INSPIRE Directive (2007/2/EC)

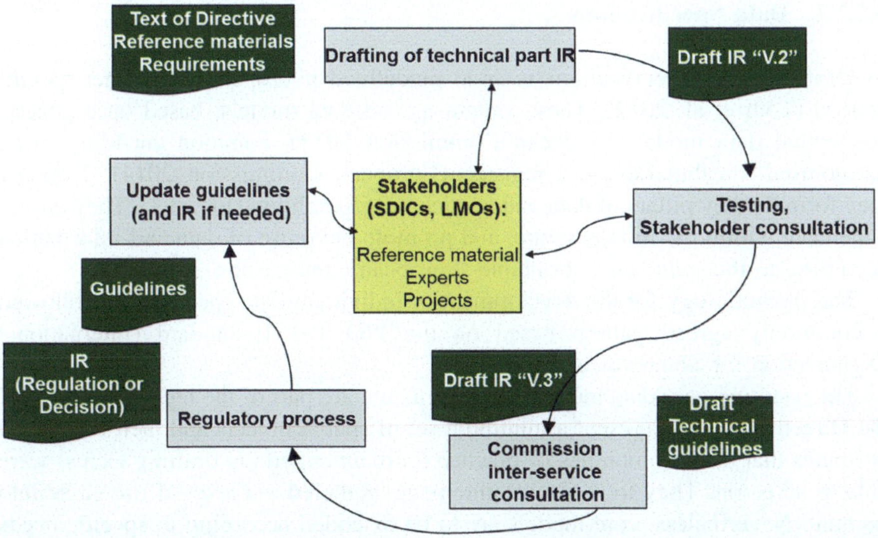

Fig. 3.2 Development cycle of implementing rules and technical guidelines. *Source* EC, Joint Research Centre

sharing are complemented by 'non-legally binding' technical guideline documents. These guidelines explain possible technical approaches for implementing the legal requirements and embed additional recommendations that may help in their implementation for a range of use cases.

3.2.2.1 Metadata

Member States ensure that metadata are created for the spatial data sets and services corresponding to one or more of the INSPIRE themes. Those metadata have to be kept up-to-date in accordance with the data and services (European Commission 2007a, b). INSPIRE metadata are based on the well-established EN ISO 19115 and EN ISO 19119 standards. Two sets of metadata elements are defined:

- discovery metadata,
- metadata elements for interoperability (sometimes also referred to as evaluation and use metadata).

In addition, some theme-specific requirements for the discovery metadata elements may occur. Those are described within the data specification of the related theme.

Metadata should be created and published through the network services in order to make the actual data discoverable. Data access is then established through view and download services which are documented in the metadata as well.

3.2.2.2 Data Specifications

In INSPIRE data specifications refer to predefined interoperability target specifications (Tóth et al. 2012). They contain agreed data models, based on a generic conceptual data model (European Commission 2013), common encoding rules, harmonised vocabularies, and registers (European Commission 2014). Together they form the key pillars of data and service interoperability (Fig. 3.3). They ensure coherence within the infrastructure, and promote the reuse of data and information according to the '*once-only*' principle (European Commission 2016a, b).

The methodology for the development of individual data specification followed a commonly agreed pattern based on the ISO 19131 standard (International Organization for Standardization 2010).

The resulting core data models for each theme are part of the legal provisions of the Directive. They represent a minimum set of spatial objects and their associated attributes that the communities of practice (domain experts in drafting teams) were able to agree on. They took also requirements gathered via a set of use cases into account. Nevertheless, core models are to be extended according to specific needs of a given domain, national use cases or applications that a user is confronted with.

The presence of controlled vocabularies is another essential pillar of interoperability. The spatial object characteristics (properties) are described by the commonly agreed semantics, which are expressed in the form of enumerations or code lists with precisely defined values and terms. All the semantics included in the legal text have been translated into 24 languages and are now accessible via the INSPIRE registry service (see Sect. 3.2.2.4). This central infrastructure component currently

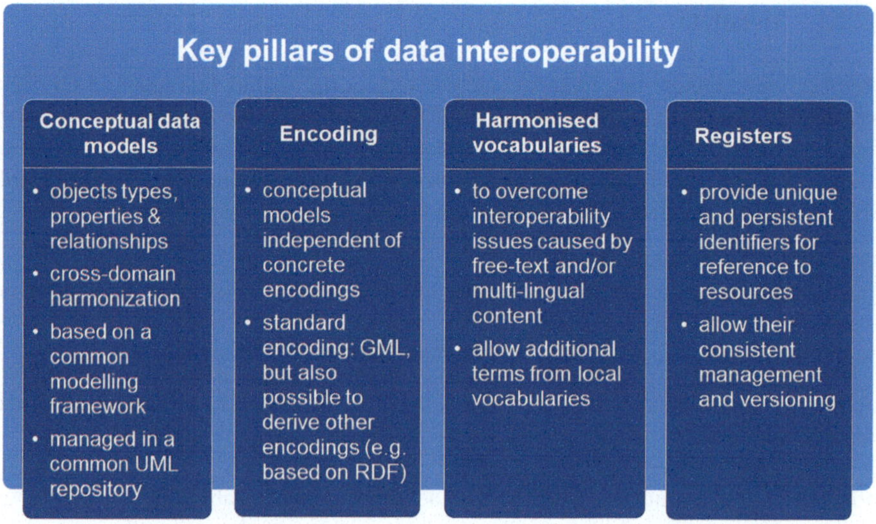

Fig. 3.3 Pillars of data interoperability in INSPIRE. *Source* EC, Joint Research Centre

contains seven different registers, including a code list register that further facilitates data and service interoperability.

Figure 3.4 illustrates how the real-world is modelled through consensus-based data specifications and an associated data model.

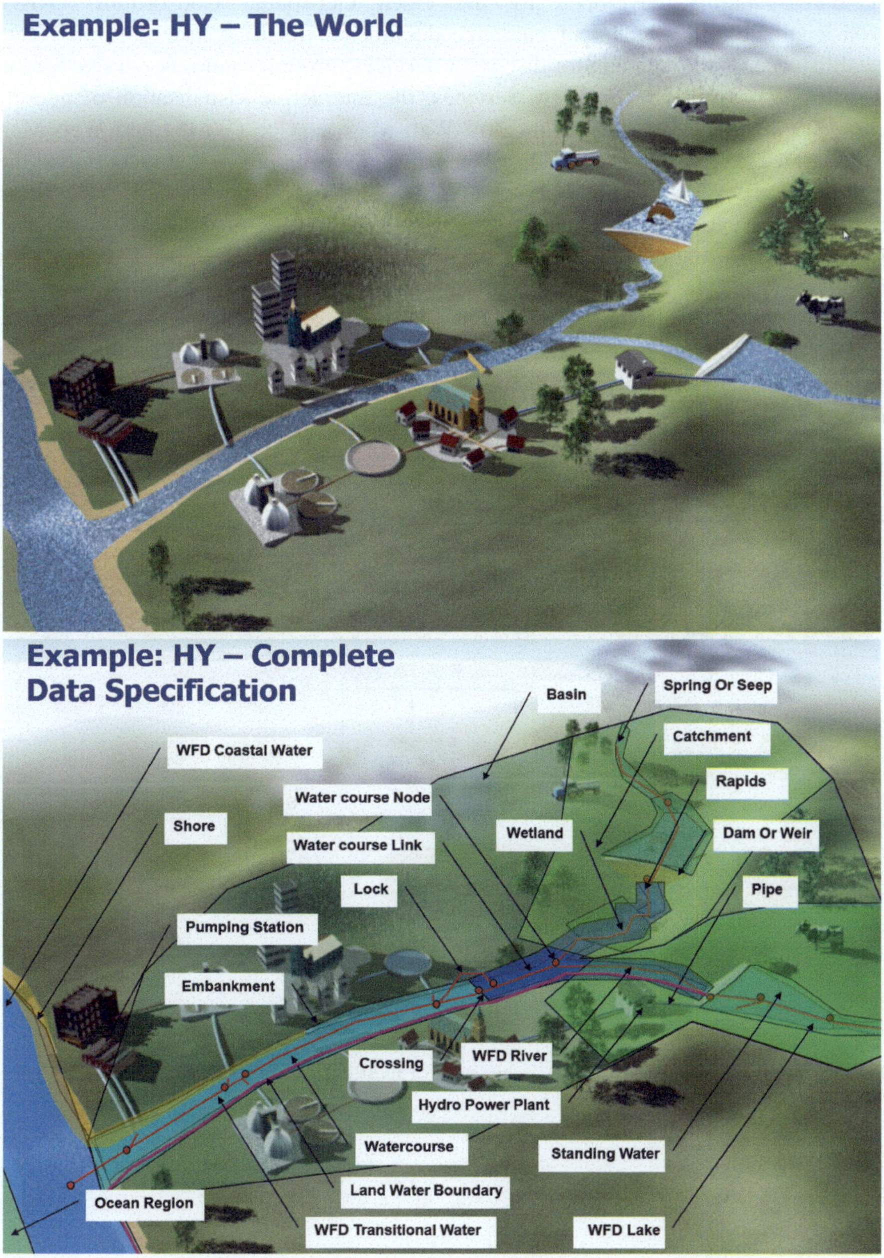

Fig. 3.4 Abstraction of the real world in INSPIRE. Example from the "Hydrography" data theme. *Source* EC, Joint Research Centre

3.2.2.3 Network Services

Data, functionality and metadata are shared through web-based services, referred to as Network Services (European Commission 2009), based on a Service-Oriented Architecture (SOA) approach (Fig. 3.5). These are based on well-established standards, mainly developed by the Open Geospatial Consortium (OGC). Non-legally binding technical guidance documents illustrate how data providers establish access to metadata for discovery services through catalogue service for the web (CSW). Similarly for View Services, the interactive visualisation of georeferenced content, involves guidance on Web Map Service (WMS) and Web Map Tile Service (WMTS). Download services also have guidelines that recommend the use of Atom, Web Feature Service (WFS), Web Coverage Service (WCS) and Sensor Observation Service (SOS), for appropriate types of data. There are also various transformation services defined, which can support coordinate and data transformations. In addition, there are generic services (registry and other spatial data services), that are implemented on a national as well as European level.

Deadlines for the establishment of network services precede those that address data harmonisation. Geospatial data is made available 'online first', thus unlocking it as is for further use (see Fig. 3.7).

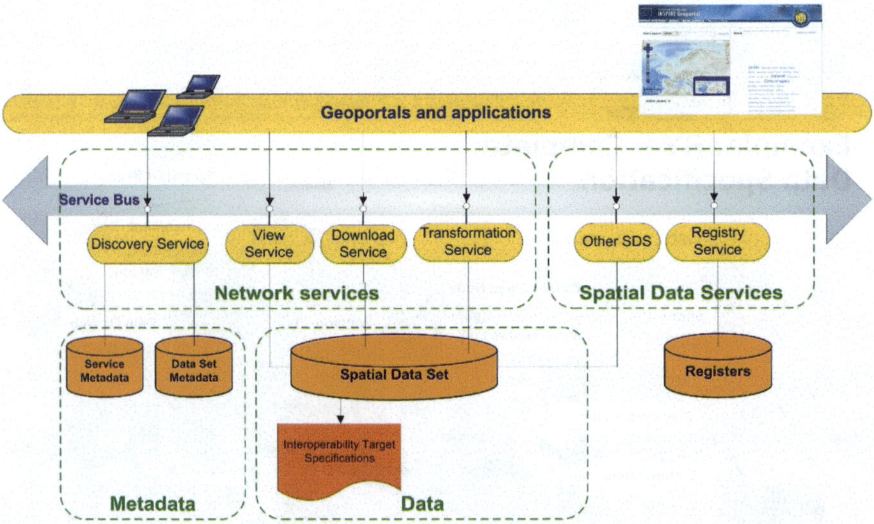

Fig. 3.5 Distributed service-oriented architecture of INSPIRE. *Source* EC, Joint Research Centre

3.2.2.4 Central Components

As shown in Fig. 3.5, several central components are illustrated that either establish interoperability on a pan-European scale or support the use of the infrastructure. Alongside national geoportals, the pan-European INSPIRE geoportal[1] serves as a central access point to data and services from organisations in the EU Member States and EFTA countries.[2] It enables not only cross-border data discovery, visualisation and use, but also metadata, data and service validation.

Another important central component of infrastructure is the INSPIRE Registry,[3] an online system to provide access to several centrally-managed registers. The content of these registers are based on the Directive, implementing rules and technical guidelines, especially for metadata and data and service interoperability. These registers facilitate semantic interoperability by offering common, persistent and clear descriptions of infrastructure items, referenced through unique identifiers. Furthermore, the content is neutral from a natural language point-of-view. Examples for such items include (i) data themes, (ii) code lists and (iii) application schemas. Registers provide a means for data providers to assign identifiers to items and their labels, definitions and descriptions (in different languages). Therefore it follows common reference material that uses controlled semantics. It can also be extended depending on the application or policy requirements.

3.2.2.5 Data-Sharing

Under the terms of INSPIRE Directive, each Member State shall adopt measures for the sharing of and enabling access to spatial data sets and services between its public authorities for the purposes of public tasks impacting the environment. It requires that such measures "preclude any restrictions likely to create practical obstacles" (European Commission 2010), occurring at the point of use, e.g. sharing of spatial data sets and services. The arrangements for sharing of spatial data sets and services shall be open to public authorities of other Member States and to the institutions and bodies of the European Union. They also "shall be open, on a reciprocal and equivalent basis, to bodies established by international agreements to which the Community and Member States are parties, for the tasks impacting the environment". These provisions are particularly important for supporting the European contribution to the global 2030 Agenda for sustainable development (United Nations 2015) and the United Nations Sustainable Development Goals.

Available guidance documents give examples of data and service sharing.[4] They include non-mandatory sample agreements which can be modified and adopted by

[1]http://inspire-geoportal.ec.europa.eu.

[2]Iceland, Lichtenstein, Norway and Switzerland.

[3]http://inspire.ec.europa.eu/registry.

[4]Guidelines and good practice documents on data sharing. http://inspire.ec.europa.eu/data-and-service-sharing/62.

public authorities. The proposed approach follows a general understanding to obtain a higher level of harmonisation with positive implications on national and European levels.

However, a number of organisational and legal challenges remain. There is still a diversity of license conditions between countries and in some cases even among national public authorities. At the same time the increasing convergence to 'open data' is an evident trend on European scale. It was recognised in the Member States reports on monitoring the implementation of INSPIRE within the last three years.[5] The evolution of this movement is fully in line with the principles of INSPIRE and its main requirement to make data accessible (see Sect. 3.2.1). Adopting this data policy approach is often a consequence of collaboration established between public administrations that decide to share the costs of implementing data services and interoperability requirements. At the same time, they have to agree on common sharing arrangements under their national Open Data policy agenda.

3.2.3 Main Actors and Stakeholder Engagement

Beyond these legal and technical aspects it was recognised that appropriate coordination mechanisms would be needed on multiple levels. Stakeholders were engaged from the outset, even before the Directive was published. The inclusiveness was important especially in the light of the decentralised architecture, where themes addressed data under the governance of different Member State ministries and at different levels of administration. The multi-disciplinarily nature of the domains, combined with the rapidly evolving ICT landscape, required the full and inclusive involvement of hundreds of stakeholders and experts from all across Europe (Fig. 3.6). Two groups of stakeholders were identified: Legally Mandated Organisations (LMOs) and Spatial Data Interest Community (SDIC). They had an active and fundamental role in proposing, developing and reviewing all technical aspects (Fig. 3.6). Member states were represented by the INSPIRE committee and appointed national contact points, who regularly provide information about implementation progress in their countries.

The breadth of activity in the Member States was also reflected by the European Commission bringing together different actors. Directorate General Environment is in the role of the policy master and the Joint Research Centre (JRC) as the technical coordinator. Eurostat and the European Environment Agency support with use-case specific developments.

Considering the large number of stakeholders involved on a voluntary basis, the development of the infrastructure and the engagement of stakeholders can be considered as a leading practice in policy and technical development.

Furthermore, after adoption of the legal frame, this spirit was sustained in the implementation and maintenance phases. A Commission Expert Group for the

[5]https://inspire.ec.europa.eu/Monitoring-and-Reporting/Monitoring-and-Reporting/69.

Fig. 3.6 INSPIRE stakeholder involvement 2007–2013. *Source* EC, Joint Research Centre

Maintenance and Implementation of INSPIRE (MIG)[6] is in place to agree a multi-annual work programme (MIF 2016) to address technical issues that need to be solved in a harmonised way for all Member States. Therefore it supports interoperability across borders, coherence between policies and the effectiveness of the results. Collaboration is also reflected in the development and sharing of reusable tools to facilitate implementation. The use of thematic platforms helps to share implementation experiences among implementers, developers and provide more information to citizens.

[6]http://inspire.ec.europa.eu/inspire-maintenance-and-implementation/.

3.2.4 Implementation and Maintenance

The deadline for Member States to transpose the Directive into their national legislation was 15 May 2009. The implementation process started immediately after, following an implementation roadmap that treats each component individually through a stepwise approach (Fig. 3.7).

The first important milestone was in December 2013, when Member States were obliged to provide their data as-is. This step established metadata and exposed data

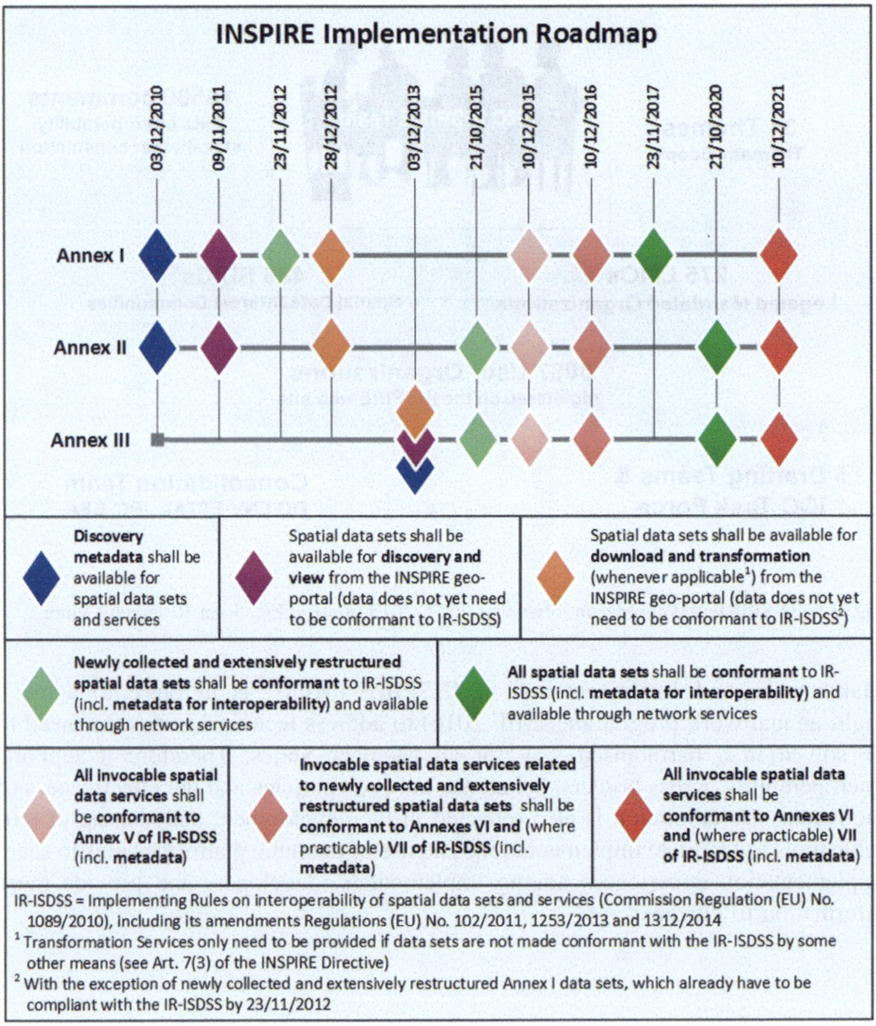

Fig. 3.7 INSPIRE implementation roadmap. *Source* EC, Joint Research Centre

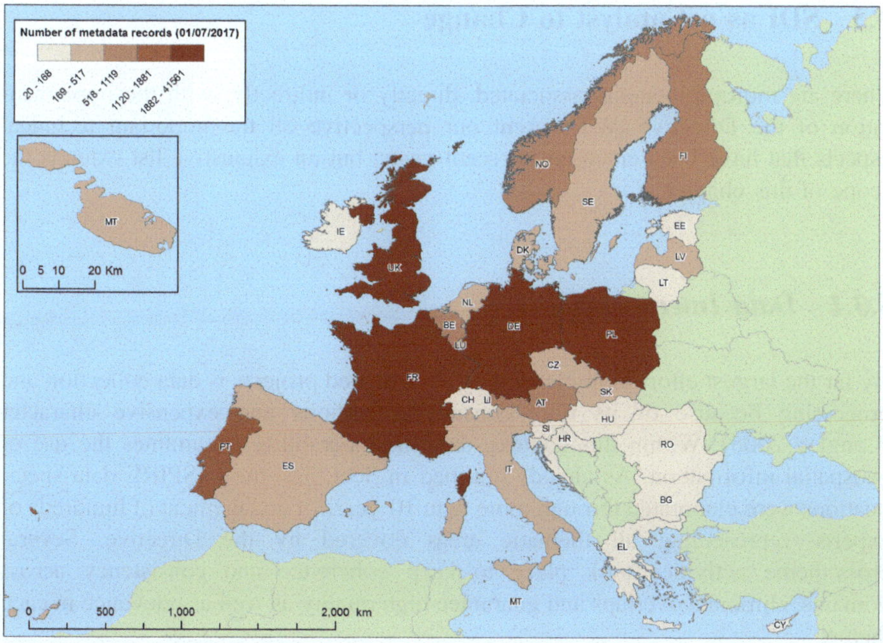

Fig. 3.8 Number of metadata records discoverable through the INSPIRE Geoportal (as of 01 July 2017). *Source* EC, Joint Research Centre

through network services. Consequently, by December 2017, datasets that fall under the scope of Annex I are expected to be in place and interoperable. Similarly, by the end of 2020, data for Annex II and III should also be conform to the Directive's requirements.

The mid-term evaluation of the status of the Directive's implementation was carried out in 2014. The results showed that a satisfactory evolution of the issues addressed by INSPIRE has been taken place. Increasing volumes of geospatial data are becoming available through the infrastructure. Figure 3.8 gives an overview of the metadata that is currently made available within the infrastructure by using discovery (CSW) services. The evaluation highlighted some ongoing challenges that include the need for coordination on multiple levels, as well as a more profound role of the private sector (European Environment Agency 2014).

Beyond this formal monitoring, collaboration can also be seen through many occasions where public authorities share their experiences, lessons learned[7] and work together to make the most out of the infrastructure.

[7]https://inspire.ec.europa.eu/document-tags/inspire-your-country.

3.3 SDI as a Catalyst to Change

There are multiple benefits associated, directly or indirectly, with the implementation of the Directive. We present our perspective on the important technical aspects that have been emerged in recent years, but an exhaustive list extends the scope of this chapter.

3.3.1 Data Interoperability

By far the largest effort in implementing data-related projects is data collection and processing because of its time consuming, tedious and expensive character (Longley 2005). Within this context, data interoperability streamlines the use of geospatial information. As already outlined in Sect. 3.2, the INSPIRE data specifications were elaborated through more than 10 years of engagement of hundreds of experts, representing all thematic areas covered by the Directive. Several cross-theme activities took place to keep coherence and consistency across domains, eliminate overlaps and guarantee high quality as well as relevance in each theme.

Interoperability of spatial data allows cross border and -thematic usage. As such INSPIRE could be seen as the "lingua franca" for geospatial data in Europe. The core data models as contained in the legal framework can be extended, following the rules defined by the generic conceptual model (European Commission 2013) and the documented semantics. This is one aspect of flexibility that preserves interoperability when applying a data model for a specific use case. It is also worth noting that on a technical level, many of the data and feature types are reused in different data models. This helps to ensure consistency across the infrastructure. Furthermore, adopting INSPIRE facilitates cross-domain interoperability, i.e. data from multiple domains are easily combined and used together. Examples include analyses in a cross-cutting setting, such as environmental impact assessments (Vanderhaegen and Muro 2005), natural hazards and disaster reduction (Tomas et al. 2015) or exposure science and spatial epidemiology (Elliott and Wartenberg 2004).

3.3.2 Streamlined Data Governance

INSPIRE is conceptualised in a way that it does not hamper processes internal to individual organisations. Instead, it helps to break silos of existing information within the public administration and streamlines data related processes on several levels. Besides the obvious gains from the improved interoperability with others, implementing the Directive offers additional benefits. They are associated with an optimised internal data management in public administration, such as (i) operating an inventory of available resources through metadata, (ii) avoiding duplication of data-gathering between organisations (once-only principle), (iii) using services for internal purposes, (iv) establishing rigid identification patterns for spatial objects based on Uniform Resource Identifiers (URI), (v) making data and information available for the private sector and citizens, including reinforcing e-Government initiatives and (vi) supporting open data developments that can help to release more data beyond the scope of the Directive.

When extended to the national level, the streamlined data governance approach leads to the establishment of an efficient governance structure. In some cases, such as the Dutch "public geo datasets at your service—PDOK" (Kruse 2016), the obligation to transpose and implement INSPIRE on a national level started a process through which heterogeneous actors were assigned clear roles according to their own data responsibility. This helped to avoid duplications in data storage and created a collaborative environment with different parties. From a governance perspective, the implementation effort has also aided coordination between public sector actors which includes different (sub-national) levels of government. Many issues are addressed in such contexts: (i) joint decision-making on specific data sets within the infrastructure, (ii) how to make data available in a most efficient manner, (iii) under which conditions data will be shared, (iv) how to maintain the infrastructure and its data, and (v) how to reach the users inside and outside public administrations.

INSPIRE is likely to be the most complex and geographically extensive SDI initiative in the world nowadays. The lessons learned from its data harmonisation activities in Europe are highly relevant to the global agenda. They have been recognised, for example, by the United Nations Committee of Experts on Global Geospatial Information Management (UN-GGIM CoE 2014) and the Global Earth Observation System of Systems (GEOSS, Butterfield et al. 2008) as notable European contributions to efforts on this topic.

3.3.3 Service-Oriented Architecture (SOA)

As outlined above in Sect. 3.2.2.3 INSPIRE utilises the SOA approach, where data are exposed in a distributed manner through web (network) services. These services allow to process existing geospatial data and are based on well-established standards. There is a growing number of readily available client applications (web, mobile, desktop) that can be used in order to consume the available data. Furthermore, services are reusable by definition. Geospatial data and metadata are independent from a particular use case and can be easily accessed.

Considering the diversity of different organisational settings and types of data within the infrastructure, a high-level objective is defined: data should be kept and exposed where it can be managed most efficiently. The implementation of services also facilitates the distribution of the most up-to-date information. Through the establishment of network services, INSPIRE is already triggering a paradigm shift that implies the establishment of architectures for 'pulling' geospatial data in contrast with the traditional approach for 'pushing' data to a centralised repository (see e.g. Kotsev et al. 2015).

There is no universal solution to fit all possible use cases regarding the establishment of services. In most cases, data are served close to the data provider, i.e. on a national or sub-national level. At the same time there is an increasing number of cases where the national coordination body is supporting data providers through a common technical infrastructure (e.g. Malta, Finland, and Czech Republic).[8] Similarly, this 'aggregation' approach is followed on a European level by thematic umbrella organisations, such as EuroGeoSurveys.[9] In some cases the distributed SOA is supplemented by a central infrastructure component, which has been put in place for performant access to pan-European data. Examples of this approach include monitoring invasive alien species through the European Alien Species Information Network—EASIN (Katsanevakis et al. 2015), industrial emissions (EEA 2017), air quality (Schleidt 2013 and Kotsev et al. 2015), and geology.[10]

From our perspective, the benefits from the exposure of data on multiple levels are manifold. Data models offer inspiration and a starting point for data users to create their own outputs. Apart from the value added, due to the implementation of network services, there are several challenges to be addressed. They are mentioned in Sect. 3.4.

3.3.4 Flexible Technical Development

The Directive and implementing rules are neutral from a technological and application point of view. They do not require a particular technology to be adopted. Existing technical documents only provide guidance on how particular requirements may be approached by using selected technological solutions. There are many options for data providers to choose from, depending on their particular needs. The described technological flexibility allows the uptake of emerging trends from relevant domains such as linked and open data, Internet of Things and cloud computing. This adaptability for future requirements can be seen as characteristic to make the infrastructure future-proof. At the same time, considering how fast the technological scenery is changing, the question on which technology should be adopted, remains open. There is clearly a trade-off between restricting the technological choices in order to ensure stability and interoperability (i), and adapting the infrastructure to new trends (ii). From our perspective, preference should be

---

[8]https://inspire.ec.europa.eu/Monitoring-and-Reporting/Monitoring-and-Reporting/.

[9]http://www.europe-geology.eu/—EGDI—European Geological Data Infrastructure.

[10]http://minerals4eu.brgm-rec.fr/—EU-MKDP—European Union Minerals Knowledge Data Platform.

given to emerging technologies that not only satisfies the requirements of the Directive, but also provides additional value to providers and users of geospatial content.

3.3.5 *Community of Practice*

Over the years, it has become clear that stakeholders, who participated in the development and implementation process, have a sense of stewardship of the results (Dawes 2010; NRC 1994). They ensure that the data specifications are aligned to particular needs, reused and semantics preserved. Furthermore, these processes have led to the establishment of a European SDI community of practice. The European Commission is supporting this community by coordinating the INSPIRE Thematic Clusters Platform.[11] It provides an online collaborative space that (i) facilitates the sharing of implementation experience, (ii) optimises the infrastructure, (iii) supports the policy objectives and (iv) increases INSPIRE's usability across thematic domains.

As practical implementers of INSPIRE, the community of practice not only maintains the infrastructure but also caters for its evolution, reflecting emerging technological trends and societal requirements.

3.3.6 *Innovative Apps and Value-Added Services*

A key aspect of INSPIRE, as other SDIs, is that it is not an end in itself. The data infrastructure is a means to mobilise efforts and resources around guaranteed, consistent, high quality and trustworthy data-sharing across Europe. Third party solution providers that have not necessarily participated in the establishment of INSPIRE are provided with a unique opportunity to build value-added applications and services. They no longer need to invest in costly data discovery and processing. Instead, developers can concentrate on innovation and improving their products. At the same time, some of the software developed to support the implementation and use on national and European level (e.g. the INSPIRE Registry[12] and Find Your Scope tool[13]) are released as reusable free and open source software tools. Such an approach can be seen as example of reuse of public goods.

In addition, public authorities in their attempts to improve and optimise the services which they provide to citizens are increasingly involved in the development of new online services. Many of them have been built on top of geospatial

[11]https://themes.jrc.ec.europa.eu.

[12]https://inspire.ec.europa.eu/registry/.

[13]http://inspire.ec.europa.eu/portfolio/find-your-scope.

data provided through the infrastructure, while (appropriately) not necessarily recognising the efforts to get such content to them. That is how INSPIRE is reinforcing the agenda related to e-government (recently referred to as digital government).

The following example highlights value added modifications in the Austrian geoinformation management. Austria defines a digital agenda for its e-government (eGFS 2017) and therefore authorities are motivated to go beyond an infrastructure on the basis of minimum consensus. The first step, but not the easiest one, is collaboration in data collection e.g. production of orthoimageries (OI) or Airborne Laser Scanning (ALS). At least each partner should make use of products under their own license model. Even the distribution in Open Government Data[14] should be enabled for all partners. Nevertheless, the main important aspects of infrastructural maintenance activities within authorities are transparency to all users, using responsible competences and enhancing quality in order to broaden the field of application. The technological framework of Service-Oriented Mapping supports these aforementioned aspects and results in economic, organisational and technological benefits.

This approach could be demonstrated trough federal maintenance of the data theme addresses. Addresses are core data because of their extensive use and key-role for the integration of geoinformation (geocoding). In Austria about 2100 municipalities collect the federal addresses. The central register of addresses instantly adds the geocoding on the basis of the Austrian cadastre and it includes several plausibility checks for quality control. In 2016 the quality of addresses has been enormously enhanced (Eurogeographics 2017) by using cross-linked geo-web services to the Austrian routes network.[15] A new workflow establishes a link from addresses to the nationwide transportation graph (GIP) which now enables the ability to route with these data. The address point is moved towards the route graph, but still stays within the parcel. The main aim for all stakeholders is to describe the access to the parcel. All involved geo-web services are autonomous, which means that the collection and initial quality control is not influenced by requesting "external" geo-web services at the GIP infrastructure. Asynchronous and periodical service mechanisms are installed to ensure a consistent dataset. Once more this example shows that collaboration leads to a win-to-win situation for all producing stakeholders: an enormous quality enhancement for addresses on one hand and precise routing functionality for the transportation graph on the other. The implementation as service-oriented architecture allows for flexible and productive linkage, which is almost independent from organisational structures. For the security of the system several specific mechanisms had to be enabled.

[14]https://www.data.gv.at/.

[15]http://www.gip.gv.at/gipat-en.html.

3.4 Discussion and Conclusions

Precisely 10 years after the adoption of the INSPIRE Directive, in this chapter we summarise some of the lessons learned, and discuss selected emerging trends that benefit the evolution of this pan-European SDI.

- INSPIRE, although not yet fully implemented, is effectively contributing to a European political priority of creating a Digital Single Market, by boosting the economy through the 'unlocking' public sector data.
- The Open Data movement and INSPIRE are mutually beneficial both in terms of development and outlook.
- INSPIRE represents a paradigm shift from 'data push' to 'data pull' that enables the reuse of data. These data are being made available from where it is best managed. This not only helps to build innovative apps, but also facilitates an evidence-based policy-making process.

At the same time, technological developments and societal challenges associated to digital transformation are quickly evolving. From our perspective, the inclusive and flexible approach used for the development of INSPIRE should continue to be applied in addressing them, in collaboration with other actors beyond the initial scope of the Directive. Sharing such experience across sectoral domains and disciplines is fundamental to reap from the investments done. Several technological challenges are addressed to ensure the sustainability and evolution of the spatial data infrastructure.

The full potential for production of innovative applications and services based on the content of the INSPIRE Geoportal[16]—a single point of access to all MS data and services—is yet to be reached. The recent JRC activity of developing "thematic views" on the Geoportal content (e.g. priority datasets for eReporting) follows the simplification of use direction defined by the Better Regulation principles.

In general the discoverability of geospatial data on the web remains a challenge. Being based on OGC standards, data in INSPIRE is accessed through the interaction with different services (CSW, WFS, WMS, etc.). Their use requires specific knowledge and often at least some understanding of the underlying infrastructure. This may be relatively clear to GIS users but limits the use of the data and services by others—non-expert users that prefer e.g. common search engines to discover content. This issue is addressed by a recent study (Geonovum 2016) and is being explored further at the JRC.

The infrastructure should adapt to changing technological trends. In that respect, new guidance may be produced that add value to the infrastructure while ensuring compliance with the legislation. Some of the questions to be addressed include e.g. the establishment of RESTful web services, use of asynchronous event-based architectures for data retrieval, encoding of data through GeoJSON, or JSON-LD (JavaScript Object Notation for Linked Data, JSON 2017).

[16]http://inspire-geoportal.ec.europa.eu.

Common semantics for spatial objects across Europe, documented in the central INSPIRE Registers, has been one of the most important interoperability achievements of INSPIRE that must be preserved and updated as requirements may evolve in the future. Common semantics achieved in INSPIRE also facilitate integration of data from different communities as well as sources (e.g. use the INSPIRE objects for eReporting (Schleidt 2013).

Emerging trends related to the use of cloud computing would optimise the performance and scalability of the service-oriented architecture and underlying data. Furthermore, they provide an opportunity for data analytics to become an inseparable part of the data infrastructure that INSPIRE does not directly address yet.

A great deal of experience has been gained by public administrations through INSPIRE's implementation, bringing innovation at all levels of public administration. Consequently, more official data is available online across Europe. It is also in line with other legislation such as the Aarhus Convention (UNECE 2017) giving the public better access to information about the conditions of their environment.

As outlined, the implementation, due to its ambition and wide scope, follows a progressive approach, according to different National and European levels of priority. It involves transversally public administration at all levels. Therefore it requires efficient and effective national coordination and monitoring. This is done by the coordination body set-up in each country. Similarly at European level, coordination is guaranteed across related policies, using the Commission instruments, following the Better Regulation's principles (European Commission 2015).

The collaboration between experts in Member States' public and private sectors, software solution providers, open source communities, international standardisation organisations, professional umbrella organisations and European institutions, has proven to be achieving wide benefits that are holding up INSPIRE as an example to follow. Similarly, INSPIRE goals, achievements as well as difficulties represent unique knowledge to be considered in recent ICT policy developments in Europe, in line with the ambitions of the Digital Single Market.

Unlocking' geospatial data remains challenging, being the technological aspects the easiest to solve. Continued interaction between all parties will ensure that investments and ongoing implementation will create a sustainable and extensible infrastructure for European and global gains.

References

Butterfield ML, Pearlman JS, Vickroy SC (2008) A system-of-systems engineering GEOSS: architectural approach. IEEE Syst J 2(3):321–332

Craglia M (2010) Building INSPIRE: the spatial data infrastructure for Europe. Arc News, pp 5–7. http://www.esri.com/news/arcnews/spring10articles/building-inspire.html

Commission Regulation (EU) No 1253/2013 of 21 October 2013 amending Regulation (EU) No 1089/2010 implementing Directive 2007/2/EC as regards interoperability of spatial data sets and services (2013). http://eur-lex.europa.eu/LexUriServ/LexUriServ.do?uri=OJ%3AL%3A2013%3A331%3A0001%3A0267%3AEN%3APDF

Dawes SS (2010) Stewardship and usefulness: policy principles for information-based transparency. Gov Inf Q 27(4):377–383. https://doi.org/10.1016/j.giq.2010.07.001

eGFS (eGovernment Factsheets) (2017) National interoperability framework observatory—eGovernment factsheets, Austria. https://joinup.ec.europa.eu/sites/default/files/ckeditor_files/files/eGovernment_in_Austria_March_2017_v_4_00.pdf. Accessed 16 Aug 2018

Elliott P, Wartenberg D (2004) Spatial epidemiology: current approaches and future challenges. Environ Health Perspect 112(9):998

Eurogeographics (2017) Annual report of eurogeographics member organisations. http://www.eurogeographics.org/sites/default/files/EURO6180_AnnualReport_Web_%28Spreads%29_190517_RH.pdf. Accessed 16 Aug 2017

European Commission (2007a) Directive 2007/2/EC of the European Parliament and of the council of 14 March 2007 establishing an infrastructure for spatial information in the European community (INSPIRE). http://eur-lex.europa.eu/legal-content/EN/ALL/?uri=CELEX:32007L0002

European Commission (2007b) INSPIRE metadata implementing rules: technical guidelines based on EN ISO 19115 and EN ISO 19119. http://inspire.ec.europa.eu/documents/Metadata/MD_IR_and_ISO_20131029.pdf. Accessed 01 Aug 2017

European Commission (2008a) Definition of annex themes and scope (D 2.3, Version 3.0). http://inspire.ec.europa.eu/reports/ImplementingRules/DataSpecifications/D2.3_Definition_of_Annex_Themes_and_scope_v3.0.pdf

European Commission (2008b) Methodology for the development of data specifications: baseline version (D 2.6, Version 3.0). http://inspire.ec.europa.eu/reports/ImplementingRules/DataSpecifications/D2.6_v3.0.pdf

European Commission (2009) Commission Regulation No 976/2009 of 19 October 2009 implementing Directive 2007/2/EC of the European Parliament and of the council as regards the network services

European Commission (2010) Commission Regulation (EU) No 268/2010 of 29 March 2010 implementing Directive 2007/2/EC of the European Parliament and of the council as regards the access to spatial data sets and services of the member states by community institutions and bodies under harmonised conditions

European Commission (2013) INSPIRE generic conceptual model. http://inspire.ec.europa.eu/documents/Data_Specifications/D2.5_v3.4rc3.pdf. Accessed 01 Aug 2017

European Commission (2014) Guidelines for the encoding of spatial data. https://inspire.ec.europa.eu/documents/guidelines-encoding-spatial-data. Accessed 01 Aug 2017

European Commission (2015) Better regulation. http://ec.europa.eu/smart-regulation/guidelines/ug_chap1_en.htm

European Commission (2016a) eGovernment action plan 2016–20120. https://ec.europa.eu/digital-single-market/en/european-egovernment-action-plan-2016-2020

European Commission (2016b) Guidelines for the use of observations & measurements and sensor web enablement-related standards in INSPIRE. http://inspire.ec.europa.eu/file/1638/download?token=EtGplOtQ

European Commission Report (2015) Report from the Commission to the Council and the European Parliament on the implementation of Directive 2007/2/EC of March 2007 establishing an Infrastructure for Spatial Information in the European Community (INSPIRE) pursuant to article 23. http://eur-lex.europa.eu/legal-content/EN/TXT/?uri=CELEX%3A52016DC0478R%2801%29

European Environment Agency (2014) Mid-term evaluation report on INSPIRE implementation. Joint EEA-JRC report. http://snig.dgterritorio.pt/Inspire/documentos/arquivo/Tech172014INSPIRE.pdf

European Environment Agency (2017) EU registry on industrial sites. Data model documentation. http://cdrtest.eionet.europa.eu/help/ied_registry/documents/EU%20Registry_datamodel.pdf. Accessed 28 June 2017

Geonovum (2016) Testbed: spatial data on the web. https://www.geonovum.nl/onderwerp-artikel/testbed-locatie-data-het-web. Accessed 01 Aug 2017

International Organization for Standardization (2010) ISO 19131:2007. Geographic information. https://www.iso.org/obp/ui/#iso:std:iso:19131:ed-1:v1:en

JSON (2017) JSON for linking data. https://json-ld.org/. Accessed 16 Aug 2017

Katsanevakis S, Deriu I, D'Amico F, Nunes AL, Pelaez Sanchez S, Crocetta F, Arianoutsou M, Bazos I, Christopoulou A, Curto G, Delipetrou P (2015) European alien species information network (EASIN): supporting European policies and scientific research. Manage Biol Invasions 6(2):147–157

Kruse D (2016) Public service on the map: a case history of PDOK; the public geodata portal in The Netherlands!. In: ECEG2016-Proceedings of 16th European conference on e-Government ECEG 2016. Academic Conferences and Publishing Limited, Reading, p 335

Kotsev A, Peeters O, Smits P, Grothe M (2015) Building bridges: experiences and lessons learned from the implementation of INSPIRE and e-reporting of air quality data in Europe. Earth Sci Inf 8(2):353–365

Longley P (2005) Geographic information systems and science. Wiley, New Jersey

Masser I (2005) GIS worlds: creating spatial data infrastructures, vol 338. ESRI press, Redlands, CA

MIF (2016) Maintenance and implementation work programme 2016–2020. https://ies-svn.jrc.ec.europa.eu/documents/58. Accessed 16 Aug 2017

NRC (National Research Council), Division on Earth and Life Studies, Commission on Geosciences, Environment and Resources, Mapping Science Committee (1994) Promoting the national spatial data infrastructure through partnerships. National Academies Press, USA. ISBN 9780309051415

Rajabifard A, Feeney MEF, Williamson IP (2002) Future directions for SDI development. Int J Appl Earth Obs Geoinf 4(1):11–22

Schleidt K (2013) INSPIREd air quality reporting. In: International symposium on environmental software systems. Springer, Berlin, pp 439–450

Tomas R, Harrison M, Barredo JI et al (2015) Towards a cross-domain interoperable framework for natural hazards and disaster risk reduction information. Nat Hazards 78:1545. https://doi.org/10.1007/s11069-015-1786-7

Tóth K, Portele C, Illert A, Lutz M, Nunes De Lima M (2012) A conceptual model for developing interoperability specifications in spatial data infrastructures. Reference report EUR 25280 EN. European Commission, Joint Research Centre. http://inspire.ec.europa.eu/file/1418/download?token=rzNYULUf

United Nations (2015) Transforming our world: the 2030 agenda for sustainable development. https://sustainabledevelopment.un.org/post2015/transformingourworld/publication

UNECE (2017) The Aarhus convention. http://www.unece.org/env/pp/treatytext.html. Accessed 16 Aug 2017

UN-GGIM Committee of Experts on Global Geospatial Information Management (2014) Europe region report to UN-GGIM

Vanderhaegen M, Muro E (2005) Contribution of a European spatial data infrastructure to the effectiveness of EIA and SEA studies. Environ Impact Assess Rev 25(2):123–142

Chapter 4
Evaluating the Efficiency of Various Styles of Distributed Geoprocessing Chains for Visualising 3D Context Aware Wildfire Scenes from Streaming Data

Lauren Hankel, Graeme McFerren and Serena Coetzee

Abstract Big data refers to the ever-increasing volumes of data being generated continuously by a large variety of sensors, such as smartphones and satellites. In this chapter, we explore solutions for challenges related to the velocity characteristic of big geospatial data. The research was inspired by the Advanced Fire Information System (AFIS), which provides near real-time fire detection from earth observation satellites. Users in Southern and East Africa, South America and Europe are automatically notified of active fire detections, with the hope that timeous information may help to mitigate the impact of wildfires. This chapter evaluates the efficiency of various styles of geoprocessing chains for generating enriched notifications containing 3D fire visualisations from an intermittent stream of active fire detection data generated by remote sensing satellites. The visualisation should be ready for viewing as soon as the user investigates the notification; this implies the requirement for rapid geoprocessing, since there may be hundreds of fire detections disseminated to hundreds of parties at any satellite overpass. Geoprocessing chains were implemented in Python using open-source libraries and frameworks. This study investigated efficiencies across three dimensions: (1) software libraries, (2) tightly-coupled/serial versus loosely-coupled/distributed geoprocessing chain implementations, and (3) standardised geoprocessing web service (Web Processing Service) implementations versus bespoke software solutions. Results show that bespoke software, using specific geoprocessing libraries, implemented on a loosely-coupled messaging architecture significantly outperforms other combinations.

L. Hankel (✉) · G. McFerren
Council for Scientific and Industrial Research, Meraka Institute, Pretoria
South Africa
e-mail: lhankel@csir.co.za

L. Hankel · S. Coetzee
Department of Geography, Geoinformatics and Meteorology, Centre for Geoinformation Science, University of Pretoria, Pretoria, South Africa

J. Döllner et al. (eds.), *Service-Oriented Mapping*, Lecture Notes in Geoinformation and Cartography, https://doi.org/10.1007/978-3-319-72434-8_4

Keywords Geoprocessing · Visualisation · Computational efficiency
Architectural style

4.1 Introduction

Big data refers to the ever-increasing volumes of data continuously being generated continuously by a large variety of sensors, such as smartphones and satellites. Four characteristics distinguish big data from other data: volume, variety, velocity, and veracity (4Vs) (Saha and Srivastava 2014; Tsou 2015; Ward and Barker 2013). In many cases, big data includes a direct or indirect reference to a location on the Earth and can then be referred to as 'big geospatial data'.

The research in this chapter was inspired by the Advanced Fire Information System (AFIS), developed by the Meraka Institute of the Council for Scientific and Industrial Research (CSIR) in South Africa. AFIS provides earth observation satellite based near real-time fire detection and notification services to users in Southern and East Africa, South America and Europe. Vegetation wildfires occur regularly in many parts of the world. Wildfires range from small scale fires, causing an insignificant amount of damage, such as minimal damage to property to large scale wildfires, causing a considerable amount of damage, such as complete destruction of property or loss of life. Wildfire incidents can potentially be mitigated by the timely dissemination of information derived from earth observation data. AFIS fulfils this need.

AFIS wildfire notifications typically take the form of an SMS (short message service) or e-mail containing a limited amount of static information, such as the location and intensity of the detected fire (McFerren et al. 2013). This chapter describes research that was conducted towards enhancing the AFIS alerting component by adding an interactive 3D visualisation of the spatial context surrounding a detected wildfire to a notification message. The spatial context is represented by variables, such as topography, vegetation types and condition, wind speed and direction, population density, land cover, infrastructure and nearby services. Figure 4.1 shows such a visualisation, overlaying the terrain with the landcover variable; it also provides values for several variables at the point of the fire.

When an interested or affected party receives a wildfire notification, the receiver should immediately be able to display the visualisation resource through a web connected device, thus, the visualisation should be available on demand. To achieve this low latency, streams of wildfire events need to be processed rapidly in relation to large datasets of contextual variables. Failure to do so would result in processing backlogs and unavailability of 3D visualisations in "demand-time". Since the broader wildfire geoprocessing system is based on Open Geospatial Consortium (OGC) service interfaces, research was required to determine if the 3D visualisation component could be similarly implemented.

The remainder of the chapter is structured as follows: in Sect. 4.2, background is provided about wildfire detection and alerting, OGC web processing and loosely

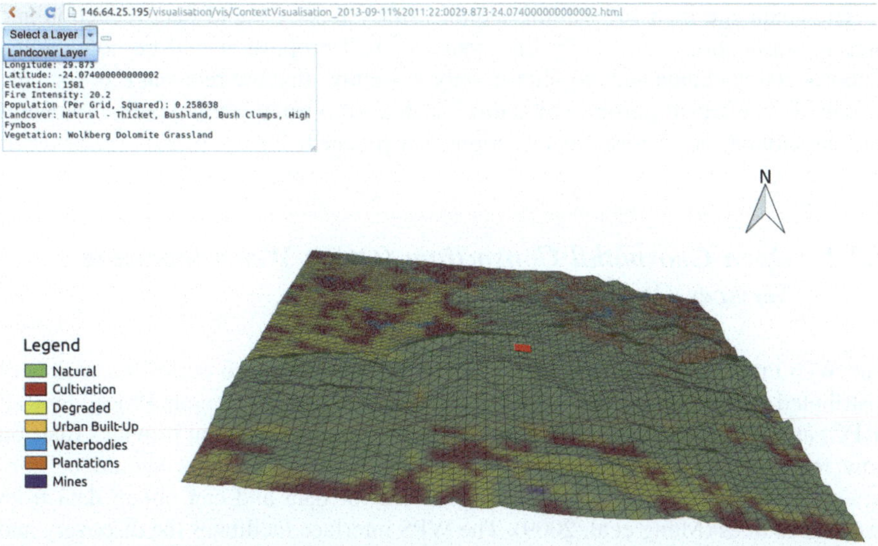

Fig. 4.1 3D context visualisation of a vegetation wildfire

coupled messaging. Section 4.3 describes the research design. Results are presented and discussed in Sects. 4.4 and 4.5 respectively. Section 4.6 concludes.

4.2 Background

4.2.1 Wildfire Detection and Alerting

"Geographical data has been broadly utilised to comprehensively describe the spatial and descriptive semantics of fire hazards and their surrounding features" according to Kassab et al. (2010). In this chapter, vegetation wildfires are characterised as intermittent data streams of geospatially referenced active fire detection events generated by geostationary and polar-orbiting earth observing satellites.

In the geostationary case, active fire events are detected by the Spinning Enhanced Visible and Infrared Imager (SEVERI) sensor aboard the Meteosat 8 satellite, which can yield events at 15 min intervals (McFerren and Frost 2009). The maximum number of events generated for a 15 min period for our study area in our time-series database is 8362. Higher spatial resolution active fire events are detected by the Moderate Resolution Imaging Spectroradiometer (MODIS) sensor aboard the Terra and Aqua polar-orbiting satellites (McFerren and Frost 2009). Events can get yielded at approximately six hour intervals for the area of study, peaking in our time-series at 48,606 events for a single overpass.

The challenge then for wildfire alerting subsystems is rapid data reduction and precise notification—of 48,606 fires, only a small proportion will be noteworthy. Geoprocessing chains need to filter in only those fires that are relevant (i.e. spatially proximal) to affected parties. Once this is achieved, visualisation of wildfire events and dissemination of these visualisations can proceed.

4.2.2 Open Geospatial Consortium (OGC) Web Processing Services (WPS)

The Web Processing Service (WPS) provides a mechanism to perform a variety of distributed web-based processing on geospatial data using a Remote Procedure Call (RPC) architectural style. The WPS is defined as an interface that provides rules on how requests and responses of arbitrary geoprocessing services should be constructed. A WPS acts as a middleware service for data and can obtain data from external sources (Meng et al. 2009). The WPS interface facilitates the discovery and publishing of geospatial processes (Open Geospatial Consortium 2007). In the design of the broader wildfire system, WPSs require consideration, for they match the architectural style of other components, e.g. Web Map Service (WMS) and Web Feature Service (WFS).

The main advantages of using the WPS are interoperability (Open Geospatial Consortium 2007) and software implementation abstraction. Geoprocessing can take place regardless of the software or hardware on a user's computer. Multiple web service access approaches are supported by a WPS. These include standardised calling approaches like Hypertext Transfer Protocol (HTTP) POST using Extensible Markup Language (XML), HTTP GET using Key Value Pair (KVP) arguments, and Simple Object Access Protocol (SOAP) requests (Kiehle et al. 2007). Geoprocessing can be highly distributed due to the fact that it can occur anywhere on the internet, on demand (Geoprocessing.info n.d). Software implementations of a geoprocessing component on the server-side of a WPS can change, but it will have no impact on the WPS client—the calling interface remain unchanged. Furthermore, WPSs enable clients to gain access to the most current data and processing implementations, without a need to change client implementations. Processes are re-usable in multiple applications (Open Geospatial Consortium 2007). WPSs can also be exploited for cloud computing, grid computing or other forms of high performance geoprocessing.

WPSs can be chained to form higher level processes. Process chaining can be executed by passing the output of one process as the input of another process (Kiehle et al. 2007). This feature is desirable for building the wildfire notification geoprocessing chains.

Implementers of WPS solutions should consider where the data utilised in the service reside. Michaelis and Ames (2007) note that it can be computationally expensive to move large datasets to a WPS as inputs—several encoding/decoding cycles need to occur and data are often increased in size in the process. It may be more feasible for data to reside locally to the WPS, if the use case allows this. It may be problematic for a client to wait for a WPS response; implementations may need to consider asynchronous responses, whereby a client is notified of process completion and then fetches output from a given location.

4.2.3 Loose Coupling Through Enterprise Messaging

To achieve the loose coupling experimental goals, this research used the OASIS Advanced Message Queuing Protocol (AMQP). AMQP enables applications to communicate over messaging middleware servers, referred to as brokers (AMQP Working Group 2012; Aiyagari et al. 2008).

An AMQP model consists of components that publish messages (known as producers), routers (known as exchanges) that move messages to queues according to a variety of messaging patterns (e.g. publish/subscribe) and consumers that act on messages.

The components of such a system are loosely-coupled, thus, they do not need to have any knowledge about each other (Eugster et al. 2003). This architecture allows flexible and scalable systems to be built. The AMQP approach allowed concurrency capabilities and distributed geoprocessing to be built into this research system.

4.3 Research Design

4.3.1 Aims, Objectives and Hypothesis

The main aim of this study was to determine if WPSs are effective for on-demand geoprocessing of intermittently high velocity streaming geospatial data for the purpose of generating 3D visualisations. To reach this aim, specific objectives had to be determined.

The first objective was to analyse the functional requirements and processing flow characteristics of the geoprocessing chain.

The second objective was to implement the geoprocessing chain components using various libraries to ascertain the fastest implementation topology (i.e. which library should be used for executing a given process).

The third objective was to determine the architectural style that delivered fastest throughput. The process chaining was executed using two completely different approaches, a tightly-coupled approach using nested method calls and a loosely-coupled approach (an enterprise bus), using a message broker and a set of producer and consumer actors.

The fourth objective was to compare the two implementation styles with and without the use of a WPS, in order to characterise the cost of web service interface interactions.

This research hypothesizes that moderately high velocity streaming geospatial data cannot be processed rapidly enough to generate and deliver notifications with 3D visualisation payload in "demand-time", by deploying a WPS-based solution on a single moderate computer resource.

4.3.2 Experiment Design

Recall that the aim of this research was to investigate the utility of the WPS to handle the geoprocessing of sporadic datastreams of geolocated events into notifications, with visualisations, sent to appropriate users in "demand time" (i.e. as rapidly as possible).

The first step of the experiment design was to determine the design of the geoprocessing chain. After the design was determined, the geoprocessing chain was implemented in four alternative ways, scripted in the Python programming language. A battery of timing tests were conducted on each implementation to ascertain which style of geoprocessing component offered the fastest performance. WPSs were then set up for each stage of the geoprocessing chain using the fastest component implementations. Timing evaluations were performed against the WPSs in two configurations: (a) tightly-coupled using method call chaining; and (b) loosely-coupled as AMQP consumers/producers. Finally, the geoprocessing components were set up without WPS facades, also using the tightly-coupled and loosely-coupled configurations, and subjected to the same test battery. The experiment was aimed at showing which component implementations, in which architectural style offered the fastest throughput of events to notifications with a visual payload.

4.3.3 Geoprocessing Chain Design

The first link of the geoprocessing chain—the structure of it illustrated in Fig. 4.2—is an input parsing process, responsible for parsing streaming data (lists of tuples) representing fire detection events into a JavaScript Object Notation (JSON) format in order to standardise the data format.

The second link of the chain is a check to determine if a given fire event occurred within a set of area of interest geometries (as defined in a notification subscription database). If so, the event data are passed along to the next step of the geoprocessing chain. If a fire event did not occur within an area of interest that fire event data is simply discarded.

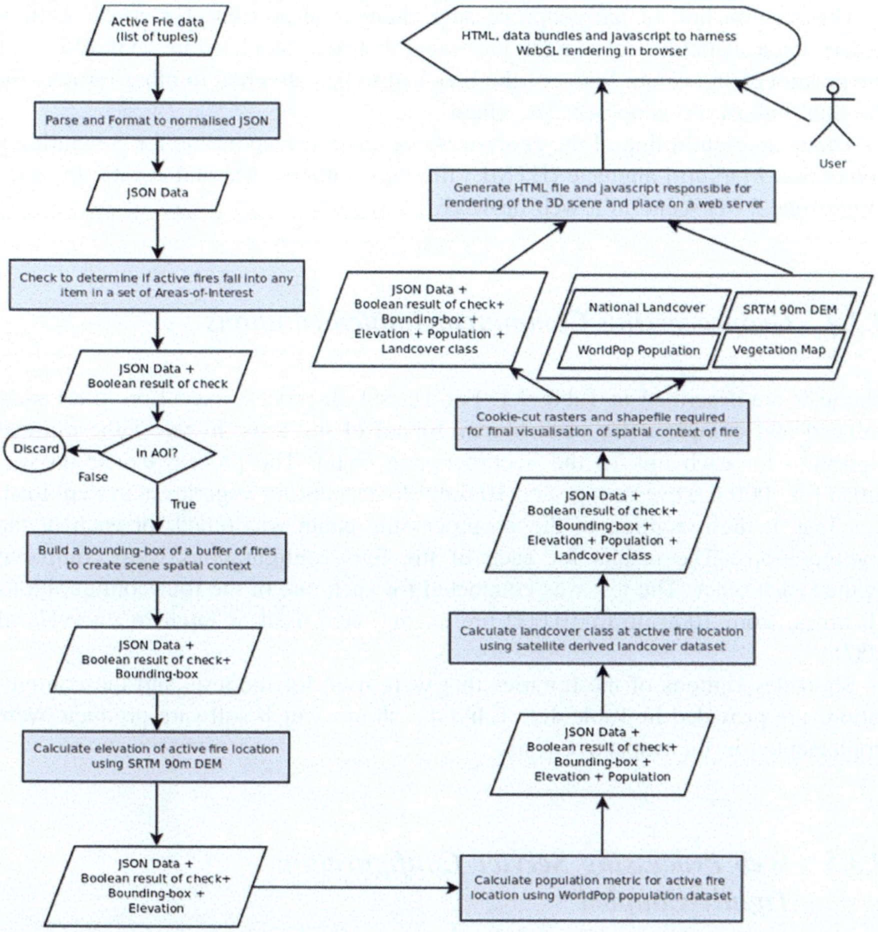

Fig. 4.2 Geoprocessing chain flow diagram

The third link of the chain derives an axis-aligned bounding box from a buffer of the event point location. The axis-aligned bounding box is ultimately used in extracting enough contextual data around an event to generate a meaningful 3D scene visualisation.

The fourth link of the chain calculates the elevation at the fire scene from a Digital Elevation Model (DEM) and attaches this new piece of data to the event data.

The fifth and sixth link of the geoprocessing chain determine the population density and land cover type around the location of a fire from a population surface GeoTIFF raster and a land cover raster respectively and attach these data to the event data.

The seventh link of the geoprocessing chain is responsible for cookie-cutting rasters and a shapefile according to the extent that was calculated in the third link of the geoprocessing chain. Some of the files had to be converted to other formats for the final link of the geoprocessing chain.

The final (eighth) link of the geoprocessing chain is responsible for generating a HyperText Markup Language (HTML) file that contains data and JavaScript code to generate a 3D scene in a web browser.

4.3.4 Geoprocessing Component Implementations

The tests are described in Table 4.1. For Tests 1–3, process execution times were collated and compared for each library for all of the tests, to select the optimal algorithm for each link of the geoprocessing chain. The process execution was timed for 1000 fire events repeated 10 times to account for vagaries in system load. For Test 4, the execution of the geoprocessing chain was timed for each of the configurations. The results for each of the four configurations were compared against each other. The test was conducted for each one of the four configurations 10 times, from 1000 up to 10,000 fire events were used as input in intervals of 1000.

Short descriptions of the libraries that were used for the tests and the configurations are provided in Table 4.2. Table 4.3 shows which software products were implemented in the respective tests.

4.3.5 Web Processing Service Configuration—Tightly-Coupled

We utilised PyWPS as the implementation of the WPS (Cepicky 2013). A tightly-coupled WPS chain class was created. Several were implemented. Each method invoked a different WPS that performed an operation on the input data and

Table 4.1 Description of tests

Test 1	Accepts fire data in a JSON format, extracts the point location of the fire and checks if the point falls within an area of interest
Test 2	Accepts fire data in a JSON format, extracts the point location of the fire, places a 2.5 km buffer around the fire point, and calculates the axis-aligned bounding box of the buffer
Test 3	Accepts fire data in a JSON format, extracts the point location of the fire, and queries the elevation from a DEM at the fire location
Test 4	The entire geoprocessing chain described in Sect. 4.3.3 is executed. Various combinations of process chaining and WPS configurations were compared

Table 4.2 Software packages used in the tests

Rtree	Python package providing advanced spatial indexing features (Python Community 2012)
GDAL/ OGR	C++ translator library for geospatial raster and vector data formats (OSGeo 2013a)
Fiona	Python package providing Python interfaces to OGR functions (Python Community 2013a)
Shapely	Python package for manipulation and analysis of geometric objects (Python Community 2013c)
PostGIS	Spatial database extender for PostgreSQL. Allows queries on geographic objects in SQL (OSGeo 2013b)
RabbitMQ	Open source enterprise messaging system that is based on the AMQP standard (Pivotal Software, inc n.d.)
Pika	Python implementation of the AMQP protocol (Python Community 2014)
PyWPS	Implementation a Web Processing Service (Cepicky 2013)

returned a result. Urllib was used to invoke a WPS. Urllib provides a high level interface to retrieve data across the internet (Python Community 2013b). The chaining was designed so that each method called the following method in the geoprocessing chain. Every result was passed along the geoprocessing chain by nesting method calls. A class instance was created and the first method to parse the data from the fire detection source was called. This initiated the geoprocessing chain. The result of a process (returned result of a method), was used as input parameter (input data) for the next process that was executed within the geoprocessing chain. Nested method calls were used.

4.3.6 Web Processing Service Configuration—Loosely-Coupled

The second approach involved loosely-coupled chaining. RabbitMQ, an enterprise messaging system based on the Advanced Message Queuing Protocol (AMQP) (Pivotal Software, inc n.d.), was used for the chaining of the processes. Figure 4.3 illustrates this chain.

JSON-formatted data are received (consumed) as inputs from the fire detection source (fire data producer). From this point forward, WPS 1 (Process 1) is invoked; the data are parsed, reformatted and published on a messaging queue. In this case, the consumer also acts as a producer. The consumer listens for messages on the Firehose (data ingest) queue.

After the consumed data has been parsed, it is published (produced) onto the Formatted Data queue. After the JSON data is received (consumed), WPS 2 (Process 2) is invoked and an area of interest check is conducted based on the Well

Table 4.3 Brief description of tests conducted

Software	Test 1	Test 2	Test 3	Test 4	
				Tightly-coupled	Loosely-coupled
Rtree	(1) Created indices that represent the area of interest (2) Constructed fire point location (3) Checked if the point intersected with an index Refer to Fig. 4.4 for results	Not applicable	Not applicable	Not applicable	Not appliable
GDAL/OGR	(1) A shapefile (area of interest) was loaded into memory (2) Constructed fire point location (3) Checked if the geometry of the shapefile contained the point Refer to Fig. 4.4 for results	(1) Constructed fire point location (2) Created a buffer around the point (3) Calculate extent around the buffer Refer to Fig. 4.5 for results	(1) Loaded the GeoTIFF into memory (2) Extracted a pixel value at the specific location of the fire Refer to Fig. 4.6 for results	Not applicable	Not applicable
Fiona and Shapely	(1) Loaded a shapefile (area of interest) (2) Determined the bounds of the area of interest	(1) Constructed fire point location (2) Placed a buffer around that point	Not applicable	Not applicable	Not applicable

(continued)

Table 4.3 (continued)

Software	Test 1	Test 2	Test 3	Test 4	
				Tightly-coupled	Loosely-coupled
	(3) Constructed fire point location (4) Checked if the point was contained within the bounds (optimised with prepared geometries) Refer to Fig. 4.4 for results	(3) Calculate extent around the buffer Refer to Fig. 4.5 for results			
PostGIS on local machine	(1) Imported the shapefile (area of interest) to database (2) Constructed fire point location (3) Connected to database, checked if geometry of point intersected with the geometry of the shapefile (area of interest) Refer to Fig. 4.4 for results	(1) Constructed fire point location (2) Created a buffer around the point (3) Calculate extent around the buffer Refer to Fig. 4.5 for results	(1) Imported the DEM into a database using PostGIS raster (2) Constructed fire point location (3) Selected value from database where the point intersects with the DEM Refer to Fig. 4.6 for results	Not applicable	Not applicable

(continued)

Table 4.3 (continued)

Software	Test 1	Test 2	Test 3	Test 4	
				Tightly-coupled	Loosely-coupled
PyWPS	Not applicable	Not applicable	Not applicable	Executed the geoprocessing chain that was chained in a tightly-coupled style, by using a WPS (refer to Sect. 4.3.5) Refer to Fig. 4.7 for results	Executed the geoprocessing chain that was chained in a loosely-coupled style, by using a WPS (refer to Sect. 4.3.6) Refer to Fig. 4.7 for results
No PyWPS	Not applicable	Not applicable	Not applicable	Executed the geoprocessing chain (refer to Sect. 4.3.3) that was chained in a tightly-coupled style, without utilising a WPS Refer to Fig. 4.7 for results	Executed the geoprocessing chain (refer to Sect. 4.3.3) that was chained in a loosely-coupled style, without utilising a WPS Refer to Fig. 4.7 for results

Any reference to bounds or bounding box refers to an axis-aligned bounding box

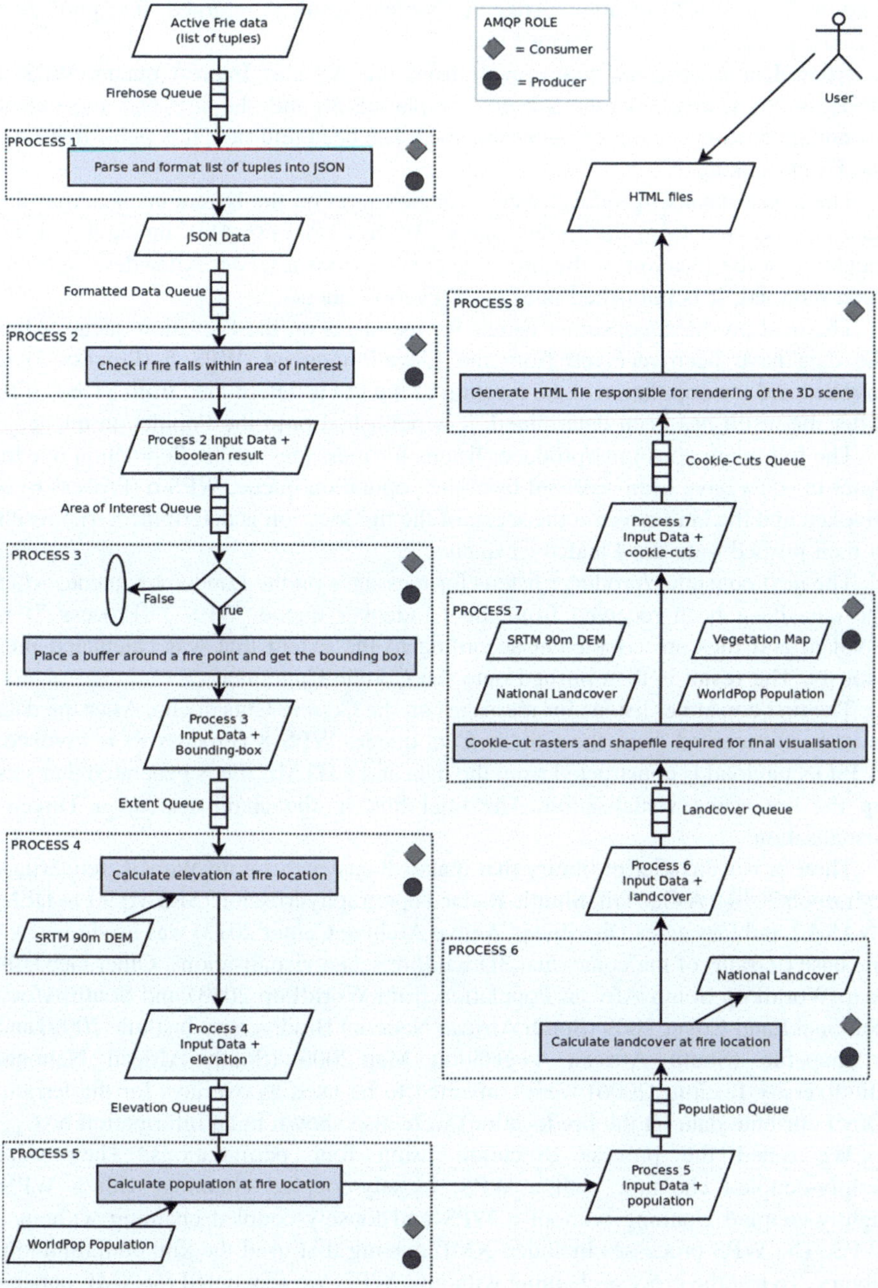

Fig. 4.3 Geoprocessing chain flow diagram

Known Text (WKT) location. After the boolean result is returned, it is published (produced) on the Area of Interest queue.

After data is received (consumed) from the Area of Interest queue, WPS 3 (Process 3) is invoked and a buffer is placed around the fire and the extent (bounding box) is calculated. After the result has been returned, it is published onto the Extent queue.

The next consumer/producer listens for messages on the Extent queue. After the data are received from the Extent queue, WPS 4 (Process 4) is invoked and the elevation at the location of the fire is calculated from a DEM. After the result has been returned, it is published onto the Elevation queue.

The next producer/consumer listens for messages on the Elevation queue. After the data have been received from the Elevation queue, WPS 5 (Process 5) is invoked and the population at the location of the fire is determined from a GeoTIFF. After the result has been determined, it is published onto the Population queue.

The following consumer/producer listens for messages on the Population queue. After the data have been received from the population queue, WPS 6 (Process 6) is invoked and the land cover at the scene of the fire location is determined. The result is then pushed onto the Landcover queue.

The next consumer/producer listens for messages on the Landcover queue. After the data have been received from the Landcover queue, WPS 7 (Process 7) is invoked and files are cookie-cut according to the extent that was calculated previously. The result is then pushed onto the Cookie-Cuts queue.

The first consumer listens for messages on the Cookie-Cuts queue. After the data have been received from the Cookie-Cuts queue, WPS 8 (Process 8) is invoked. A JSON payload is constructed from the data and a HTML file is generated that sets up the fire scene visualisation. The final link in the chain sets up a Three.js visualisation.

Three.js is a JavaScript library that makes it easier to set up WebGL renderings (Mr.doob 2014). A GeoTiff, Shuttle Radar Topography Mission (SRTM) 90 m DEM (NASA Land Processes Distributed Active Archive Center 2013) was used to create the base (terrain) of the contextual 3D wildfire scene visualisations. Other GeoTiffs (e.g. WorldPop South African Population from WorldPop 2013) and South Africa National Land Cover 2009 (South African National Biodiversity Institute 2009) and a shapefile (South African Vegetation Map 2006 (South African National Biodiversity Institute 2006) were converted to be used as overlays for the terrain. Other attribute data (at the fire location) were also shown in an information box.

We timed the process execution using four permutations. They were tightly-coupled chaining with a WPS, loosely-coupled chaining with a WPS, tightly-coupled chaining without a WPS and loosely-coupled chaining without a WPS. The WPS processes included XML parsing that used the xml.dom.minidom library. To test the process chaining without a WPS, we accounted for XML parsing by writing our result into XML and also parsing the XML with the xml.dom.minidom library.

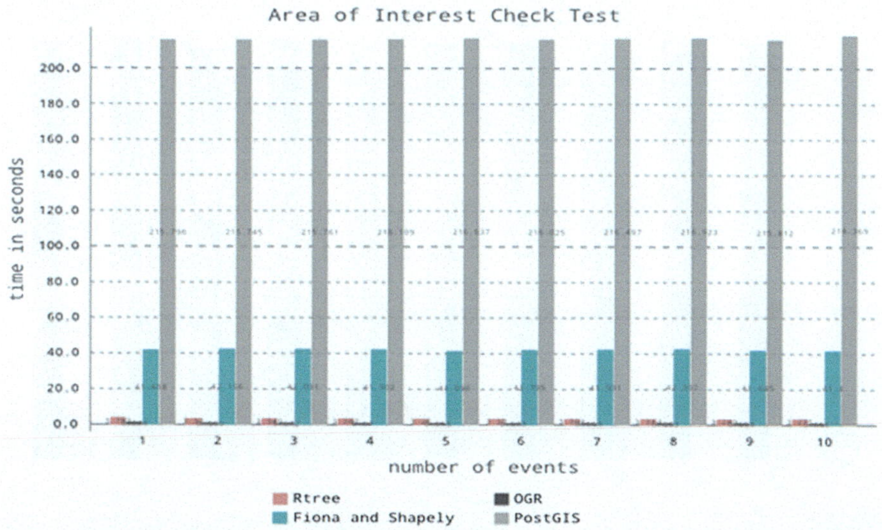

Fig. 4.4 Area of interest check test results

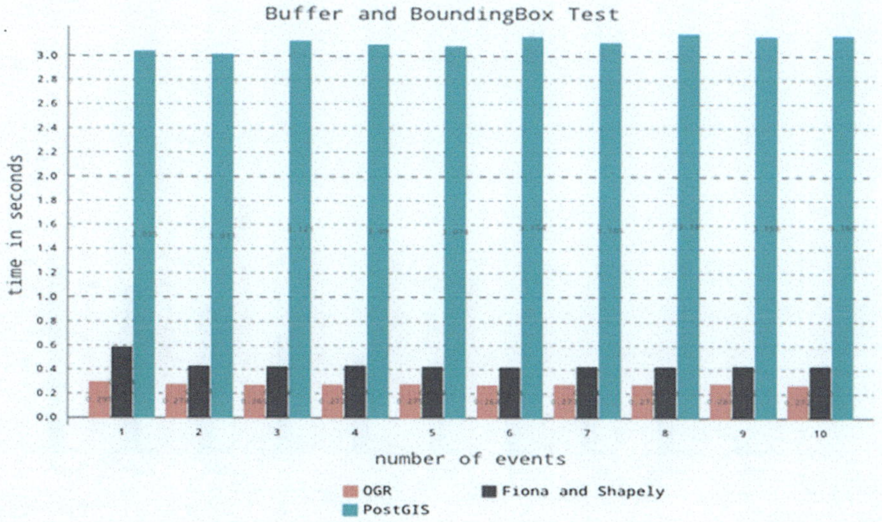

Fig. 4.5 Buffer and boundingbox test results

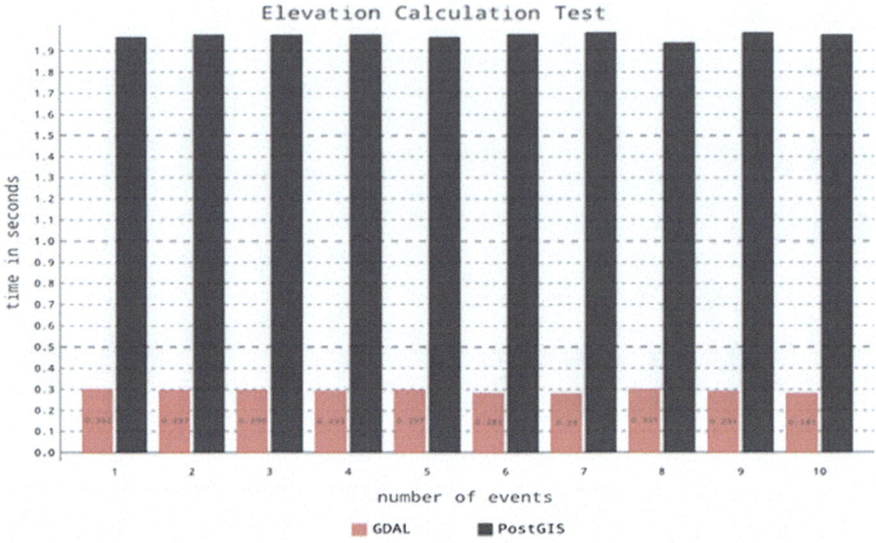

Fig. 4.6 Elevation calculation test results

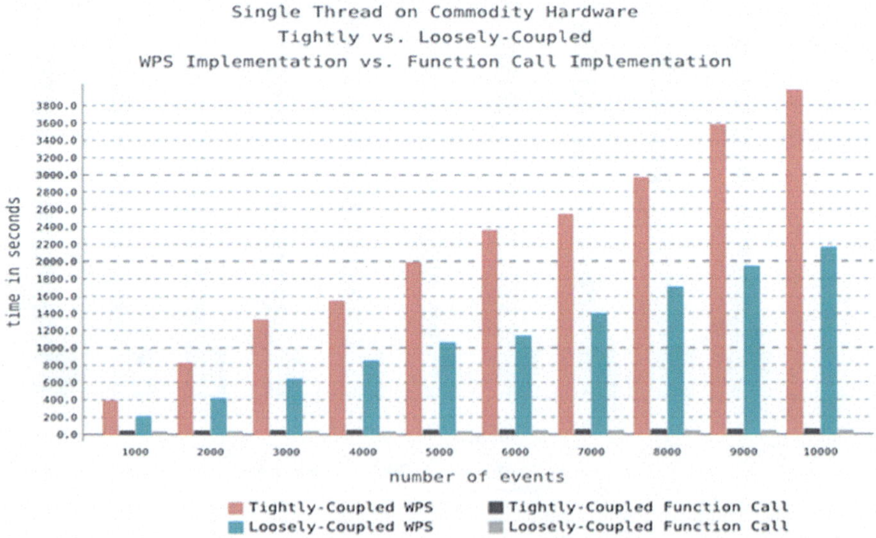

Fig. 4.7 Tightly-coupled versus loosely-coupled|WPS implementation versus function call implementation test results

4.3.7 Environment

The tests were executed on a laptop with an Intel Core i7, quad core Central Processing Unit (CPU) with four multi-threaded cores, therefore, eight cores running at 2.10 GHz with 8 GB of Random Access Memory (RAM).

4.4 Results

All of the tests were set up to be able to time process execution from start to finish. Shorter execution time is better.

The use of the OGR library with the area of interest test proved to be the fastest option, as observed in Fig. 4.4. PostGIS exhibited the poorest performance by a significant margin.

In Fig. 4.5 it can be observed that OGR performed the best in the buffer and axis-aligned bounding box test. Fiona and Shapely offered similar performance as in the area of interest test whilst PostGIS approach underperformed.

GDAL strongly outperformed PostGIS in the elevation calculation test, see Fig. 4.6. Other links in the geoprocessing chain (getPopulation and getLandCover) used the same method, so results for those links are likely represented by the elevation extraction.

Referring to Fig. 4.7, it can be observed that the loosely-coupled process chaining approach is significantly faster than the tightly-coupled process chaining approach. It can also be noted from this illustration, that the approaches that do not utilise the WPS to facilitate geoprocessing performed far better than the approaches with a WPS. Illustrated times represent the interval between fire event ingest and 3D scene rendering output.

4.5 Discussion

Results from the first three sets of tests were unexpected. It was expected that Rtree would perform best in the area of interest test due to its specialised spatial indexing functionality, whilst Shapely/Fiona and PostGIS were expected to rival, if not surpass OGR in the other tests, due to the use of the underpinning specialised geometry engine (GEOS) software (OSGeo 2014). Rtree performed worse than OGR in the area of interest check process. A number of optimisation techniques to improve the performance of Rtree indexing are possible, such as streamed data loading, which improves performance through pre-sorting before indexing (Python Community 2012). If the size of the index is known a priori, Rtree performance gains can be made by presizing the index to save on storage.

The OGR library allows performant spatial indexing of shapefiles through support for the .sbn and .sbx indexes (GDAL n.d.). The floating point comparisons in Rtree are a likely explanation for the slower performance of Rtree compared to OGR (Python Community 2012), which uses integer comparisons (Rouault 2012). The variation in behaviour of Rtree and OGR was clearly demonstrated in the spatial filtering step, i.e. area of interest check process.

PostGIS also supports indexing through the use of Rtree-over-Generalised Search Tree (GIST) schemes (OSGeo 2013b). This advanced indexing feature was expected to deliver good performance, but the connection setup and query planner overhead are suspected to be slowing the process substantially. This merits further investigation. Raster tile sizes in the PostgreSQL database can similarly influence the performance of the processes. Given the unpredictable, episodic, yet long running nature of the fire event data stream, it was deemed necessary to re-establish spatial database connections everytime a new fire event was processed. This unavoidably caused a processing overhead leading to underperformance of the PostGIS solution.

The result of the Fiona and Shapely implementation could be explained quite easily. Fiona utilises Python objects which impact negatively on the performance, compared to OGR which uses C pointers (Python Community 2013a), which are less memory intensive. Fiona's documentation states that it "trades memory for simplicity and reliability". Fiona copies features from the data source to Python objects. The performance of the Fiona and Shapely implementations improved after using prepared geometries.

By comparing the two chaining approaches, the loosely-coupled chaining approach provided a better performance than the tightly-coupled chaining approach. The process execution time halved with the loosely-coupled chaining approach that utilises the enterprise messaging solution, compared to the tightly-coupled chaining approach where the process execution time doubled. This is due to the fact that with the tightly-coupled chaining approach, one event has to be processed from start to finish before the next one can start (serial geoprocessing). With the loosely-coupled chaining approach, one event does not have to be fully processed for the next one to start, because of multiple consumers and producers in the system (parallel geoprocessing).

A significant variation in performance can be noted from the process chaining with a WPS and without a WPS. The processing time from 1000 to 10,000 events without the WPS for tightly-coupled and loosely-coupled chaining was almost constant. The processing time with the WPS from 1000 to 10,000 events increased linearly. These results require further investigation. The PyWPS implementation itself might be slow, but perfomance likely suffers due to (a) the time it takes to construct and destroy WPS instances in a web server environment such as Apache; and (b) serialisation and deserialisation of WPS request payloads. Differences between passing geoprocessing payloads via POST payloads or GET URL key/value pairs do not appear to be a significant contributor to performance results.

This chapter describes a wildfire scenario and the research is informed by this industry use case. However, the architecture and components are reusable for other applications where the visualisations are required. Situational awareness, asset management and emergency response applications are examples of applications where 3D visualisation of spatial context may be beneficial.

4.6 Conclusion

This chapter shows that careful consideration needs to be given to the design of geoprocessing chains for working with streaming geodata in a 'demand time' case. Loosely-coupled, lightweight processes in a distributed architecture are more likely to meet the 'demand-time' constraint than WPS-based systems. Surprising performance differences exist among geoprocessing chain implementations, depending on underlying software libraries and styles of retrieving data; in particular, the set of experiments that compare SQL request to database versus cached file accessed via Application Programming Interface (API) suggest that the database could be a bottleneck. This research indicates that 'demand time' geoprocessing is difficult to achieve in a non-distributed architecture—10,000 responses in $\sim$30 s is reasonable, but is not quite reaching a level of performance that matches the requirement to have a prebuilt web resource ready as soon as a user is alerted to a wildfire, for example.

4.7 Further Recommendations and Future Work

We are performing further investigations in this area of research. It is particularly important to understand why the PostGIS process implementations performed poorly, since they would be expected to deliver decent performance and offer intriguing possibilities for analysis of different spatial data types e.g. tiled rasters in the database. Newer WPS implementations may bridge the performance gap to bespoke geoprocessing services, so they need consideration. The most important investigation that will be undertaken in the future, will be to determine if distributed (especially cloud-based) computing can speed up geoprocessing, by allowing horizontal scaling (addition of more cloud computing instances, i.e. nodes or virtual machines) to address the workload. Protocols and architectures for loose coupling other than AMQP will be tested to see how they will affect the results of the research.

References

Aiyagari S, Arrott M, Atwell M, Brome J, Conway A, Godfrey R, Greig R, Hintjens P, O'hara J, Radestock M, Richardson A, Ritchie M, Sadjadi S, Schloming R, Shaw S, Sustrik M, Trieloff C, van der Riet K, Vinoski S (2008) Advanced message queueing protocol specification. Version 0.9.1. OASIS

AMQP Working Group (2012) Advanced message queueing protocol specification. Version 1.0. OASIS

Cepicky J (2013) Welcome to PyWPS. http://pywps.wald.intevation.org. Accessed 21 Nov 2013

Eugster PTH, Felber PA, Guerraoui A, Kermarrec AM (2003) The many faces of publish/subscribe. ACM Comput Surv 35(2):114–131

GDAL (n.d.) OGR: Simple feature library. http://www.gdal.org/org. Accessed 21 Nov 2013

Geoprocessing.info (n.d.) Geoprocessing. http://geoprocessing.info/wpsdoc. Accessed 21 Nov 2013

Kassab A, Liang S, Gao Y (2010) Real-time notification and improved situational awareness in fire emergencies using geospatial-based publish/subscribe. Int J Appl Earth Obs Geoinf 12(6):431–438

Kiehle C, Greve K, Heier C (2007) Requirements for next generation spatial data infrastructures-standardized web based geoprocessing and web service orchestration. Trans GIS 11(6):819–834

McFerren G, Frost P (2009) The South African fire information system. In: Landgren J, Jul S (eds) Proceedings of the 6th international ISCRAM conference, Gothenburg, Sweden, 10–13 May 2009, pp 1–6. http://hdl.handle.net/10204/3621. Accessed 25 July 2014

McFerren G, Swanepoel D, Lai C (2013) Wide area alerting system for wildfires & other nasties. FOSS4G 2013, Nottingham, United Kingdom, 17–21 Sept 2013. E-learning for the open geospatial community conference series. http://elogeo.nottingham.ac.uk/xmlui/handle/url/214. Accessed 15 Apr 2014

Meng X, Bian F, Xie Y (2009) Geospatial services chaining with web processing services. In: Proceedings of the international symposium on intelligent information systems and applications, Qingdao, China, 28–30 Oct 2009, pp 7–10

Michaelis CD, Ames DP (2007) Evaluation of the OGC web processing service for use in a client-side GIS. OSGEO J 1:1–8

Mr.doob (2014) mrdoob/three.js GitHub. https://github.com/mrdoob/three.js. Accessed 20 Aug 2014

NASA Land Processes Distributed Active Archive Center (LP DAAC) (2013) SRTM Version 3. USGS/Earth Resources Observation and Science (EROS) Center, Sioux Falls, South Dakota

Open Geospatial Consortium (OGC) (2007) OpenGIS® web processing service version 1.0.0. OGC 05-007r7. http://www.opengeospatial.org/standards/wps. Accessed 20 Aug 2014

OSGeo (2014) GEOS—geometry engine, open source. http://trac.osgeo.org/geos/wiki. Accessed 20 June 2013

OSGeo (2013a) GDAL/OGR info sheet. http://www.osgeo.org/gdal_ogr. Accessed 21 Nov 2013

OSGeo (2013b) PostGIS—spatial and geographic objects for PostgreSQL. http://postgis.net/. Accessed 21 Nov 2013

Pivotal Software, inc (n.d.) RabbitMQ—messaging that just works. http://www.rabbitmq.com/. Accessed 21 Nov 2013

Python Community (2012) Rtree 0.7.0: Python Package Index. https://pypi.python.org/pypi/Rtree/. Accessed 21 Nov 2013

Python Community (2013a) Fiona 1.0.2: Python Package Index. https://pypi.python.org/pypi/Fiona/. Accessed 21 Nov 2013

Python Community (2013b) Open arbitrary resources by URL: Python Package Index. http://docs.python.org/2/library/urllib.htm/. Accessed 21 Nov 2013

Python Community (2013c) Shapely 1.2.18: Python Package Index. https://pypi.python.org/pypi/Shapely/. Accessed 21 Nov 2013

Python Community (2014) Pika 0.9.14: Python Package Index. https://pypi.python.org/pypi/pika/. Accessed 14 Aug 2014

Rouault E (2012) Geo tips and tricks: GDAL/OGR using Shapefile native .sbn spatial index. http://erouault.blogspot.co.za/2012/06/gdalogr-using-shapefile-native-sbn.html. Accessed 18 November 2013

Saha B, Srivastava D (2014) Data quality: the other face of big data. In: Proceedings of the international conference on data engineering, pp 1294–1297

South African National Biodiversity Institute (2006) Vegetation Map 2006. http://www.bgis.sanbi.org/vegmap/project.asp. Accessed 20 Aug 2014

South African National Biodiversity Institute (2009) National Land Cover 2009. http://www.bgis.sanbi.org/landcover/project.asp. Accessed 20 Aug 2014

Tsou MH (2015) Research challenges and opportunities in mapping social media and big data. Cartography Geogr Inf Sci 42(sup1):70–74

Ward JS, Barker A (2013) Undefined by data: a survey of big data definitions. arXiv:1309.5821

WorldPop (2013) WorldPop population map. http://www.worldpop.org.uk/data/summary/?contselect=Africa&countselect=South+Afriaca&typeselect=Population. Accessed 20 Aug 2014

Chapter 5
Service-Oriented Map Production Environments: The Implementation of InstaMaps

Rafael Roset, Marc Torres, Wladimir Szczerban, Jessica Sena
Victor Pascual, Montse Ortega, Isaac Besora and Sergio Anguita

Abstract To overcome paper map limitations and promote the use of its own digital cartography the ICGC (Institut Cartogràfic i Geològic de Catalunya, the national mapping agency) created, developed and maintains a SaaS (Software as a Service) open platform starting from 2013 that allows anyone to combine all sorts of geodata and publish it online as a new (thematic) map.

Keywords Map production use cases · Geoinformation management
Best practice examples · InstaMaps

5.1 Introduction

The Institut Cartogràfic de Catalunya (ICC, cartographic institute of Catalonia) was established in 1982 as the office in charge to carry out all works on cartography and geodesy in Catalonia (Spain). In 2014 its name changed to Institut Cartogràfic i Geològic de Catalunya (ICGC, cartographic and geological institute of Catalonia) as a result of the merger with all activities related to knowledge, information and prospecting of the soil and subsoil. It is an agency of the Generalitat de Catalunya (local government) and part of the Departament de Territori i Sostenibilitat (DTES, Department of Territory and Sustainability).

The online presence of the ICGC started back in 1998, when the first website was deployed. The site then hosted the catalogue of printed products of ICGC, and evolved into a catalogue of digital services, from WMS to full resolution download of aerial images among all other products. These digital services were targeted to

R. Roset (✉) · M. Torres · W. Szczerban · J. Sena · V. Pascual · M. Ortega · I. Besora
Geostart Unit, Institut Cartogràfic i Geològic de Catalunya, Barcelona, Spain
e-mail: rafael.roset@icgc.cat

S. Anguita
Head of Geotechnologies Area, Institut Cartogràfic i Geològic de Catalunya, Barcelona, Spain

J. Döllner et al. (eds.), *Service-Oriented Mapping*, Lecture Notes in Geoinformation and Cartography, https://doi.org/10.1007/978-3-319-72434-8_5

and were used by professionals and businesses, but lacked ease of use for all those without deep knowledge of core GIS operations or programming expertise.

In this context the need arose to provide all users with a platform to create their own maps by aggregating base maps from ICGC along with users own geodata, thus facilitating usage of ICGC's digital products as well as encouraging the use of maps as a means to disseminate geospatial information.

To create such a platform ICGC decided to gather a group of professionals, already working in-house in other areas of the business, into a multidisciplinary task force, called Geostart, dedicated, amongst other things, to create, develop, deploy and maintain InstaMaps, a Software as a Service (SaaS) online platform open to everyone. Another example of using maps and the Internet, allowing for easy access to geospatial information as well as allowing for customisation of the map is presented by Estella and Gibson (2018) in Chap. 10 in this volume.

5.2 InstaMaps, a 'Glocal' Vision (Think Global, Act Local)

One of the goals in creating InstaMaps was to be a showcase for everyone to use the cartographic basemaps and orthophotos of the ICGC. Most of these users were already accustomed to consuming geo-applications, such as Google Maps (Miller 2006) and Open Street Maps (Haklay and Weber 2008).

Among the benefits of Google Maps we would like to highlight the sound style of the map, its focus on location and the innovative use of the pseudo-Mercator projection (Battersby et al. 2014) to display the whole globe in web environments.

When you have spent over 30 years working with large scale and regional reference systems -UTM areas -, such as we have with the basemaps of the ICGC, it is not always easy to understand that we must think globally to implement a local product. This approach lead to a discussion of usable projections and the creation of InstaMaps. The new maps were created out of the existing products of the ICGC and were designed for location, not for metrics.

The products were reprojected and 'pyramids' were created from the ICGC basemaps using the pseudo-Mercator reference system, leaving aside the more classic scaling factors that were derived from paper on which the ICGC had worked on so far (Schmidt and Jörg 2012). Once these products were finalized they were tailored with other global suppliers such as OSM (2018).

The result is a global map of the world where the ICGC created data were seamlessly matched and still maintaining its "local" characteristics derived from our best knowledge of our own territory. But at the same time can be consumed globally along with other global suppliers such as OSM and Google Maps. In a small way the ICGC has improved the overall vision of the world by adding the best local knowledge.

5.3 Design

5.3.1 *Simplicity Versus Complexity*

One factor for the acceptance of InstaMaps is the simple design of the graphical interface according to the user experience. Three years ago when the ICGC launched InstaMaps, its flat design surprised many, especially those in the government for whom design was never a priority.

The landing page, as well as the dashboard, have been created along the lines of the so-called flat design (flatification) (Hanseth 2010). These pages are designed for optimum Search Engine Optimization (SEO) in our website and show the basic potential of InstaMaps: "explore", "create", "view" and "share" as shown in Fig. 5.2.

The visualization is adaptive (responsive) and has been designed to be easily available on different screens, terminals, operating systems and resolutions.

From the landing page, two main actions are promoted: access to the public gallery of maps and access to map making.

The final version of the toolbar for the InstaMaps editor was achieved after over 30 iterations as seen in Fig. 5.1, until we agreed to a good one. This process from the "cartography" approach to a "basic user" centered approach is illustrated in Fig. 5.1.

Currently the inclusion of too many functionalities in InstaMaps has rendered the tool less intuitive. The ICGC is in the process of redesigning the interface so

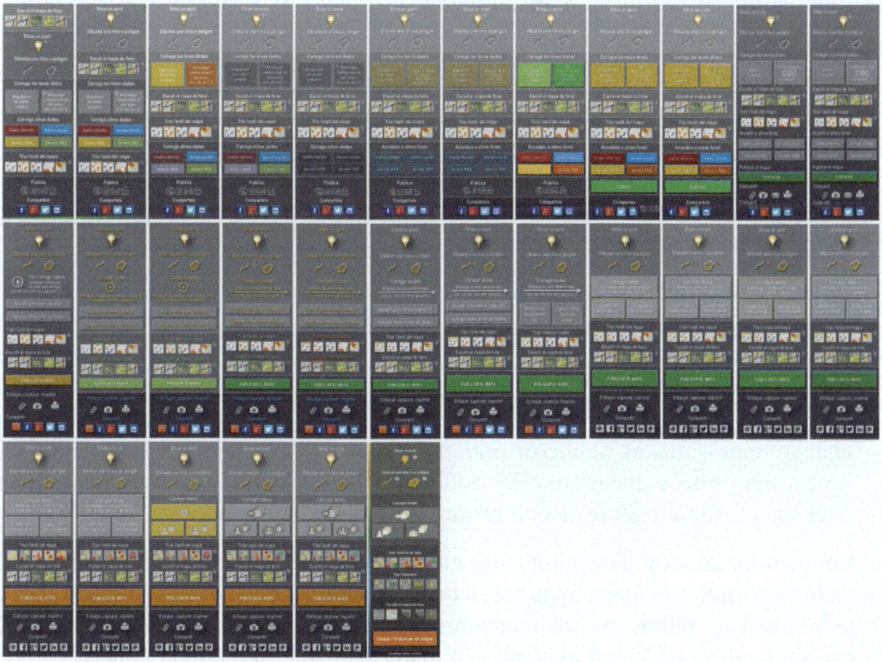

Fig. 5.1 Iterations until the final toolbar was approved

that InstaMaps does not scare users (Hassan-Montero Y, Martín-Fernández FJ (2003)). This is done by prioritizing the visibility of the tools that are more frequently used and by hiding advanced tools, such as GIS functions, aimed at advanced users. Complexity has to be kept to a minimum. InstaMaps has to adapt to the user, not the other way round.

5.3.2 Features of InstaMaps

InstaMaps is a platform for creating and sharing maps online quickly and easily. This allows users to create their own geo-information, either by drawing from scratch or by uploading their own files or by accessing information from other data sources, such as open data portals or agencies, which are available online.

InstaMaps is a service that allows users to quickly create, share and host their own maps online. The creation of a online map through InstaMaps can be done in four simple steps:

- manually or automatically add owned or third party information;
- select a map style;
- choose a base map that highlights the message to convey;
- publish and share the newly generated map.

Features of InstaMaps. The features of InstaMaps that allows for easy map creation are discussed in more detail below:

Access
Access to InstaMaps can be done by using an ICGC free user account or using your own credentials on the major social networks such as Twitter, Facebook, Google + or LinkedIn.
InstaMaps has the option to remember the password.
Upon logging in, the user is redirected to its own private map gallery.
From the private map gallery one can:

- make a new map;
- browse any map in visor mode;
- edit any existing map;
- get the link to the map;
- delete a map;
- change map status as public or private;
- turn a map into collaborative by adding up to 10 editors, sending invitations over email. (new feature)-clone or duplicate an existing map.

Public map gallery. The public map gallery is available to anyone on the Internet and allows sorting the maps by name, date or creator. The gallery harvests all those maps marked as public by their creators from private map galleries. Figure 5.2 shows the landing page and examples of maps available in the map gallery.

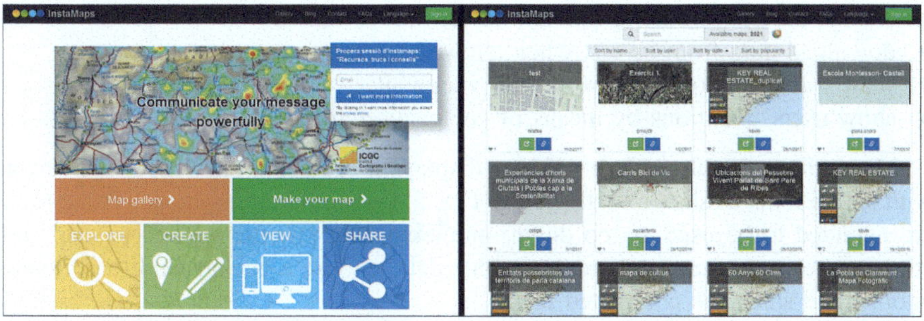

Fig. 5.2 Left: InstaMaps homepage. Right: InstaMaps public map gallery

InstaMaps top toolbar. The top tooblbar of InstaMaps located on the black strip allows to:

- log into the app;
- open the blog page (http://betaportal.icgc.cat/wordpress/category/instamaps/). It is a meeting point with all sorts of resources to solve open InstaMaps questions;
- get in touch with the team developing InstaMaps via email (instamaps@icgc.cat);
- access the FAQ (Frequent Asked Questions) section (http://betaportal.icgc.cat/wordpress/faq-dinstamaps);
- change the language between Catalan, Spanish and English.

Map viewer of InstaMaps

The map viewer as shown in Fig. 5.3 displays the map as it was created. It is both a tool to visualize maps and to interact with their contents, but it does not allow any editing:

- turn on and off selected layers;
- search for geographical places;

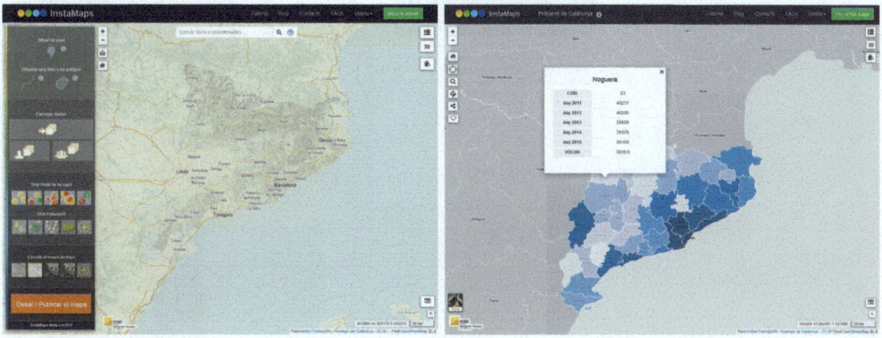

Fig. 5.3 Left: InstaMaps editor. Right: InstaMaps viewer

- change basemaps among different styles and products;
- download data from layers, when allowed by map creator;
- share a map in social networks and via email;
- show users' positions by means of geolocation and tracking (see Fig. 5.4).

Map editor dashboard. InstaMaps displays a toolbar at the left of the screen with all the tools as shown on the left in Fig. 5.3 when in editing mode. These tools are intended to be used as top down, mimicking the process applied to map creation, namely add data, style and combine data, choose a background (base map), publish and share the completed map.

Placing geometries. Adding and drawing geoinformation on top of the basemaps in InstaMaps has different options:

- add points, change marker and edit style;
- add lines and style their color and weight;
- add polygons, choose their outline color, fill color and transparency;
- present style options for any new type of geometry added by clicking on the plus sign next to drawing tools;
- all elements added to a map can be edited afterwards. They can also be moved and geometry can be changed;
- geometries added to the map are assigned to a new layer, or can be moved to any existing layer of the same kind. Layers are displayed on a layer drawer to the

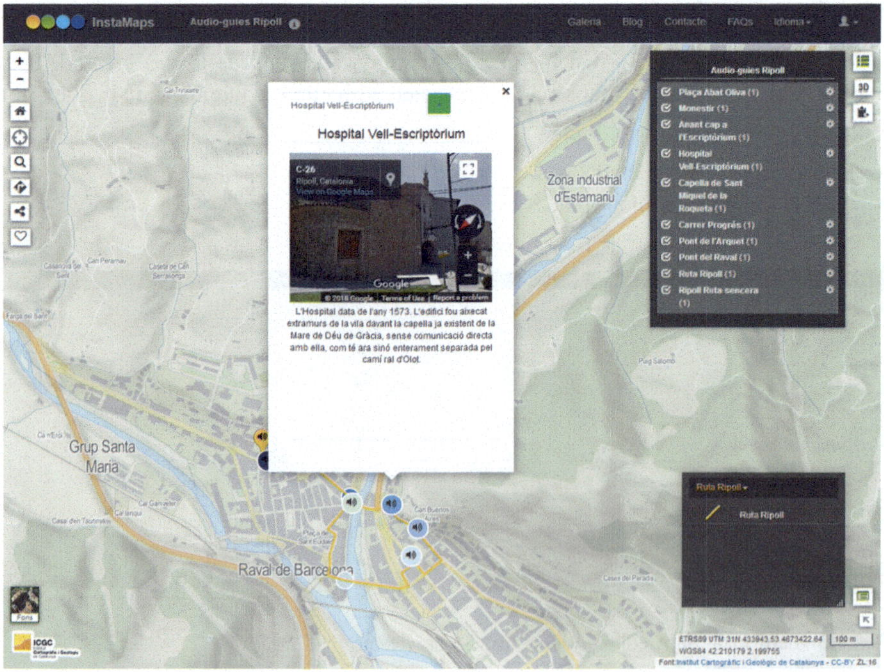

Fig. 5.4 A map displayed in InstaMaps

right of the map. From there layers can be swaped, deleted and their contents can be downloaded in different formats (kml, gpx, shp(zip), dxf, geojson) with different reference systems (SRS).

Loading user's own files. Geometries can be added to the map simply by dragging any file on top of the InstaMaps window. Currently different SRS are recognized for different formats: KML, KMZ, TXT, CSV, XLS, XLSX, SHP (zipped), DXF, GeoJSON and GML.

- GPX data: InstaMaps can read multigeometry GPX files;
- Both KML and KMZ are supported, but a different file is needed for each type of geometry: point, line, polygon;
- GeoJSON is recognized with no multigeometry;
- GML has its own specification but the same restrictions as GeoJSON apply;
- to load a shapefile (ESRI SHP) all files belonging to the project (dbf, prj, sbn, sbx, shp) must be zipped in a single file;
- TXT and CSV files must be encoded in UTF-8. The first row must necessarily have column names;
- XLS and XLSX: only the first book of Excel files will be added to InstaMaps. The first row must contain column names.

There are also three different ways to geolocate data that have been added from files of type TXT/CSV/XLS/XLSX:

- **by coordinates**: two separate columns must exist, X and Y. The reference system of the coordinates must be selected by its EPSG code.
- **by address**: right now it only works for addresses in Catalunya and uses the ICGC geocoder.
- **by code**: applies only to Catalunya. A column with the official statistical codes for regions must be present.

File loading is carried out by using the open source software GDAL (Warmerdam 2008). Thus the number of file formats supported in InstaMaps is likely to expand.

Files up to 500 MB can be loaded into InstaMaps, but performance will suffer due to web browser limitations as the number of geometries increases. Files bigger than 50 MB are ingested and then converted to a WMS service. This conversion disallows a further adoption of styles to the layers but increases the overall performance.

All the data ingested into InstaMaps is saved for later use in the private cloud, which is linked to the profile of the data provider. This private cloud is maintained for free at the ICGC data processing center. This data can be reused by anyone. The assumption is that the original data provider grants reuse privileges to the uploaded data.

InstaMaps is also able to recognize and upload data from a multitude of external data sources, such as:

- From open data portals, such as the Generalitat de Catalunya open data portal http://www.dadesobertes.cat.
- From social networks: geotagged content from Twitter, Panoramio and Wikipedia can be turned into maps.
- From WMS services: almost any existing WMS service can be added to a map in InstaMaps as a data layer and queried, provided it supports EPSG:3857 requests.
- From cloud files: all files with geodata that have been made publicly accessible on platforms such as Google Drive, Github, Dropbox and others can be directly loaded in InstaMaps. Such files can be loaded just once or can be dynamically linked so that the map is updated when a data change is detected.
- From JSON services: any JSON compliant data stream can be added to InstaMaps.

Metadata that describe geometries, which were loaded in InstaMaps, can be presented as a table to be edited (only while in editor mode) and can be exported as a GeoJSON, Excel, CSV or TXT file.

Styling of geographical elements in InstaMaps. Different styles can be applied to the geoinformation loaded in a map:

- Basic: the standard uniform symbolization of elements.
- Categories: elements in a layer can have their style changed triggered by a field in the associated data. This is helpful for thematic mapping.
- Size: the style changes based on a numerical field of the data. Useful for making bubble maps.
- Concentration: data density is turned into a new colorful layer like a heat map.
- Clustering: automatic grouping of data depending on density and map scale.

Applying basic GIS operations to data layers. A limited list of basic GIS operations is available in InstaMaps:

- Union by column: allows to join two data sets. Useful for combining geometries and alphanumerical data that share a common column.
- Points inside a polygon: a spatial filter that is used to restrict data belonging to a given zoning.
- Filter: the standard alphanumeric filter.
- Union: creates a new geometry out of two existing ones.
- Buffer: draws an outline to points.
- Intersection: creates a selection of two matching zones.
- Tag: transfers alphanumeric information from polygons to points.
- Centroid: creates a center point for a given area.

All these operations are usually observed in GIS systems. This functionality offers some additional flexibility to advanced users in InstaMaps. The downsize is the limits imposed and the degraded performance due to web browser constraints.

Base maps. A selection of basemaps exists that can be used as background for the creation of thematic maps in InstaMaps:

- Topographic map of Catalunya: the standard topographic map of the ICGC, which is made by combining topographic maps at different scales (depending on zoom level). Outside Catalunya, OpenStreetMap with a pastel coloured style created ad hoc by the ICGC is the basemap.
- Light topographic map: a very light pastel style was applied to the ICGC topographic database aiming at providing a basemap that would not interfere with users content while providing geographical context.
- Orthophotomap: the orthophotomap series of the ICGC at different scales (depending on zoom level). Outside Catalunya the global image coverage provided by ESRI is used.
- Hybrid: a blend of ICGC orthophotomap with a selection of geometries from the topographic base at 1:25,000. Outside Catalunya images and maps from ESRI are used.
- Administrative gray map: a simple gray map with administrative divisions used for thematic mapping of statistical data.
- Other map bases: gray-scale orthophotomap, topographic maps in different colors/hues, historical orthoimages from 1945 or 1956, historical topographic map from 1936, black and white elevation maps, shaded relief map from the 5 m Digital Terrain Model.

Publishing the resulting map. The last step to make a map live on the Internet is publishing:

- Give the map a title, description and some tags that will help to locate and index it.
- The initial map view will take into account the selected zoom level, visible layers and geographic area.
- The map with all these settings can be published as a private map (only the author will have access to it) or as a public map (available in the public map gallery).
- Data layers to be published will have to be defined as reusable and thus downloadable in this step.
- Selecting a legend covers the decision of being static or dynamic for one or every data layers used. Furthermore the default value upon initial map loading for an open or closed legend has to be set.

The final outcome of the publishing process is displayed in a new window that contains:

- the link to the map;
- the necessary HTML code to embed it in web pages (websites, personal blogs, content management systems);
- the link to the WMS service created ad hoc for each map;
- a link to visually check the map;
- links to share the map on social networks and over email.

Other features. InstaMaps offers a set of tools that aim at providing hard copies of the maps:

- a screenshot of the current view of the map as a PNG file;
- send to printer the current map view;
- export the current map view as a GeoPDF file, with all layers;
- export the current map view as a GeoTIFF file;
- export all vector data layers as a GeoPackage file.

A geolocation aware button allows InstaMaps to work as a moving map by keeping track of the current location of the client.

5.4 Software

5.4.1 Free and Open Source Software: FOSS, FOSS4G

From the beginning InstaMaps was intended to be a platform based on a scalable architecture, and that's why all InstaMaps components are built on open-source and free software tools. The main FOSS4G projects and components that are reused in InstaMaps are PostgreSQL and PostGIS, GDAL, MapServer, LeafletJS and Cesium.

Unfortunately InstaMaps has not yet been released as FOSS in our own Github, but the team is actively collaborating in the projects providing translations and bug corrections.

5.4.2 Architecture

The InstaMaps platform has a distributed and scalable 3-tier architecture (Liu et al. 2005):

Presentation tier. Communicates with the other 2 tiers by outputting the results to the browser/client tier. Servers in this layer provide HTTP communication between server and clients. We chose Nginx over Apache because it provides a most needed reverse proxy capability.

Application tier. This tier controls an application's functionality by performing detailed processing. This layer's servers contain the Java HTTP Web Server and the Java EE web application. They also contain the geospatial libraries required for data translation and processing.

The logical layer was programmed in Java (struts2), and we used *Hibernate* to map objects between the applications and the database. Fortunately, the architecture was available from a previous project which shortened the time frame for a release.

However it resulted in not using the latest available technology which would have improved the performance of this tier.

In this tier GDAL is used for file conversion and the import/export of data, along with Mapserver (because of its ability to deal with hot config files) to create WMS services out of layers that are too big to be displayed in real time and Tomcat as a web server for the logical layer.

Data tier. The data tier includes the data persistence and the data access mechanisms. This tier is implemented in a server that has the Database Management System and the user access control service. The database is replicated and synchronized on a second server using internal procedures. In this tier we use PostgreSQL as a database along with PostGIS extension to provide support for geographical objects and spatial functions.

All the data is stored in 36 tables using a relational database schema. Some other tables exist in a non-relational schema that add flexibility to deal with the different data structures as well as improving the speed and simplicity in performing certain types of queries that would take longer using a relational schema. The layers in this tier illustrated in Fig. 5.5 are as follows:

- Client layer (Geocat 1): two Nginx web servers behind a load balancer;
- Logic layer (Geocat 2): two Tomcat web servers behind a load balancer;
- Data layer: two database servers, one acting as master for read and write operations and the other acting as slave for backup purposes.

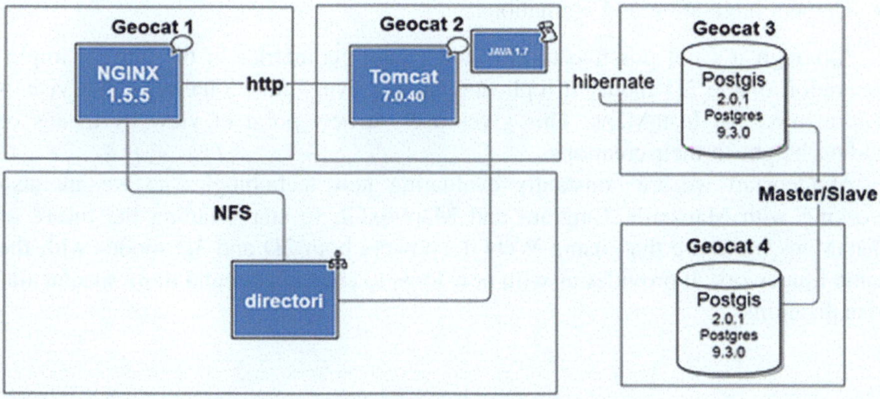

Fig. 5.5 Schematics of the server architecture

5.4.3 Front-End: The Web Client Application

The web client was programmed using Bootstrap framework with HTML5 and CSS3 and links more than 55 JavaScript libraries. Yet it is lightweight and responsive. For 2D mapping Leaflet and for 3D mapping Cesium has been used.

Leaflet is designed with simplicity, performance and usability in mind. It works efficiently across all major desktop and mobile platforms, can be extended with lots of plug-ins, has a beautiful, easy to use and well-documented API and a simple, readable source code that is a joy to contribute to.

Cesium is an open-source JavaScript library for world-class 3D globes and maps.

While Bootstrap allows developers to create responsive applications, the initial design of InstaMaps was created for desktop environments. We are now rethinking the design to solve some issues and to improve its usability in mobile clients. Leaflet was chosen as the best open source library to interact with maps because it was lightweight and offered a high degree of flexibility for loading plug-ins as well as customizing available features.

Cesium libraries enpower InstaMaps'3D mode. Cesium provides a lot of geographically-related 3D functionalities out of the box. The most prominent are the following ones:

- Ability to load and visualize the high-resolution terrain with high performance;
- Ability to use the tile-based layer imagery;
- Ability to draw vector formats, such as KML and GeoJSON, and clamp these layers to the terrain;
- Camera navigation and animations.

Although it's not possible to edit and create geometries in the current implementation of the 3D mode, it replicates almost every functionality of the visualization mode of InstaMaps. This gives users a new point of view by means of adding height to their creations.

At Geostart we are constantly evaluating new technologies so we are also working with Mapzen's Tangram and MapboxGL to start shaping the future of InstaMaps. A future that, using WebGL, powers both 2D and 3D modes with the same framework. It provides us with new tools to create better and more spectacular visualizations.

5.5 Team

The task force assembled to develop InstaMaps is multidisciplinary with a huge IT background and high programming skills for both the front-end and back-end. But it also integrates other technical profiles not necessarily in the IT field. This mix of approaches to mapping ensures that all aspects are covered when developing new contents.

It is worth mentioning that this group of 7 people who created and maintain InstaMaps are individually involved in the geogeek community in Barcelona and in other social initiatives around geographical information in their spare time. This is relevant when it comes to innovating the concept of maps, the dissemination of geoinformation, combining user generated content with official information and other aspects that impact daily activities of a national mapping agency like ICGC.

This group has been situated around a table where the persons sit facing each other, so that all of us can participate in any conversation. This setting establishes a better engagement at the workplace and the project (Becker and Steele 1995). This has been really useful for solving doubts and problems, but also for checking design proposals and discussing any aspect of programming. It added some noise to the whole work ambiance but proved fruitful for the outcome. Although it brings some chaos in, it also has created a collaborative group that shares a different approach working in government offices.

The team does also answer all email queries related to InstaMaps and Betaportal prototypes, since the interaction with users and first hand knowledge of issues are key in solving errors. Every now and then members of the team speak at conferences. They give lectures on InstaMaps and mobile apps. They visit universities to spread the word in order to determine the requirements that should be fulfilled using InstaMaps. This contact between the user base and the developers helps in steering InstaMaps in the right direction.

5.6 Ecosystem

In the end InstaMaps is an innovative tool that aimed at changing the procedure for the development of a web mapping IT environment and the way cartography is offered to the users. As a result the project InstaMaps has been very well received. It has become a key factor for helping people in getting real benefit by adding their own information to maps.

As a natural consequence, the tool itself starts to grow. Around this environment other tools and services are being developed. These other tools follow the same philosophy and complement each other. Linking these tools with InstaMaps creates an ecosystem of geocentered applications.

5.6.1 Ecosystem and Other Apps

The current InstaMaps ecosystem encompasses different tools and services that are developed and maintained by the same team that created InstaMaps as seen in Fig. 5.6:

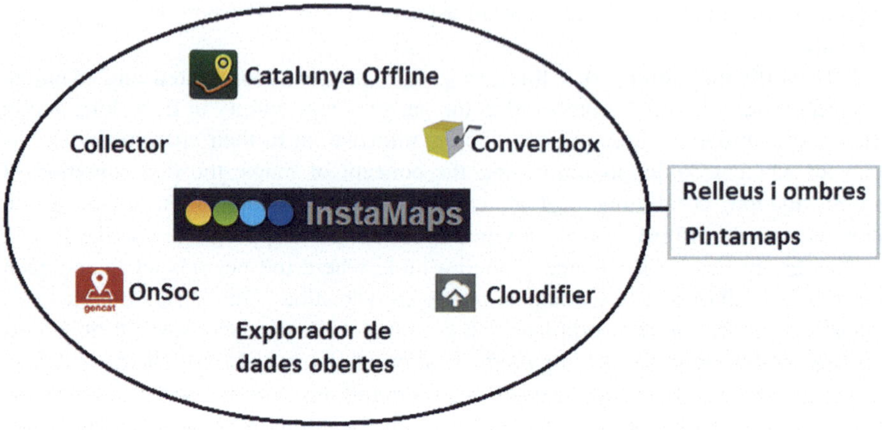

Fig. 5.6 Current InstaMaps ecosystem

- **Convertbox**. An interactive online web application that converts coordinates to and from different spatial systems and converts also certain file formats. It uses GDAL and is now completely integrated into InstaMaps.
- **Cloudifier**. A web application which turns any kind of geoinformation into a WMS/TMS service which can be loaded as a layer in InstaMaps. Developed initially to provide raster capabilities to InstaMaps.
- **Open Data Explorer (Explorador de dades obertes)**. A small set of tools which query data from Socrata (a data management environment used in government) and adds it to InstaMaps.
- **Catalunya Offline**. A mobile app for Android and iOS for hikers to get outdoors, record GPS tracks and way points and display maps even offline without a data connection. These outdoor recorded tracks can then be transferred to InstaMaps as new content.
- **Collector**. A small mobile application aimed at collecting all sorts of geodata and transfer it to InstaMaps.
- **OnSoc**. A mobile app to recover the current location and create a new map in InstaMaps.
- **Storymap**. An online web application that combines user provided Instamaps maps and HTML content to create interactive presentations.

In the meantime other tools have been developed which are currently not tied to InstaMaps, but which can be considered as part of the ecosystem. The ultimate objective is to add these tools as core functionality to InstaMaps:

- **Pintamaps/Pintamaps World**. A prototype that uses vector tiles and JSON styles to completely customize the looks of the topographic base map of ICGC.
- **Relleus i ombres**. Another prototype that helps users to understand and interact with the relief of Catalunya by using the map which has been generated from the 2 m DEM.

All of these prototypes are available online at the Betaportal (http://betaportal.icgc.cat), a web site created and maintained by the same team to help disseminate these new tools that explore new uses for existing cartography.

The ultimate goal is to expand this ecosystem, increase the number of interconnected InstaMaps services and provide augmented value to the maps of ICGC.

5.6.2 Raster to Vector

Within the InstaMaps ecosystem, the mobile environment is a different playground that requires some rethinking of the way we offer and consume geospatial information. This aspect directly relates to cartography.

In this scope speed and agility are key aspects. For these reasons the change from raster (image) to vector is very important.

Traditionally the consumption of cartography worked with raster formats, which means that "pre-made" images (image tiles) that were styled in a certain manner, projection and with a considerable load on the system will be consumed. In contrary the vector format offers a direct data consumption: it reduces the system load and makes the system more agile. In addition the rendering and visual customisation of the data are made on the client side (Gaffuri 2012).

This opens up a huge range of possibilities, which in our case have not yet been explored nor exploited. For instance the vector tile format, which allows to have all the cartography of Catalonia on your mobile device (e.g. for offline use), becomes an option. The system will not suffer from cellular data usage, the mobile coverage or the storage space. A recent example of the topographical base map of Catalunya in vector tiles has been published in the Betaportal (in catalan here: http://betaportal.icgc.cat/wordpress/topografic-5000-vector-tiles/).

In the case of InstaMaps, the idea is to finally work with vector tiles. In the same way as the tool Pintamaps/Pintamaps world allows users to create their own stylization and even to upload their own style files, InstaMaps will implement it.

Thus, the data vector and vector tiles should be the main way of geographical data consumption. All those who work in the geoinformation sector should keep it in mind.

5.6.3 *API*

Once seeing all the existing ecosystems around InstaMaps, it is entirely plausible to observe this as a natural evolution towards API (Application Programming Interface) functionality.

Providing an API would untie all internal functions and services developed so far from the client side. Thus its use and reuse is facilitated in external tools.

This would not only give more visibility and recognition of our work, but ultimately would allow us to increase the number of users of our own tools.

5.7 Conclusion

Mapping agencies in the 21st century must master a new set of skills beyond map-making in order to ensure that their products reach the market.

In InstaMaps there's still room for improvement. Bugs have to be addressed and new features need to be developed. For instance an isochrone calculation and viewshed analysis could easily be added. But right now the focus must be on two aspects: usability, which provides user retention, and geodata, which provides more opportunities for users to design the perfect map for their use case.

Acknowledgements InstaMaps as a project was granted an innovation award from Generalitat de Catalunya (local government) on March 16th 2017 (La innovació sovint apareix en els marges. II Jornada d'Innovació. https://innogent.blog.gencat.cat/2017/03/16/la-innovacio-sovint-apareix-en-els-marges-ii-jornada-dinnovacio/. Accessed 23 Mar 2016).

References

Battersby SE, Finn MP, Usery EL, Yamamoto KH (2014) Implications of web mercator and its use in online mapping. Cartographica Int J Geogr Inf Geovisualization 49(2):85–101

Becker FD, Steele F (1995) Workplace by design: mapping the high-performance workscape. Jossey-Bass, San Francisco

Gaffuri J (2012) Toward web mapping with vector data. In: International conference on geographic information science. Springer, Berlin, pp 87–101

Haklay M, Weber P (2008) Openstreetmap: user-generated street maps. IEEE Pervasive Comput 7 (4):12–18

Hanseth O (2010) From systems and tools to networks and infrastructures–from design to cultivation: towards. In: Industrial informatics design, use and innovation: perspectives and services: perspectives and services, p 122

Hassan-Montero Y, Martín-Fernández FJ (2003) Guía de evaluación heurística de sitios web. No Sólo Usabilidad. http://www.nosolousabilidad.com/articulos/heuristica.htm. Accessed 10 Mar 2016

Kuparinen L, Silvennoinen J, Isömaki H (2013) Introducing usability heuristics for mobile map applications. http://icaci.org/files/documents/ICC_proceedings/ICC2013/_extendedAbstract/424_proceeding.pdf. Accessed 17 June 2016

La innovació sovint apareix en els marges. II Jornada d'Innovació. https://innogent.blog.gencat.cat/2017/03/16/la-innovacio-sovint-apareix-en-els-marges-ii-jornada-dinnovacio/. Accessed 23 Mar 2016

Liu X, Heo J, Sha L (2005) Modeling 3-tiered web applications. In: 13th IEEE international symposium on modeling, analysis, and simulation of computer and telecommunication systems, 2005, pp 307–310

Miller CC (2006) A beast in the field: the google maps mashup as GIS/2. Cartographica Int J Geogr Inf Geovisualization 41(3):187–199

OSM (2018) Explanation of OSM. https://en.wikipedia.org/wiki/OpenStreetMap, visited Jan 2018

Rosenfeld L, Morville P, Arango J, Morville P (2015) Information architecture for the World Wide Web. O'Reilly, Beijing

Schmidt M, Jörg W (2012) basemap.at—creating a harmonised web basemap for Austria. In: Jobst M (ed) Service oriented mapping, pp 143–150. ISBN 978-3-9502039-2-3

Warmerdam F (2008) The geospatial data abstraction library. In: Open source approaches in spatial data handling. Springer, Berlin, pp 87–104

Chapter 6
Depiction of Multivariate Data Through Flow Maps

Alberto Debiasi, Bruno Simões and Raffaele De Amicis

Abstract Flow maps are graphical representations that depict the movement of a geospatial phenomenon, e.g. migration and trade flows, from one location to another. These maps depict univariate spatial origin-destination datasets, with flows represented as lines and quantified by their width. One main feature of these maps is the aggregation of flows that share the same origin. Thus, visual clutter is reduced and the readability improved. This chapter describes a novel technique that extends flow maps to visualize multivariate geographical data. Instead of a univariate color scheme, we interpolate strategic color schemes to communicate multivariate quantitative information. Additionally, our approach crystallizes on augmenting flows with pie charts. To evaluate the relevance and impact of our approach, three case studies are presented.

Keywords Map production use cases · Flow maps · Origin-destination data

6.1 Introduction

An increasing amount of spatial datasets are now freely available on the Internet, and this number is expected to increase as sensor networks are becoming a ubiquitous part of our environment. A substantial subset includes origin-destination data describing the movement of phenomena between locations. Phenomena can involve both tangible (e.g. bank notes, people, and goods) and intangible objects (e.g. reputation, ideas, and energy). Static media, such as paper maps, and dynamic media, such as electronic devices can be used to transport geoinformation (Debiasi et al. 2014a, 2015a, b). Compared to the latter, static media impose certain requirements on the map production, especially when multivariate data must be

A. Debiasi (✉)
Fondazione Bruno Kessler, Trento, Italy
e-mail: adebiasi@fbk.eu

B. Simões · R. De Amicis
Fondazione GraphiTech, Trento, Italy

J. Döllner et al. (eds.), *Service-Oriented Mapping*, Lecture Notes in Geoinformation and Cartography, https://doi.org/10.1007/978-3-319-72434-8_6

communicated. Since the user cannot manipulate the mapping information in a static context, it must be encoded in such a way that is easily understandable for the viewer.

This chapter is focused on the development of new static representations to depict spatial origin-destination data with multiple attributes. The requirements fulfilled by our approach can be clustered in constraint and analytical requirements.

Constraint requirements:

- CR1. The mapping must be static.
- CR2. The information must be depicted on a limited space (Mazza 2009) (e.g. the computer screen or a page on a newspaper) in such a way that it is clearly understandable by the viewer.
- CR3. The number of visual primitives must be minimized to reduce the cognitive overload.

Analytical requirements:

- AR1. Possibility to reason about the geographic patterns.
- AR2. Must enable, simultaneously, the visualization of the flow structure and its multivariate patterns.
- AR3. Possibility to analyze the outliers and commonality between each destination.
- AR4. Possibility to analyze the outliers and commonality for groups of moving entities.

Parallel coordinates are popular visualization methods for displaying multidimensional data. They consist of several vertical axes, each representing one data dimension. A data entity consists of several data points, each located on one of the axes. *Star plot*, also called radar chart, is a similar method. The axes are equally spaced around a circle, originating from the center. The values of neighboring axes are connected by a line, resulting in a star-like shape. Although parallel coordinates and star plots are powerful methods, which do not require any interactive media, they do not preserve the geographic information, i.e. the position of each location (the requirement AR1 is not satisfied). Flow map representations are more suitable to communicate univariate spatial origin-destination data. They depict the movement of phenomena between geographic locations (Slocum et al. 2009). The curves in a flow map, called flow lines, describe the movement of a phenomenon from one location to another. Flow maps offer a simple and effective instrument to depict geographical movement. However, its visualization without visual clutter and cognitive overload can be challenging (Fig. 6.1a) (Debiasi et al. 2015c, 2016). An extension to tackle those problems is the Sankey Flow drawing technique (Schmidt 2008). Flow lines sharing the same origin are aggregated together and their width is defined proportional to the sum of the flows' magnitude that they represent (Fig. 6.1b).

There are three approaches that extend Sankey flow maps so they can be used to depict multivariate data:

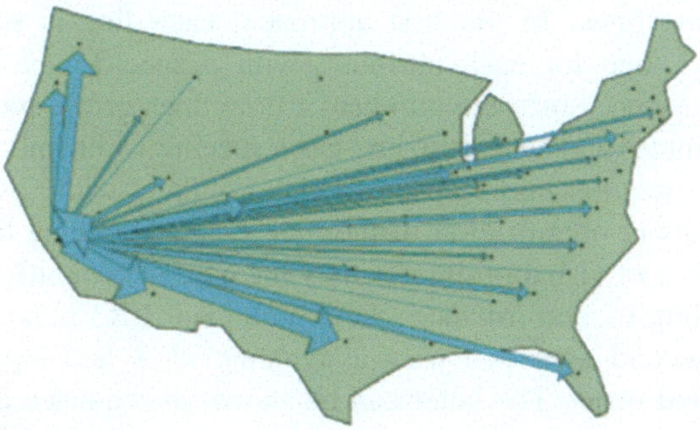

(a) Simple Flow map.

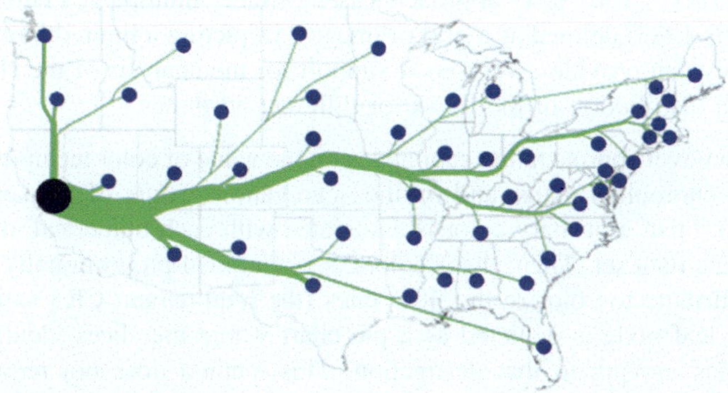

(b) Sankey Flow map.

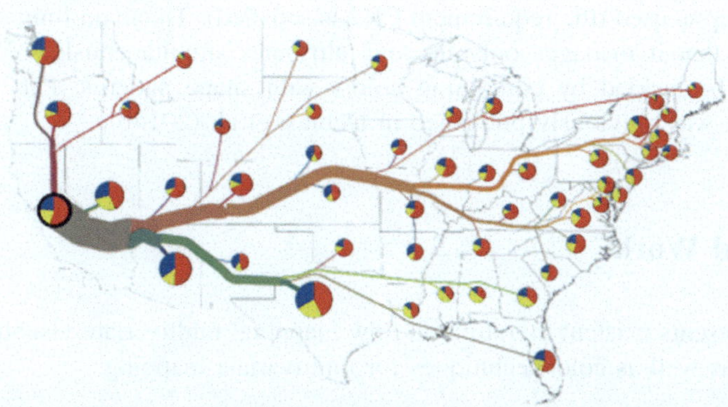

(c) Sankey flow map enriched with information about the ethnicity of migrations: red are Caucasian people, yellow are African Americans people and blue are Asian people. Compared to (a) and (b), (c) communicates more information without heavily increasing the visual complexity.

Fig. 6.1 Single-origin flow maps depicting immigration from California

- Subdivision of flow lines: In the first approach, each line is subdivided in "parallel channels", one for each attribute, with a specific color and width proportional to its value. Since the number of flow lines grows proportional to the number of attributes and each flow line has a specific width and margin, then this approach also requires more space (which conflicts with CR2). Because space is limited, this re-organization of the map is a challenging task. Further, the increased number of visual primitives may increase the cognitive load of the user, thus, decreasing its performance (the requirement CR3 is not satisfied).
- Animation: The second approach uses a dynamic view and represents each attribute in separated maps. The latter can be shown in sequence in predefined time intervals, or interactively using triggers to switch between the separated maps (the requirement CR1 is not satisfied).
- Small Multiples: The last approach uses small multiples (Tufte and Graves-Morris 1983) defined as a grid of images. Depicting a limited number of attributes, they can provide a very good support for the analysis of the changes of the overall distribution of the flows for different attributes.

We propose a novel approach that capitalizes on the value of color techniques, as already done in choropleth maps, and on the aggregation feature of Sankey flow maps (Fig. 6.1c). First, a primary color is associated with each numerical attribute of the multivariate data set. Then, these colors are combined proportionally to the value of their attribute to colorize the flow lines (the requirement CR3 satisfied). Moreover, each leaf node is depicted as a pie chart where the slices identify the amount of entities moving to that destination. This method does not require an interactive platform, thus it can depict a map as a static image (the requirement CR1 is satisfied). Compared to small multiples, this method requires less screen space because one map is used (the requirement CR2 is satisfied). The main limitation of our method is that it manages only up to 3 attributes simultaneously. Yet, this limitation can be tackled by combining colors with shape patterns. Part of the presented work was previously published in Debiasi et al. (2014b).

6.2 Related Work

This section presents existent literature on flow mapping, multivariate visualization and mapping, as well as color techniques for multivariate mapping.

6.2.1 Flow Mapping

There are different algorithms that are designed to automatically aggregate and bundle flow lines for flow maps.

In edge bundling, the links of a graph are bundled together if certain conditions are met. Holten and Wijk (2009) presented a force-directed algorithm in which links are modeled as flexible springs that can attract each other while node positions remain fixed. This algorithm is extended to separate opposite-direction bundles, emphasizing the structure of the graph (Selassie et al. 2011). Cui et al. (2008) described a mesh-based link-clustering method for graphs. Control mesh generation techniques were used to capture the underlying edge patterns and to generate informative and less cluttered layouts. Ersoy et al. (2011) created bundled layouts of general graphs, using skeletons as guidelines to bundle similar links. Pupyrev et al. (2012) proposed a link routing algorithm based on ordered bundles. With this technique, the links are placed in parallel channels to avoid overlap. The peculiarity of the edge bundling is that it is used to understand the overall patterns but not the single links of the graph because the connections become harder to read.

In Phan et al. (2005), flow maps are automatically generated aggregating links and making them curves. The authors defined a binary hierarchical clustering to formulate the layout in a simple and recursive manner. Then, in a post creation phase, users are given the possibility to alter the shape of the flow lines by moving their control points. Verbeek et al. (2011) introduced a method to create flow maps using directed angle-restricted Steiner trees of minimum length, also known as spiral trees. They are designed to avoid crossing nodes, as well as user-specified obstacles. To straighten and smoothen the tree, a cost function is defined. At each iteration, a new position of the intermediate nodes is computed and the cost function is calculated. Then, they apply the steepest descent to minimize the global cost function, which is the sum of the cost function for each intermediate node. The layout is updated according to the magnitude of the flows. Nocaj and Brandes (2013) proposed an approach based on a new drawing style for the generation of flow maps called confluent spiral drawings. Confluent spirals consist of smooth drawings of each bundle in which edge directions are represented with smooth appearance. A sequence of spiral segments is used to represent an edge between the origin and the target node, and the vortex of each spiral corresponds to a target node. When an obstacle overlaps a branching point, it is branched out in an earlier or later phase of the parent spiral to miss that obstacle. An exhaustive survey on edge bungling and flow map techniques is provided by Zhou et al. (2013).

6.2.2 *Multivariate Visualization and Multivariate Mapping*

Different methods have been designed to visualize multivariate datasets with no spatial content. The basic tools are tables, charts, histograms, and scatter plots (Harris 1999). More advanced approaches include, for example, parallel coordinate plots (Inselberg 1985), scatterplot matrices (Andrews 1972) and glyphs (Pickett et al. 1995; Tominski and Schulz 2012). For a detailed list of the key visualization methods used in this context the surveys Keim et al. (2004) and Chan (2006) are recommended.

In the area of Air Traffic Control, Klein et al. (2014) have presented a set of visualization techniques for the exploration of movement data. The trails are colored using a directional hue color map to discover close-and-parallel, opposite-direction, and flight paths.

Multivariate mapping has long been an interesting research problem. An approach that has been successfully applied to mapped data is the Chernoff faces (Chernoff and Rizvi 1975), where different variables are related to different facial features to form a face icon for each data object. Similar glyph-based techniques were presented by DiBiase et al. (1994), Gahegan (1998), Wittenbrink et al. (1995), Tominski et al. (2005). These approaches typically require the viewer to interpret each icon by memorizing its configuration and constructing its meaning on the fly.

Andrienko and Andrienko (2001) proposed an effective approach to reason about spatial patterns through dynamic linking between a geographic map and non-spatial multivariate representations. However, this approach cannot generate a visualization that shows the distribution of multivariate patterns in a limited screen space.

6.2.3 Color Techniques for Multivariate Mapping

The use of colors was first introduced in the Golden Age of data graphics (1850–1899) (Friendly and Denis 2008). Since then cartographers have been using it to highlight specific aspects of the map.

A choropleth map is a thematic map in which areas are shaded or patterned in proportion to the value of one or more variables being displayed, as described in detail in Robinson (1955). A color scheme is built to accommodate all possible magnitude combinations of the variables. These maps facilitate the recognition of patterns, as well as an easy interpretation of the variables as individual entities, and as a set (Leonowicz 2006). The chosen colors are an important aspect of the visualization: they should be distinguishable and the transition of the colors must be smooth and visually coherent (Eyton 1984). Depending on the number of properties to depict, different variations of choropleth maps exists.

A univariate choropleth map (e.g. Vanderbei n.d.) displays data for only one variable. To differentiate between the variable values different shades of the same color are used. Bivariate choropleth maps enable the portrayal of two independent variables. The color schemes for bivariate choropleth maps usually take the form of a square. Since univariate maps are considered effective tools to recall patterns, they are preferred to the use of two distinct univariate maps (Eyton 1984). A trivariate choropleth can be used only for three attributes that add up to 100% (Slocum et al. 2009). In such case the color scheme takes the form of a triaxial graph (Byron 1994).

In single-origin flow maps, existing color techniques only represent the magnitude of the flow line together with, or as an alternative to the use of the width property (Holten et al. 2009). If the phenomenon has multiple starting points, the color can identify the origin (Phan et al. 2005; Verbeek et al. 2011). In such case the color does not provide additional information. Instead, it emphasizes the geographical location of the moving entities. Flow maps can also depict qualitative data characterized by a continuous attribute, such as the average age of the migrants. The color gradient that encodes the attribute is used to color each line (Guo 2009). As limitation, no aggregation method is applied, thus the presence of cluttered areas is more probable and the aesthetically result may vary.

6.3 Proposed Approach

We extend the algorithm for the automatic generation of Sankey Flow map described in Debiasi et al. (2014c), to depict more than one attribute. Compared to simple Sankey Flow maps, the expressiveness of nodes and lines are enriched by a continuous color scheme (see Sect. 6.3.1) applied to the lines (see Sect. 6.3.2) and to the nodes represented by circular diagrams (see Sect. 6.3.3). The geographic dataset is represented as a tree t where the origin is the root node r and the n geographical destinations correspond to the leaves $l_1, l_2, \ldots, l_n$, respectively with flow magnitudes $magn(l_1), magn(l_2), \ldots, magn(l_n)$. Each destination (represented by a leaf) can have associated complementary attributes, i.e., their union is equal to the whole magnitude. An example of supported attributes is the number of males and the number of females of migrating people. Another example is the motivation that led them to travel: family, job and study. Let $attr_{x_i}(l) \in [0, 1]$ the value of the characteristic x (e.g. x can identify the motivations) of the attribute with index i (e.g. i can identify the family, job or study) assigned for each destination l, such that:

$$\sum_{i=1}^{h} attr_{x_i}(l) = 1$$

where h can be at most 3. Our approach can be applied to datasets with the following data characteristics:

- Case 1: Two attributes $attr_{x_1}$ and $attr_{x_2}$, which compose the whole amount, i.e. for each leaf l, $attr_{x_1}(l)$ and $attr_{x_2}(l)$ sum up to 1.
- Case 2: Three attributes $attr_{x_1}$, $attr_{x_2}$, and $attr_{x_3}$ which compose the whole amount, i.e. for each leaf l, $attr_{x_1}(l)$, $attr_{x_2}(l)$, and $attr_{x_3}(l)$ sum up to 1.
- Case 3: Two complementary attributes from distinctive characteristics $attr_{x_1}$ and $attr_{y_2}$, i.e. for each leaf l, $attr_{x_1}(l)$ and $attr_{y_2}(l)$ do not necessarily sum up to 1.
- Case 4: Three complementary attributes from distinctive characteristics $attr_{x_1}$, $attr_{y_2}$, and $attr_{z_3}$, i.e. for each leaf l, $attr_{x_1}(l)$, $attr_{y_2}(l)$, and $attr_{z_3}(l)$ do not necessarily sum up to 1.

6.3.1 Color Models

The color is used to convey information about the attributes of the geographical data depicted in a flow map. There are many color models to specify colors in maps. Models that are based on hardware specifications, such as RGB, are frequently used because of their long history and their consequent familiarity. Although RGB color spaces reflect color mixing operations in color printers and computer monitors, they are not recommended as reference model of color for human beings. In fact, pigment mixing in early childhood art training makes use of subtractive color models that differ from RGB the model. For such reason, our work employs a technique designed by Gossett and Chen (2004); inspired by paint mixing using a subtractive color space with Red, Yellow and Blue as primary colors. To obtain an effective color coding it is fundamental to have color mapping function that is invertible, i.e. every attribute value (or every group of data values) is associated with exactly one color, and vice versa (Tominski et al. 2008). Moreover, a color compositing technique is only useful if it allows the viewer to discern the individual components that form the visualization (Gossett and Chen 2004).

For each case described at the beginning of this section, different attributes are associated to the color mapping function, thus different color spaces are used.

In Case 1, we use a pattern composed by red and blue. $attr_{x_1}$ is associated to the red channel and $attr_{x_2}$ is associated to the blue channel, see Fig. 6.2a. In Case 2, we use a triaxial graph composed by red, yellow and blue. In this case $attr_{x_1}$ is associated to the red channel, $attr_{x_2}$ is associated to the yellow channel and $attr_{x_3}$ is associated to the blue channel, see Fig. 6.2b. Only a small portion of the color space is used because the sum of the primary colors must be 255. In Case 3, we use the square pattern composed by red and blue colors. The association between attributes is the same as Case 1, see Fig. 6.2c. Finally, in Case 4, we use the entire colors space N^3, and the association between attributes is the same as Case 2, see Fig. 6.2d.

Generating many colors from a limited range of values can be challenging. If users cannot distinguish between colors then they cannot perceive the magnitude of the value it represents. Hence, we normalize each primary color interval using their minimum and maximum value, see Sect. 6.4 for an application of such method. As shown in Fig. 6.2a, for Case 1 the two attributes share the same range, therefore the formula to normalize their value is:

$$norm\_attr\_x_{\mathrm{var}}(l) = \frac{attr\_x_{var}(l) - \min(attr\_x_{var})}{\max(attr\_x_{var}) - \min(attr\_x_{var})}, \quad var = 1,2 \tag{6.1}$$

Figure 6.3 shows an example of Case 1 where the values of the attributes are normalized. For Case 2 the formula to normalize the value of each attribute is defined as follows:

For Case 3 and 4, where attributes describe distinctive characteristics, we normalize the attributes according to its minimum and maximum value.

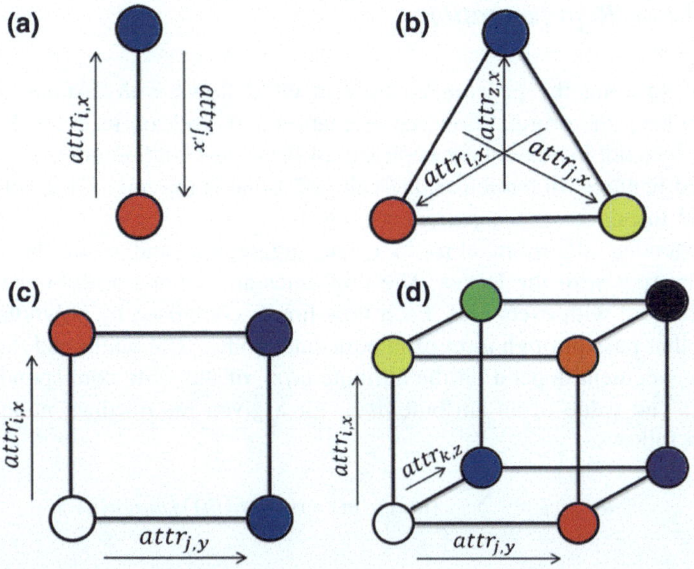

Fig. 6.2 RYB interpolation patterns used for each case

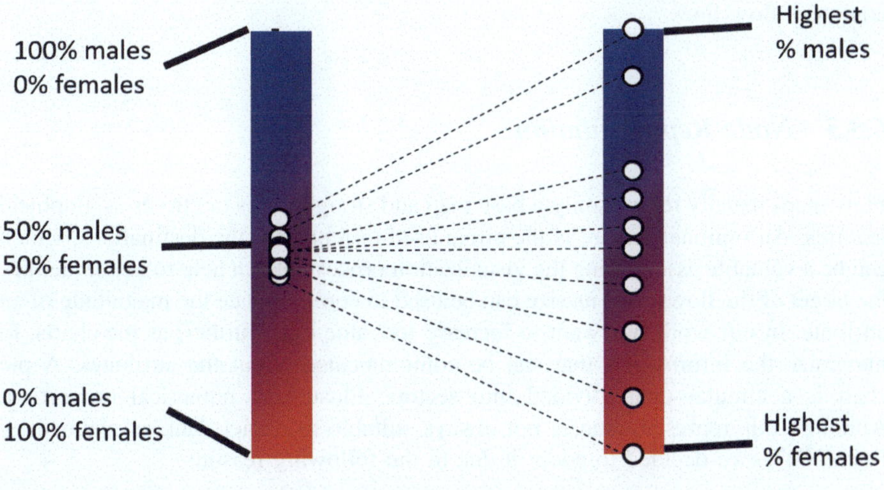

Fig. 6.3 Example of procedures to normalize the attributes in Case 1

$$norm\_attr\_x_{var}(l) = \frac{attr\_x_{var}(l) - \min(attr\_x_{var})}{1 - \min(attr\_x_1) - \min(attr\_x_2) - \min(attr\_x_3)}, \quad (6.2)$$

$$var = 1, 2, 3$$

(see Eq. 6.1)

6.3.2 Flow Representation

Instead of splitting the generalized flow in child flows with distinct colors, we decided to keep the original flow representation and work on its color. Hence, our approach does not increase the complexity of the representation in terms of number of depicted features. In particular, the color of a line is the mix of the colors of the aggregated lines.

The extended algorithm aims to create, aggregate, and color the lines that connect the root with the leaves. The flow lines are defined as cubic splines that connect the root with every leaf. Each flow line is composed by polynomial curve segments that pass through a set of intermediate nodes. The width and the color of each curve segment depend on the attribute $attr_{x_i}$ of the node corresponding to its end point. The value of an attribute $attr_{x_i}$ for a given intermediate node n is calculated as follows:

$$attr_{x_i(n)} = \sum_{m \in child} (magn(m) * attr\_x_i(m))/magn(n)$$

where the set *child* contains all the child nodes of the node n. The color of each curve segment is calculated using the attributes of its end point and the appropriate mapping function described in Sect. 6.3.1. Figure 6.4 shows an example of generation of flow lines.

6.3.3 Node Representation

Flow maps usually require only a base map and straight lines or curves as graphical features. An optional feature is the circle used to represent the destinations, which can be a valuable asset during the visualization process. It can help to better identify the target of the flows, and its size can be used to communicate the magnitude of an attribute. In our work, we want to increase its value even further, as pie charts, to maximize the information that can be communicated about the attributes. A pie chart is a circular chart divided into sectors, illustrating numerical proportion. Although this representation is not always suitable to depict data, as claimed by Few (2007), we decided to adopt it due to the following reasons:

- Nodes are usually depicted as circles. Therefore, pie charts can be easily integrated without an effecting the layout of the flow map.
- The attributes are at most three, thus this representation is not misleading.

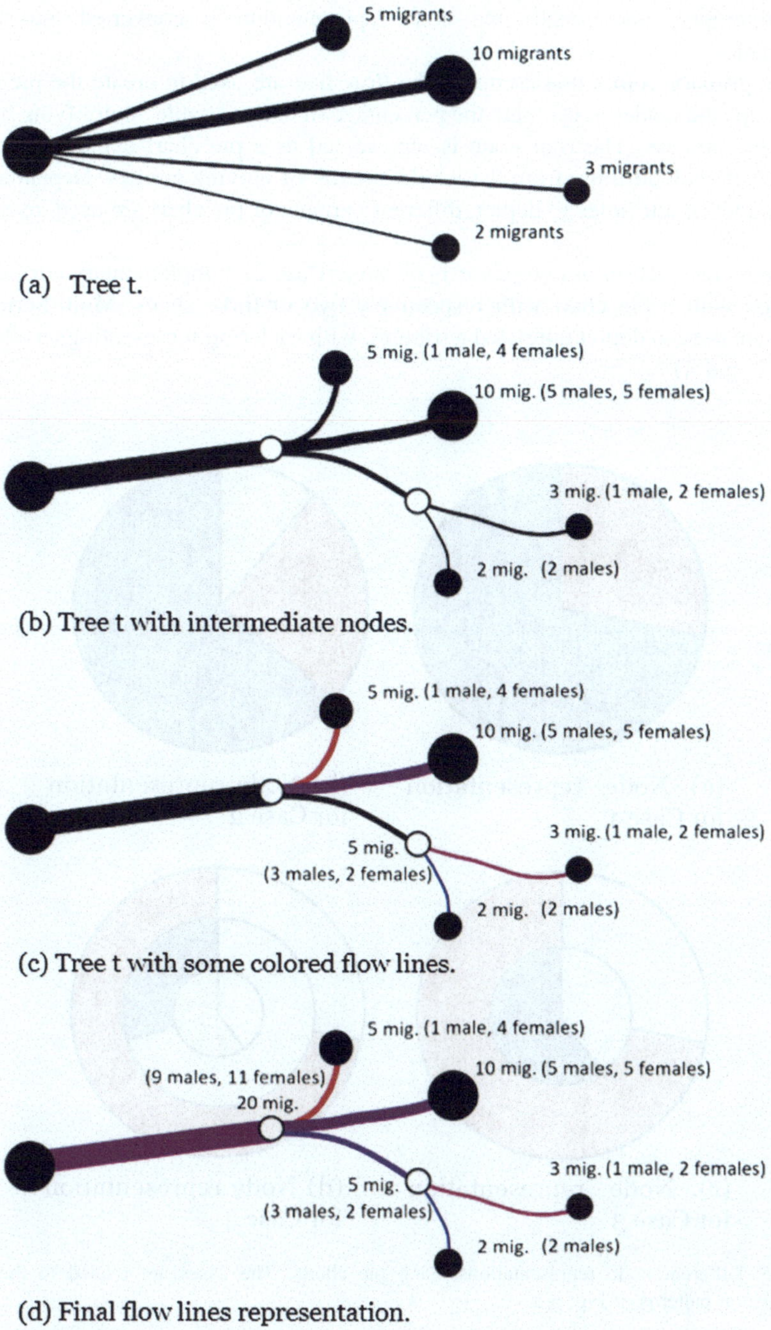

Fig. 6.4 Example of generation of flow lines. Each destination has associated the number of migrants and two complementary attributes: number of males and number of females

The empty space inside the node representation is converted into visual information.

The primary colors that compose the flow line are used to create the pie chart. This helps the reader to estimate the percentage of each variable, simplifying his/her cognitive process. The root node is represented as a pie chart depicting the percentage of each attribute from the whole amount of moving entities. Depending on the number of attributes to depict, different variants of pie chart are used, as shown in Fig. 6.5.

The representation of two (Case 1) or three (Case 2) complementary attributes is depicted with a pie chart with respectively two or three slices. Multi-Series pie charts are used to depict unrelated attributes, with each ring representing an attribute (Case 3 and 4).

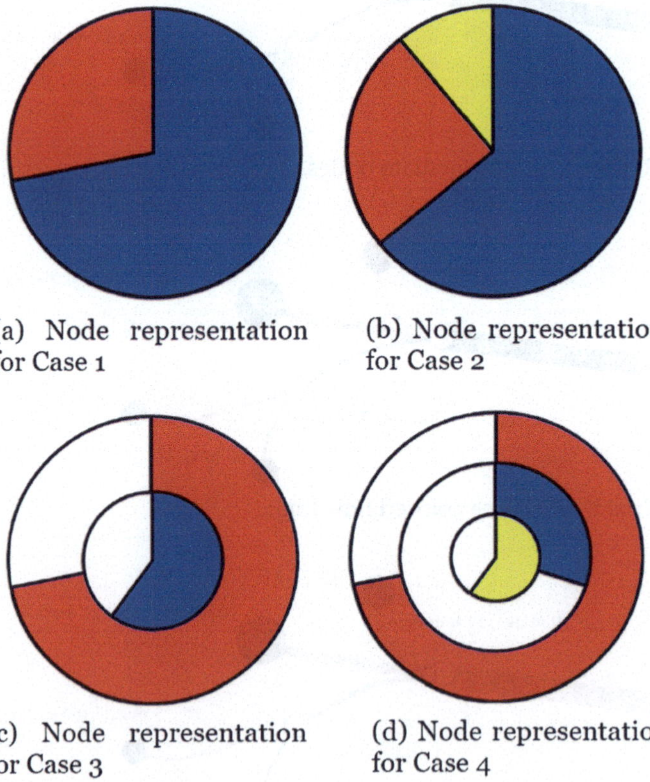

(a) Node representation for Case 1

(b) Node representation for Case 2

(c) Node representation for Case 3

(d) Node representation for Case 4

Fig. 6.5 Different node representations using pie charts. The cases are related to the RYB interpolation patterns of Fig. 6.2

6.4 Case Studies

In this section, we provided an evaluation of our approach based on the County-to-County Migration Flows dataset (Bureau 2014). It describes migration phenomena from multiple states of the USA. Although this dataset has been extensible used to generate flow maps (Ersoy et al. 2011; Phan et al. 2005; Pupyrev et al. 2012; Verbeek et al. 2011), existent algorithms depict only one attribute. As opposite, with our approach, the color of the flow lines and the pie charts enable the viewer to gain information not only about the number of people moving but also about their age group and ethnicity.

To emphasize differences between the attributes, all values are normalized following the procedure described in Sect. 6.3.1. Thus, the color of the flow lines does not identify exactly the percentages of each attribute, but communicates whether the attribute is greater, lesser or equal to the same attribute in other flow lines. For example, if a flow line is 100% red, it does not mean that the percentage of the attribute associated to red is 100%. It means that the flow line has the highest percentage of the red attribute compared to other flow lines and the lowest percentage of other attributes respectively.

In Fig. 6.6 it is possible to extract information about age groups in the migratory flows from Colorado. This attribute is subdivided into three groups: younger than 30 (red), ranging from 30 to 54 (yellow) and over 54 (blue). Pie charts depict the percentage of people for each group. Moreover, flows color encodes such information also for aggregated destinations. For example, by looking at the main split in the west of Colorado, we can conclude that the flow directed to north (area A) has

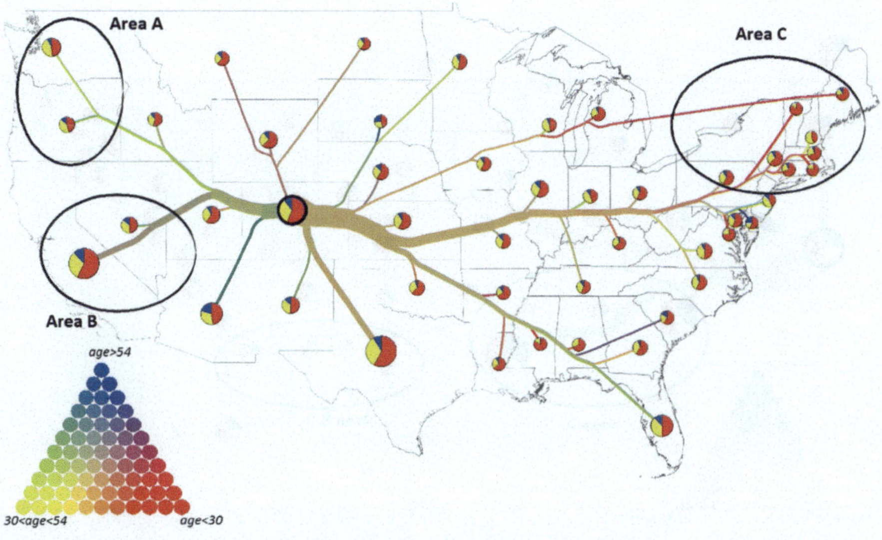

Fig. 6.6 Migratory flows from Colorado. Color identifies age: red are people younger than 30, yellow from 30 to 54 and blue the over 54

more migrants with age between 30 and 54 than those directed to California and Nevada (area B) that have more young people. The flows directed to the north east (area C) contain a high percentage of young people.

In Fig. 6.7 the flow of migrants is shown together with the information about their ethnicity. The colors red, yellow and blue represent the Caucasian, African American and Asian ethnicity, respectively. The thick flow directed to Arizona, New Mexico and Texas (Area A) is colored with a more vivid blue, which indicates a higher percentage of Asians migrants. As opposite, in the south east (area B) African Americans people are the majority of the migrants.

The flow map depicted in Fig. 6.8 shows the migration from Texas. It is enriched with information describing distinctive characteristics: the Asian people that are moving (red) with age from 30 to 54 (blue). The map can help analysis to find correlations between these two attributes. If the color tends to be redder, a high percentage of migrants are Asian people and if the color tends to be bluer, a high percentage of migrants have an age between 30 and 54. Moreover, the lightness/darkness of the color defines respectively if both attributes have low/high percentage with respect to the whole amount of people that are moving. The thick flow directed to west (Area A) is colored with a strong red, indicating that the presence of Asian is higher over the migrants directed in that zone.

In Area B the lines are colored primarily with blue, which implies a high percentage of people with ages between 30 and 54 and a low percentage of Asian, if compared with other flows of migrants.

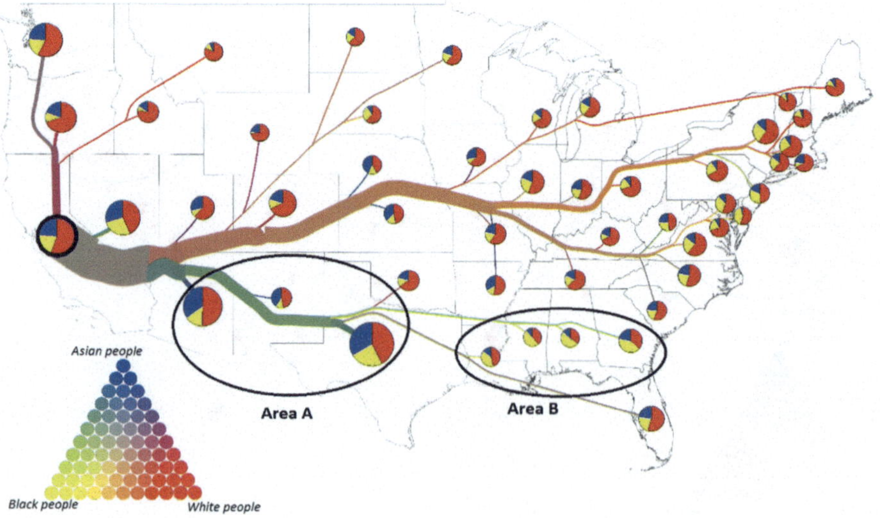

Fig. 6.7 Migratory flows from California. Color identifies the ethnicity of migrants: red are Caucasian people, yellow are African Americans people and blue are Asian people

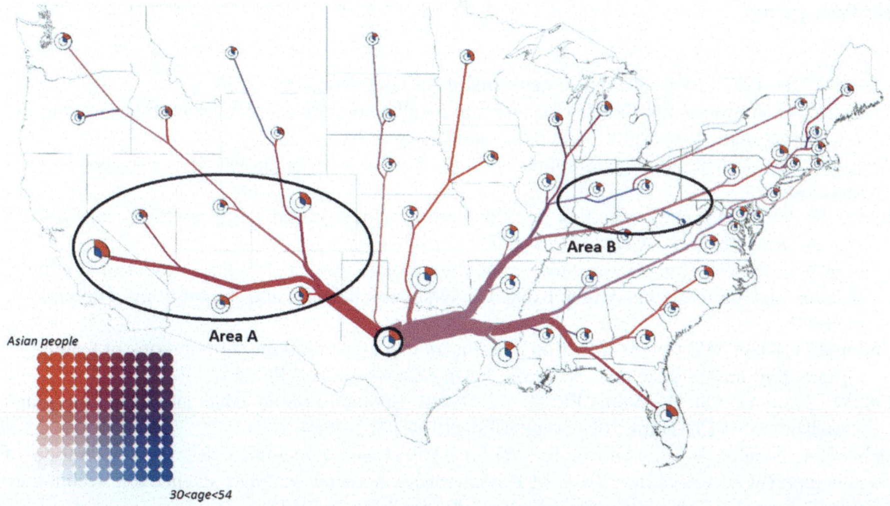

Fig. 6.8 Migratory flows from Texas. Color identifies two attributes from distinctive characteristics (ethnicity and age) of people that are moving: (Red) Asian people, (Blue) people with age between 30 and 54

6.5 Conclusion and Future Work

A color scheme blending is used in conjunction with the aggregation aspects of flow maps and pie charts, to visualize in a static representation spatial origin-destination data with multiple attributes. Compared with parallel coordinates and star plots, our approach analyses the outliers and commonality between each destination and for groups of moving entities. Moreover, it overcomes the limitations of the techniques that extend flow maps with animation, small multiples and subdivision of flow lines.

Case studies are performed through three automatically generated flow maps to communicate information not only about the number of migratory flows but also about their age group and ethnicity. However, to further validate our work a user study must be performed.

The analysis of the geographic patterns depends on the criteria used to aggregate the flow lines, i.e. different flow maps that depict the same dataset can give a different interpretation. However, the algorithm used for the automatic generation of the flow map gives the cartographer the possibility to modify the direction and the aggregation of the flow lines. As future work, we plan to reduce significantly the used colors discretizing the color scale. In fact, cartographers recommend the use of five to seven classes of color for a map (Dent 1990).

References

Andrews DF (1972) Plots of high-dimensional data. Biometrics, 125–136
Andrienko G, Andrienko N (2001) Constructing parallel coordinates plot for problem solving. In: 1st international symposium on smart graphics, pp 9–14
Bureau C (2014) Migration/geographic mobility. Retrieved from https://www.census.gov/hhes/migration/
Byron JR (1994) Spectral encoding of soil texture: a new visualization method. In: GIS/LIS Proceedings, Phoenix, Airz, pp 125–132
Chan WWY (2006) A survey on multivariate data visualization, vol 8. Department of Computer Science and Engineering, Hong Kong University of Science and Technology, Hong kong, pp 1–29
Chernoff H, Rizvi MH (1975) Effect on classification error of random permutations of features in representing multivariate data by faces. J Am Stat Assoc 70:548–554
Cui W, Zhou H, Qu H, Wong PC, Li X (2008) Geometry-based edge clustering for graph visualization. IEEE Trans. Vis Comput Graph 14:1277–1284
Debiasi A, Simões B, De Amicis R (2014a) 3DArcLens: a technique for the exploration of geographical networks. In: 2014 IEEE conference on visual analytics science and technology (VAST), pp 245–246. https://doi.org/10.1109/vast.2014.7042512
Debiasi A, Simões B, De Amicis R (2014b) Depiction of multivariate data through flow maps. In: Proceedings of the SOMAP—symposium on service-oriented mapping
Debiasi A, Simões B, De Amicis R (2014c) Supervised force directed algorithm for the generation of flow maps. In: Proceedings of the WSCG—22nd international conference on computer graphics, pp 193–202
Debiasi A, Simões B, De Amicis R (2015a) 3DArcLens: interactive network analysis on geographic surfaces. In: Proceedings of the 6th international conference on information visualization theory and applications, pp 291–299. https://doi.org/10.5220/0005255202910299
Debiasi A, Simões B, De Amicis R (2015b) GeoPeels: deformation-based technique for exploration of geo-referenced networks. In: Proceedings of the WSCG—23rd international conference on computer graphics, pp 53–62
Debiasi A, Simoes B, Amicis RD (2015c) Schematization of node-link diagrams and drawing techniques for geo-referenced networks. In: 2015 international conference on Cyberworlds (CW), pp 34–41. https://doi.org/10.1109/cw.2015.68
Debiasi A, Simões B, De Amicis R (2016) Schematization of clutter reduction techniques in geographic node-link diagrams using task-based criteria. In: Proceedings of the 7th international conference on information visualization theory and applications. https://doi.org/10.5220/0005255202910299
Dent BD (1990) Cartography: Thematic map design. WC Brown, Dubuque, IA
DiBiase D, Reeves C, MacEachren AM, Von Wyss M, Krygier JB, Sloan JL, Detweiler MA (1994) Multivariate display of geographic data: applications in earth system science. In: Visualization in modern cartography, pp 287–312
Ersoy O, Hurter C, Paulovich FV, Cantareiro G, Telea A (2011) Skeleton-based edge bundling for graph visualization. IEEE Trans Vis Comput Graph 17:2364–2373
Eyton JR (1984) Complementary-color, two-variable maps. Ann Assoc Am Geogr 74:477–490
Few S (2007) Save the pies for dessert. In: Perceptual edge visual business intelligence newsletter, pp 1–14
Friendly M, Denis DJ (2008) Milestones in the history of thematic cartography, statistical graphics, and data visualization. Seeing Science: Today American Association for the Advancement of Science
Gahegan M (1998) Scatterplots and scenes: visualisation techniques for exploratory spatial analysis. Comput Environ Urban Syst 22:43–56
Gossett N, Chen B (2004) Paint inspired color mixing and compositing for visualization. In: IEEE symposium on information visualization, 2004. INFOVIS 2004, pp 113–118

Guo D (2009) Flow mapping and multivariate visualization of large spatial interaction data. IEEE Trans Vis Comput Graph 15:1041–1048

Harris RL (1999) Information graphics: a comprehensive illustrated reference. Oxford University Press, Oxford

Holten D, Wijk VA (2009) Force-directed edge bundling for graph visualization. Comput Graph Forum 28:983–990

Holten D, van Wijk, Jarke J (2009) A user study on visualizing directed edges in graphs. In: Proceedings of the SIGCHI conference on human factors in computing systems, pp 2299–2308

Inselberg A (1985) The plane with parallel coordinates. Vis Comput 1:69–91

Keim DA, Panse C, Sips M (2004) Information visualization: scope, techniques and opportunities for geovisualization. Bibliothek der Universität Konstanz

Klein T, van der Zwan M, Telea A (2014) Dynamic multiscale visualization of flight data. GRAPP/VISAPP

Leonowicz A (2006) Two-variable choropleth maps as a useful tool for visualization of geographical relationships. Geografija 42:33–37

Mazza R (2009) Introduction. Introduction to information visualization. Springer Science & Business Media

Nocaj A, Brandes U (2013) Stub bundling and confluent spirals for geographic networks. In: Graph drawing, pp 388–399

Phan D, Xiao L, Yeh R, Hanrahan P (2005) Flow map layout. In: IEEE symposium on information visualization, 2005. INFOVIS 2005, pp 219–224

Pickett RM, Grinstein G, Levkowitz H, Smith S (1995) Harnessing preattentive perceptual processes in visualization. In: Perceptual issues in visualization. Springer, Berlin, pp 33–45

Pupyrev S, Nachmanson L, Bereg S, Holroyd AE (2012) Edge routing with ordered bundles. In: Graph drawing, pp 136–147

Robinson AH (1955). The 1837 maps of Henry Drury harness. Geogr J 121:440–450. Retrieved from http://www.jstor.org/stable/1791753

Schmidt M (2008) The Sankey diagram in energy and material flow management. J Ind Ecol 12:82–94

Selassie D, Heller B, Heer J (2011) Divided edge bundling for directional network data. IEEE Trans Vis Comput Graph 17:2354–2363

Slocum TA, McMaster RB, Kessler FC, Howard HH (2009) Thematic cartography and geovisualization. Pearson Prentice Hall, Upper Saddle River, NJ

Tominski C, Schulz HJ (2012) The great wall of space-time. In: Vision, modeling & visualization, pp 199–206

Tominski C, Schulze-Wollgast P, Schumann H (2005) 3D information visualization for time dependent data on maps. In: Proceedings of the ninth international conference on information visualisation, 2005, pp 175–181

Tominski C, Fuchs G, Schumann H (2008) Task-driven color coding. In: 12th international conference on information visualisation, 2008. IV'08, pp 373–380

Tufte ER, Graves-Morris PR (1983) The visual display of quantitative information, vol 2. Graphics press, Cheshire, CT

Vanderbei RJ (n.d.) On graphical representations of voting results. http://citeseerx.ist.psu.edu/viewdoc/summary?doi=10.1.1.140.7759. Website accessed on 15 Feb 2018

Verbeek K, Buchin K, Speckmann B (2011) Flow map layout via spiral trees. IEEE Trans Visual Comput Graphics 17:2536–2544

Wittenbrink CM, Saxon E, Furman JJ, Pang A, Lodha SK (1995) Glyphs for visualizing uncertainty in environmental vector fields. In: IS\&T/SPIE's symposium on electronic imaging: science & technology, pp 87–100

Zhou H, Xu P, Yuan X, Qu H (2013) Edge bundling in information visualization. Tsinghua Sci Technol 18:145–156

Chapter 7
Humanitarian Demining and the Cloud: Demining in Afghanistan and the Western Sahara

Peter Schmitz, Johannes Heymans, Arie Claassens, Sanet Carow and Claire Knight

Abstract Communities in war torn countries continue to face many life threatening situations long after the end of a war. These situations include the contamination of the environment by landmines and explosive remnants of war (ERW). One of the main objectives of mine action is to address problems faced by communities owing to landmine contamination. Since the removal of all landmines worldwide is improbable, the humanitarian demining sector focusses on removing landmines from areas where communities are most affected. Due to the decrease in donor funding, there is continued pressure for more effective and efficient mine action through improved and appropriate data collection, analysis and the use of the latest technologies. Proper data management, sharing of data in the collaborative cloud and improved decision support systems to prioritize areas for demining, will result in more effective mine action. This chapter will discuss humanitarian demining as one of the components of mine action and will emphasize the importance of mapping an area for demining purposes. The importance of data management for decision support systems to prioritize areas for demining is covered with specific reference to data collection, manipulation, dissemination and data quality.

P. Schmitz
CSIR Built Environment, Meiring Naude Rd, Brummeria, Pretoria, South Africa

J. Heymans · A. Claassens · S. Carow (✉) · C. Knight
Esri South Africa, Midrand, South Africa
e-mail: sanetcarow@gmail.com

P. Schmitz
Department of Geography, Geoinformatics and Meteorology, University of Pretoria, Lynnwood Rd, Hatfield, Pretoria, South Africa

P. Schmitz
Fakultät für Vermessung, Informatik und Mathematik, Hochschule für Technik, Schellingstrasse 24, 70174 Stuttgart, Germany

J. Heymans · A. Claassens
School of Geography, Faculty of Environment, University of Leeds, Leeds LS2 9JT, UK

P. Schmitz · S. Carow
Department of Geography, University of South Africa, Christiaan de Wet Road, Florida, Johannesburg, South Africa

J. Döllner et al. (eds.), *Service-Oriented Mapping*, Lecture Notes in Geoinformation and Cartography, https://doi.org/10.1007/978-3-319-72434-8_7

The important role that the collaborative cloud plays in data dissemination and sharing is expanded upon. Use cases of the collaborative cloud and humanitarian mapping are noted and the role of data security is described. The latest decision support systems for humanitarian mapping are briefly discussed. The main shortcoming of these decision support systems is the lack of a spatial analysis component. The development of a decision support tool based on a Geographical Information System is illustrated by referring to case studies in Afghanistan and Western Sahara. The successful use of the GIS based decision support system has consequently lead to the development of a spatial multi-criteria analysis tool. The spatial multi-criteria analysis tool emphasizes the importance of sharing data in the collaborative cloud and the use of quality data. This tool contributed to humanitarian demining and mapping by assisting in better and faster decision making processes at a reduced cost.

Keywords Humanitarian demining · Humanitarian mapping · Collaborative cloud Data management · Decision support systems · Multi-criterial decision support tool

7.1 Introduction

Communities in current and past conflict areas are affected by landmines, improvised explosive devices (IED) and explosive remnants of war (ERW). This chapter will discuss humanitarian demining as one of the components of mine action and will emphasize the mapping of an area for demining purposes (Sect. 7.2). This will be followed in Sect. 7.3 by a discussion on the importance of data management for decision support systems in order to prioritize areas for demining. The important role that the collaborative cloud plays in sharing data for mine action and demining is discussed in Sect. 7.4. The latest decision support systems and tools for prioritizing areas for demining are discussed in Sect. 7.5 by referring to case studies in Afghanistan and Western Sahara. From the case studies, a spatial multi-criteria decision tool had been developed and will be discussed in Sect. 7.6. Section 7.7 concludes the chapter.

7.2 Mine Action and Humanitarian Demining

Communities in war torn countries face many life threatening situations long after the end of a war. (ICRC 2012). These situations include the contamination of the environment by landmines and ERW. The main objective of mine actions are to address problems faced by communities as a result of landmine contamination. These activities aim to reduce the social, economic and environmental impact of landmine contamination on affected communities. Mine actions also determine areas where people can live safely and create an environment in which economic, social and health development can occur free from the constraints imposed by landmine contaminations. Mine action includes five complementary activities namely: mine risk education and awareness, humanitarian demining, victim assistance, stockpile destruction and advocacy against the use of anti-personnel mines (Langholtz 2014).

Humanitarian demining (also referred to as demining) includes all activates that lead to the removal of landmines and unexploded ordinance (UXO). The removal and clearance of landmines and UXO is just one step in the humanitarian demining process. The process of demining an area also includes technical surveys, mapping, clearance, marking, documentation, community mine action liaison, the handover of the cleared land (Langholtz 2014) and decision support systems (Heymans 2015). Humanitarian demining may be emergency based or developed over a longer period.

Since the removal of all landmines worldwide is improbable, the humanitarian demining sector focusses on removing landmines from areas where communities are most affected. The removal of landmines is less about mines and more about communities and their interactions with a landmine contaminated environment (Heymans 2015). The effects that landmines have on the lives of people are devastating and hard to quantify. Figure 7.1 indicates the number of casualties worldwide as monitored by the Landmine Monitor from 1999 to 2015. The number of landmine and ERW casualties worldwide for 2015 was 6461, up by 3492 casualties since 2014. The sharp increase can be attributed to more landmine and ERW casualties recorded in armed conflicts in Libya, Syria, Ukraine and Yemen (Anon 2016). Landmines are used as weapons of terror against local communities. The landmines are spread to deprive communities of access to farmlands, roads, drinking water and fire-wood and often force them to relocate elsewhere. This has a negative effect on the economy and social structures of impacted communities (Demining 2017).

According to the latest available estimates, 110 million landmines are threatening the lives of communities in 46 different countries. The countries with the

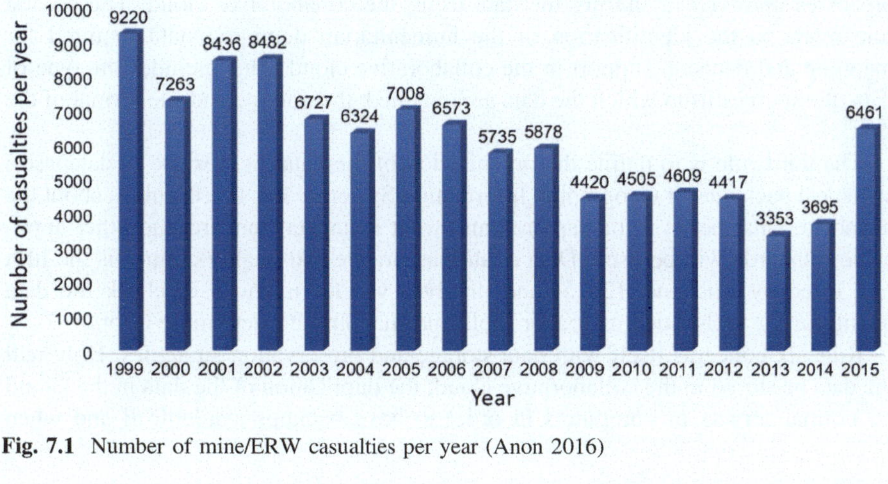

Fig. 7.1 Number of mine/ERW casualties per year (Anon 2016)

highest estimated number of landmines are Bosnia and Herzegovina, Cambodia, Croatia, Egypt, Iraq, Afghanistan, Angola, Iran and Rwanda (UNICEF n.d.). The highest number of landmine related deaths is reported in Afghanistan. Since 1997, the Convention on the Prohibition of the Use, Stockpiling, Production and Transfer of Anti-Personnel Mines and on their Destruction (also referred to as the Ottawa Convention) has put an international ban on the use of landmines (UNOG n.d.). To date, 133 states have signed the Ottawa Convention while 162 states have agreed to be formally bound by the Convention (UNOG n.d.).

The cost of obtaining a landmine is estimated at $3–$10 while the cost of removing a landmine is estimated at $300–$1000 per landmine. The clearance of field mines relies heavily on contributions from donors. The total contribution from donors during 2016 amounted to $340.1 million (Anon 2016). The Landmine Monitor 2016, reported that for the third year in a row, donors decreased their international mine action assistance. Due to the decrease in funding, there is continued pressure for more effective and efficient mine action through improved and appropriate data collection, analysis and the use of the latest technologies (Heymans 2015). Proper data management, sharing of data and improved decision support systems to prioritize areas for demining will result in more effective mine action. An overview of data management is given in the next section.

7.3 Data Management

Data management is both an action and a system and both are interlinked. Michener (2015) looked at data management as an action when the author published some rules to establish a good data management plan. Although the context is in the health and biological research environment, these rules can be adapted and applied to the spatial and humanitarian demining domain.. The first rule by Michener (2015) is to determine the requirements, with regards to data management, within the spatial and humanitarian demining context. This includes the standards and processes involved in sharing the data using the collaborative cloud. The second rule refers to the identification of the humanitarian demining data required for mapping and decision support in the collaborative cloud. This includes the type of data, the sources from which the data are obtained, the volume and file format of the data.

The third rule is to define the organisation of the data, i.e. for use in databases, statistical packages or Geographic Information Systems. The fourth rule is about the metadata, what needs to be captured and what metadata standard and other applicable standards will be used. Data quality assurance and quality control is the fifth rule listed by Michener (2015) and involves various methods to check the data quality using tools, such as scatter plots and mapping, to determine errors.

Rule six concerns itself with data storage and preservation strategies. How will the data be stored in the collaborative cloud, the duplication of the data in the Cloud on normal servers or computers in order to have backups available if and when needed as well as the backup cycle. Michener (2015:5/9) lists three questions that need to be answered, namely “How long will the data be accessible? How will data be stored and protected over the duration of the project? How will data be preserved and made available for future use?”. All these questions are relevant for the data that will be used in the collaborative cloud. Rule seven defines the data policies, such as sharing arrangements, access control, data security and restrictions on data, such as non-disclosure agreements.

Rule eight concerns itself with the manner of data dissemination. Dissemination depends on accessibility, such as broad band connectivity, collaborative cloud and the format of the data being made available, such as raw data, data that has been enhanced and made available as text, comma delimited files or as shape files or even as printed data to entities outside the humanitarian demining entity. The ninth rule by Michener (2015) determines who is responsible for what, with regards to the data, and who will be responsible for the management of the data. Rule ten is about budgeting for data management and involves the cost of collecting, storing, manipulation, disseminating data, software, hardware, Internet connectivity, renting space in the Cloud, office space and personnel. This section, and the above sections, discussed data management from an activity viewpoint; the next few sections will look at data management as a system.

There are two main components of database management namely transactional database management and analytical database management (Abadi 2009). Transactional database management is found in the banking industry, e-commerce, airline industry, hotel booking industry and railway and bus online ticketing systems. These are mainly financial database systems that have high security levels and, owing to their nature, are not well suited for cloud computing. The second component is analytical database management which is used in problem-solving, data analysis and decision support and is well suited for cloud computing (Abadi 2009). Analytical database management is applicable in the humanitarian demining environment since decision support and mapping include various forms of analysis.

In the context of cloud computing, there are three main service paradigms namely Infrastructure as a Service (IaaS), Platform as a Service (PaaS) and Software as a Service (SaaS) which will be discussed in more detail in the next section. However Agrawal et al. (2011) argue that cloud computing can also serve as Database as a Service or as a Storage as a Service. The advantage of cloud computing is the concept of scalable database management systems which allow for updating and managing heavy workloads. Cloud computing’s ability means that very large data sets can be stored and/or ad hoc analysis of the data and decision support can be made much more effeciently (Agrawal et al. 2011). The latter two analysis are of importance to the humanitarian demining community. The main advantage of using the Cloud as a data management facility is that communities, such as those active in humanitarian demining, can access the data from anywhere around the globe without having to spend large amounts of money on hardware and software in order to setup large data management systems to support demining activities (Abadi 2009).

The advent of big data in the Cloud heralded new forms of data management and data management systems, such as NoSQL (Not only SQL) and NewSQL, which have relational database management systems (RDMS) designed for big data management and analytics (Moniruzzaman and Hossain 2013). However, in comparison to big data, humanitarian demining data sets are comparatively small and therefore do not require NoSQL or NewSQL. The SQL type of relational RDMS, such as the Oracle RDMS's, MySQL or Microsoft SQL Servers, are sufficient to manage the data pertaining to demining in the Cloud. This section looked at data management as an action as well as data management as a system. Data management as a system can be used in the Cloud. The next section discusses the collaborative Cloud for data management and other applications for humanitarian demining and mapping.

7.4 Collaborative Cloud

Cloud computing (also referred to as the Cloud) is an ever evolving, and rapidly growing, technology and paradigm. The Cloud is a model that enables on demand network access, to a pool of computing resources, in a convenient and timely manner (NIST 2016).

The Cloud is used by many IT professionals, business managers and researchers. Each of these groups define the Cloud according to their own needs and depending on how they apply the Cloud in their own application fields (Armbrust et al. 2009). For the purpose of this chapter, the definition of the National Institute of Standards and Technology (NIST) from the United States Department of Commerce is used (Mell and Grance 2011):

> Cloud computing is a model for enabling ubiquitous, convenient, on- demand network access to a shared pool of configurable computing resources (e.g., networks, servers, storage, applications, and services) that can be rapidly provisioned and released with minimal management effort or service provider interaction. This Cloud model is composed of five essential characteristics, three service models, and four deployment models.

The above definition refers to the five essential characteristics, three service models and four deployment models of the cloud. The explanation and definition of the characteristics, service models and deployment modules fall outside the scope of this chapter but are summarized in Table 7.1 for clarity purposes.

Figure 7.2 illustrates the characteristics, service and deployment models of the Cloud. An organization needs to consider each one of the characteristics and models carefully before migrating data to the Cloud.

The collaborative cloud (or cloud collaboration) is a concept that has evolved from cloud computing. The same principles of the Cloud apply to the collaborative cloud. The collaborative cloud is defined as a way of sharing computer files (e.g. spatial data files) using the Cloud (Bradley 2013). The data files are uploaded to the Cloud and are shared, edited and updated by many users such as Open Street Map,

Table 7.1 Definitions related to the characteristics, service and deployment models of cloud computing (Mell and Grance 2011; Hamdaqa and Tahvildari 2012; Hassan 2011; Rong et al. 2013)

Terminology	Definition
Characteristics:	
On demand self service	A single user can facilitate computing capabilities as needed
Broad network access	Capabilities can be accessed over high speed internet connections using various platforms e.g. mobile phones, tablets, laptops and/or workstations
Resource pooling	Resources are shared among various users without knowing the exact location of the resources
Rapid elasticity	Resources can be scaled in and out as needed, often automatically. This gives the illusion of unlimited resources to the user
Measured service	Measuring services monitor the usage and health of systems. This is crucial because it assists with optimizing resources and provides transparency for both consumers and providers
Service models:	
Software as a Service (SaaS):	Applications running on Cloud infrastructure to provide services to end-users
Platform as a Service (Paas):	Tools and resources running on Cloud infrastructure to provide services to end-users
Infrastructure as a Service (IaaS):	Are the computing resources, e.g. storage, networks, servers and operating systems, used to provide services to clients
Deployment models:	
Private cloud	Is owned or rented by a single organization and the whole resource is dedicated to the single organization
Community cloud	The resources are shared among organizations or members of a closed community with similar interests. This model is very similar to the private Cloud
Public cloud	Is owned by a service provider that sells or rents resources to organizations. The resources are shared among many organizations and end-users
Hybrid cloud	This is a combination of two or more Cloud infrastructures. A hybrid Cloud provides extra resources when needed e.g. migrating from a private to a public Cloud for intensive computational tasks

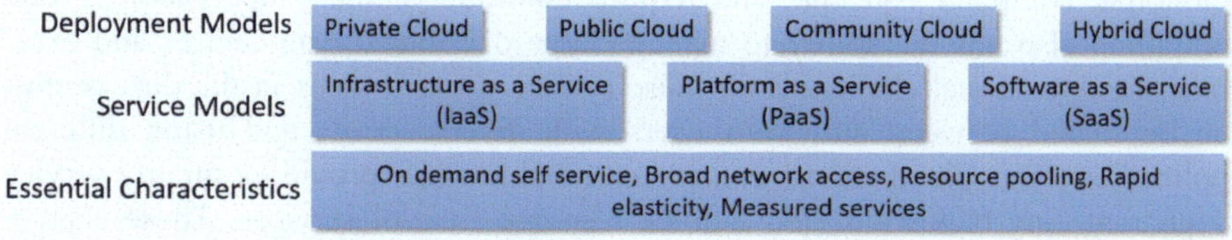

Fig. 7.2 Characteristics, service and deployment models of cloud computing

Natural Earth Data and Google maps. In terms of this definition, the collaborative cloud seems to be suitable for use in humanitarian demining. This approach will provide the opportunity to share data files within and between demining organizations. This type of deployment model is called the community cloud. Cloud collaboration in a community cloud has the potential to save considerable time and resources in terms of obtaining data sources in areas not yet or poorly mapped. The data files can be obtained from various resources, uploaded into the community cloud where they are updated and shared with various other demining organisations.

One of the apparent concerns with regards to adopting cloud technology is the security available for information stored within the Cloud. Reports indicate that there is a significant gap between the claims of vendors and the view of users when it comes to cloud security, privacy and transparency (Kshetri 2013). Since a private cloud provides an individual or organisation with more controllability and flexibility, with regards to security, consumers have greater trust in a private Cloud than a public Cloud. Consumers also need to consider that the public Cloud providers implement best practices and aim to provide as much security as possible to protect the information of all users (Hamdaqa and Tahvildari 2012).

Kshetri (2013) argues that issues related to security and privacy in the Cloud are well documented but only partially understood. Cloud users have more holistic perspectives than non-cloud IT users and are becoming more educated on how to communicate the relevant issues important to them, such as cost saving, productivity gain and security and privacy issues. An educated approach will lead to better service delivery and improvement of security by the Cloud vendors (Kshetri 2013).

While the issue of security remains, and continues to be debated, the benefits of storing information in the Cloud outweigh the security concerns. The benefits of the Cloud are well documented and discussed in the report by Hamdaqa and Tahvildari (2012). In their report, the Cloud is compared with other computer paradigms, such as grid computing, parallel computing, utility computing, autonomic computing and virtualisation technologies. The Cloud has replaced expensive supercomputers and clusters, as a means of running intensive calculations, and the cloud is making expensive supercomputers and clusters available for those users that otherwise would not have the means to access supercomputers and clusters. Computer resources in the Cloud are also provisioned fairly, easily and quickly. The challenges related to the complexity of managing large and distributed Cloud data-centres, increased resource availability and enhanced flexibility are solved by autonomic computer concepts and require minimum human supervision. Cloud computing also has the ability to manage large distributed data-centres and overcome the operational challenges of these centres. Data sources in the data centres can be shared between numerous users with diverse needs and using different applications. The Cloud can be considered reliable while providing quality service by guaranteeing bandwidth and a quick response time of services. These characteristics make the Cloud suitable for mission critical systems.

Cloud computing has proven to be the result of the development and adoption of existing technologies and paradigms. The goal of cloud computing is to assist users in benefiting from all these technologies, without the need for knowledge or expertise of any of the existing systems (Hamdaqa and Tahvildari 2012). Furthermore, when users make use of cloud computing, they should confirm the location of the data centres, background checks of employees and certifications from auditing and professional organisations to ensure the safety of their sensitive data sources (Kshetri 2013).

In the case of humanitarian demining, an organization needs to share their data sources carefully to ensure that security is not compromised. Data layers that are considered to be sensitive (e.g. the location of mines) should be saved in a local and secure location.

An internet search has indicated that numerous geospatial softwares are available for use in the Cloud. The more familiar applications include ArcGIS Online, MangoMap, GISCloud, Mapbox and CartoDB. Each one of the applications have advantages and disadvantages. It is not the purpose of this chapter to compare the different softwares, but a few aspects need to be considered when deciding which software to use for humanitarian demining. These include the analytical capabilities of the software, the availability of data sources, the ability to handle data sources and layers obtained from different sources and in different formats. The software should also be able to handle plugins and have the option to develop customized plugins for specific applications.

Various examples of the use of the Cloud or collaborative cloud in a variety of application fields have been published. The application areas include the development of a Cloud based platform for watershed analysis and management (Alencar et al. 2014), the military environment (Tutino and Mehnen 2013), healthcare environment (Hoang and Chen 2010) and local municipalities (Ali et al. 2015). An initial internet search could not provide any documented examples of using the Cloud or collaborative cloud for humanitarian demining and mapping.

It is predicted that the future development of the Cloud will focus on mobility (Kshetri 2013). More devices will link to cloud data centres including smartphones, vehicles and televisions. The Cloud is also expected to promote what is called the 'Decision support as a service' industry. Decision support as a service is based on collecting and analysing information and providing users with accurate decisions. It is expected that decision support as a service will have a significant impact on certain application fields such as medicine, through smart medicine diagnoses systems, the economy, through automatic financial market brokers and environmental science, through accurate climate-change forecasting (Hamdaqa and Tahvildari 2012). The Cloud, and specifically the collaborative cloud, could have a similar impact on humanitarian demining through supporting decision support systems for clearance operations.

This section and Sects. 7.2 and 7.3 discussed the background to humanitarian demining, data management and the collaborative cloud. The next section discusses examples of decision support and mapping in the collaborative cloud for humanitarian demining.

7.5 Using GIS for Humanitarian Demining Decisions and Mapping

7.5.1 Effectiveness of GIS as a Decision Support System (DSS) on a Global Level

Heymans (2015) conducted research to assess the effectiveness of GIS as a decision support system for mine action at a global level. This was achieved by mapping and analysing the global strategic indicators with regards to demining to inform global management decisions for mine action in terms of priorities and funding using available GIS technologies and data. The prepared data on demining activities was uploaded to ArcGIS online, Esri's GIS in the Cloud that allows users to create, manage, and store maps and applications online. After uploading the data, a web mapping application was published to compare the selected strategic indicators and mine action funding from 2000 to 2012.

For example, cleared area values are displayed as a percentage of the total area cleared for all the selected countries. The comparative values for each indicator, i.e. cleared area, casualties and funding, were obtained by normalizing the respective country values as a percentage of the total across the full reporting period. This is best practice for presenting data as 'choropleth maps', which are maps where spatial units (in this case the countries) are filled with an associated colour. For example, the casualties in Angola are represented as a percentage of all the casualties across all the countries in the study, from 2000 to 2012.

The web application is intuitive and has the basic navigation tools that allow users to pan and zoom to areas of interest. This allows decision makers without any GIS technical skills to use web application for decision making. The time slider functionality was included to allow the user to look at a specific year's data, for example, looking at the area cleared, causalities and mine action funding for 2007 to see if there are correlations between the different indicators and the funding levels. Typically, the user would expect that countries with higher funding levels would correlate with countries with higher numbers of casualties and that the area cleared would also be higher among countries that received more funding.

Based on the web map applications, the user was able to compare strategic indicators in comparison with the funding levels. After visual inspection of the three different maps, the starkest observation concerned casualties where the amount of funding for clearance were relatively high in one country while null causalities were reported. This observation illustrates the value of projecting data on a map for visual representation, while the fact that zero or null causalities were recorded for Angola in 2008 highlights the issue of data quality and completeness. This is shown in Figs. 7.3, 7.4 and 7.5.

Fig. 7.3 Area cleared for 2008

Fig. 7.4 Causalities for 2008

7.5.2 *Effectiveness of GIS as a DSS at Country Level*

Traditional planning of landmine clearance operations relies heavily on geographical and spatial analysis, GIS technology is perceived by some as playing a progressive role in humanitarian demining. A number of attempts have been made to apply GIS and cartographic methodologies in the humanitarian demining field.

However, further research including that of and the geographical contextual aspect of problems of interest to the user.

GIS allows decision makers to "integrate great quantities of information… and to provide a powerful repertoire of analytical tools to explore this data". This extends the multi-criteria decision-making process by using various different datasets as input within a GIS, applied preferences (policies) to generate various

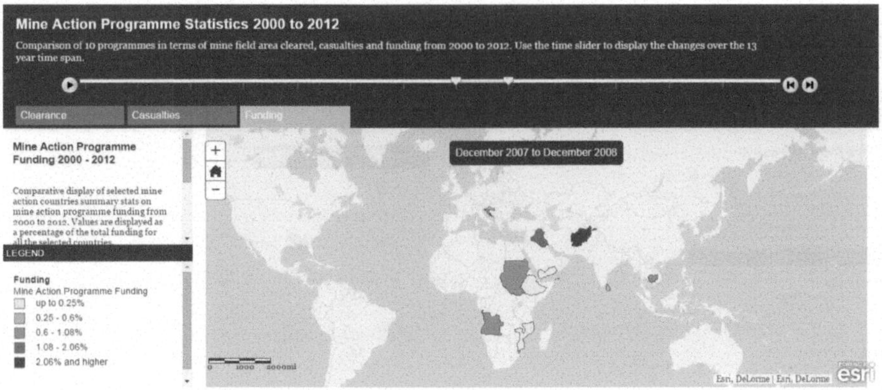

Fig. 7.5 Mine action funding for 2008

possible solutions. Ultimately, the decision maker needs to apply sound judgement in selecting the final solution appropriate to the problem.

7.5.3 *Spatial Decision Support System: Western Sahara Case Study*

Western Sahara was selected because it is a small mine action programme in comparison to other mine action programmes. No landmine impact survey has previously been conducted and, consequently, the available data in IMSMA NG, an international humanitarian demining organisation, is limited. The Western Sahara is mostly a desert which poses challenges to decision makers with regards to prioritising demining activities. The mine action database was made available for the research project, conducted by one of the authors, by UNMAS, a demining initiative by the UN, in Western Sahara.

Western Sahara has limited data available for strategic and operational planning. Based on the literature review by and spatial analysis used in Afghanistan, free data sources have been identified and used for Western Sahara. Using the community and hazard geospatial data from IMSMA NG – the national mine action database, and freely available datasets, GIS generated attributes were derived:

- Hazard and accident data from IMSMA NG was exported and prepared as a point dataset.
- ASTER 30 m digital elevation model (DEM) imagery data and ArcGIS spatial analyst tools were used to extract altitude, slope and aspect values from the DEM data.
- For land cover, GlobCover is a global land cover map available.
- For population, the LandScan global population data was used.

Table 7.2 Western Sahara classification bands for hazards

Classification	Score
High	10 and more
Medium	6–9
Low	0–5

- HydroSHEDS data (Lehner et al. 2008) is a hydrological dataset for the analysis of the river network, determining the number of rivers within 1 km buffer for each hazard.
- OpenStreetMap has a lot of free data available on infrastructure, water, land cover, land use, points of interest (POI).

Each of these datasets was then reclassified using a weight value of 0, 1, 2 or 3. This was carried out so that each of the factors under consideration contributes equally to the final Spatial Decision Support System (SDSS) score. Based on the score, each hazard was then classified as High, Medium or Low SDSS priority (see Tables 7.2 and 7.3). This was all done by using ArcGIS model builder (see Fig. 7.6).

Applying the analyses, the results from the SDSS are then used in SPSS, a statistical software package, to perform frequency analysis. From the table below it can be noted that 16.6% of the hazards have been classified as High, while 48.9% have been classified as Medium and the remainder, 34.5%, as Low SDSS priorities as shown in Table 7.4.

It is clear from this research that the application of such models can add value in the decision making process. In addition, the fact that it makes use of freely available data sources can make a substantial decision support contribution in countries where limited resources are available for obtaining expensive imagery and data. Extending these capabilities to web based or Cloud SDSS solution will extend the functionality and allowing it to be used to integrate additional data sources if required (Claassens 2016).

7.5.4 *Effectiveness of GIS as a DSS at Tactical and Operational Levels*

Building on the research analysis conducted in Afghanistan, three additional attributes were added to the dataset to the SDSS with the aim to assist decision makers on tactical and operational levels when developing operational clearance plans, (1) the number of mines found during clearance operations; (2) area in sq meters cleared during operations; and (3) clearance completion data.. Figure 7.7 shows the study area in Afghanistan. From the initial analysis of the dataset, specifically the number of mines found during clearance operations, 15% did not record any mines found. Using frequency analysis at the regional level, it was observed that in the Western region, more hazards were cleared with no mines

Table 7.3 Reclassification criteria for input datasets

Criteria	Range	Weight
Population	0	0
	1–50	1
	51–100	2
	>100	3
Land cover	200—Bare areas	1
	150—Sparse vegetation	3
Slope	>15°	1
	11°–15°	2
	0°–10°	3
Aspect	Flat, West, South, Southeast and Southwest	0
	East	1
	Northwest and Northeast	2
	North	3
Rivers/water basins	No rivers within 1 km	0
	1 river within 1 km	1
	2 rivers within 1 km	2
	more than 2 rivers within 1 km	3
Infrastructure	No infrastructure within 1 km	0
	1 infrastructure feature within 1 km	1
	2 infrastructure features within 1 km	2
	more than 2 infrastructure features within 1 km	3
POI	No POI within 1 km	0
	1 POI feature within 1 km	1
	2 POI features within 1 km	2
	more than 2 POI features within 1 km	3
Key Features	No key features within 1 km	0
	1 key feature within 1 km	1
	2 key features within 1 km	2
	more than 2 key features within 1 km	3
Accidents	No accidents within 5.5 km radius	0
	One accident within 5.5 km radius	1
	Two accident within 5.5 km radius	2
	Three or more accident within 5.5 km radius	3

recorded than hazards cleared where mines were found. The aim was to try and identify a potential set of attributes that can be used in a regression model.

Using SPSS and ArcGIS, with the available dataset, the attempt to identify linear variables that could be used to predict if mines would be found during clearance operations did not yield any significant results. None of the variables available in

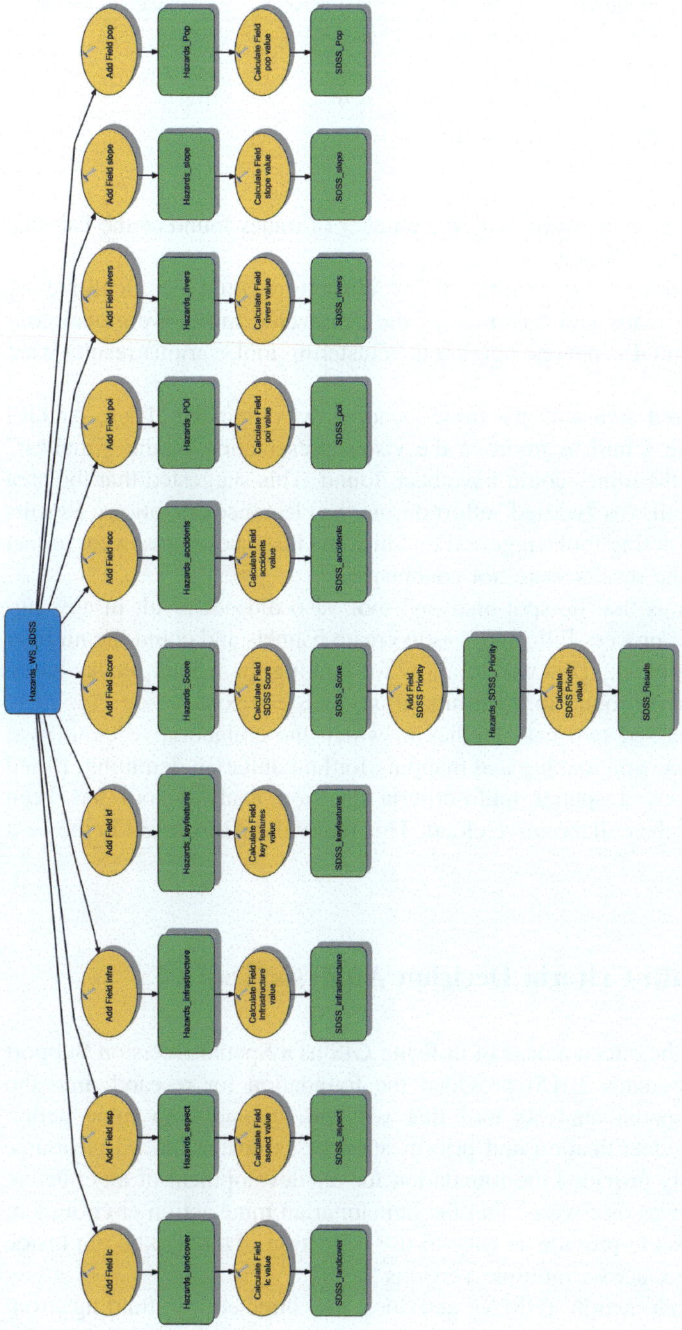

Fig. 7.6 Western Sahara spatial decision support system model

Table 7.4 Results from SDSS for Western Sahara

		Frequency	Percent	Valid percent	Cumulative percent
Valid	High	83	16.6	16.6	16.6
	Low	173	34.5	34.5	51.1
	Medium	245	48.9	48.9	100.0
	Total	501	100.0	100.0	

the dataset had a strong correlation with the number of mines found or the fact that mines were found or not found.

Further investigations to determine if any clustering could be identified by examining clearance maps and locations of the discovered mines were also conducted. Using the ArcGIS 'nearest neighbour' clustering tool, various results were observed.

The original cleared area with the mine locations was digitized. Then, ArcGIS was used to generate a hull to improve the visual presentation of the "smallest" possible area where the mines could have been found. This suggested that the area cleared outside the hull was "wasted" effort during the clearance operations. Results from the ArcGIS clustering tool suggested that in some instances mines were in fact clustered, however the results were not conclusive.

Other tools, such as the 'hotspot analysis' tool, also did not result in any significant findings. The process followed was to create fishnets and count the number of mines inside each block. This was to allow for ranking the fishnet blocks based on the number of mines found to determine if any clusters existed (see Fig. 7.8).

This section discussed two case studies in which the collaborative cloud was used in supporting decision making and mapping for humanitarian demining. Based on these case studies, a spatial multi-criteria decision analysis tool has been developed for use in the collaborative cloud. This tool will be discussed in the next section.

7.6 Spatial Multi-Criteria Decision Analysis Tool

Heymans' study on the effectiveness of utilising GIS as a Spatial Decision Support System of DSS (Heymans 2015) provided the foundation for research into the development of a spatial analysis tool that will aid humanitarian mine action organisations in the identification and prioritisation of explosive hazard clearance tasks. While the study provided the foundation for the development of an effective mine action DSS, it was discovered that the humanitarian mine action environment lacked a formal toolset to provide an easy-to-use collection of tools to be run inside GIS desktop software, across multiple versions.

Humanitarian mine action (HMA) activities are directed by funding from national government institutions and, to a large extent, international donor

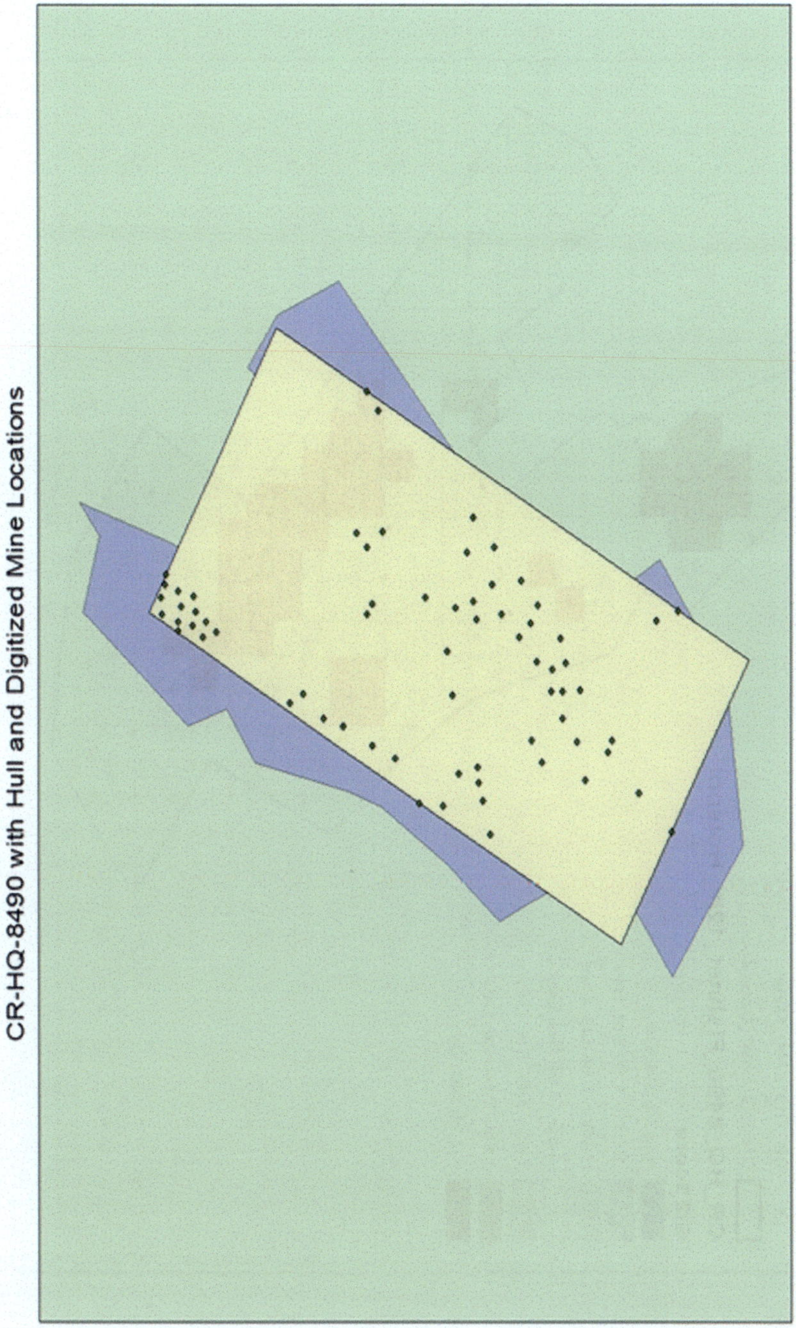

Fig. 7.7 Map of minefield CR-HQ-8490 that shows the original cleared area in purple, rectangular bounding hull in yellow and points of mine locations founded

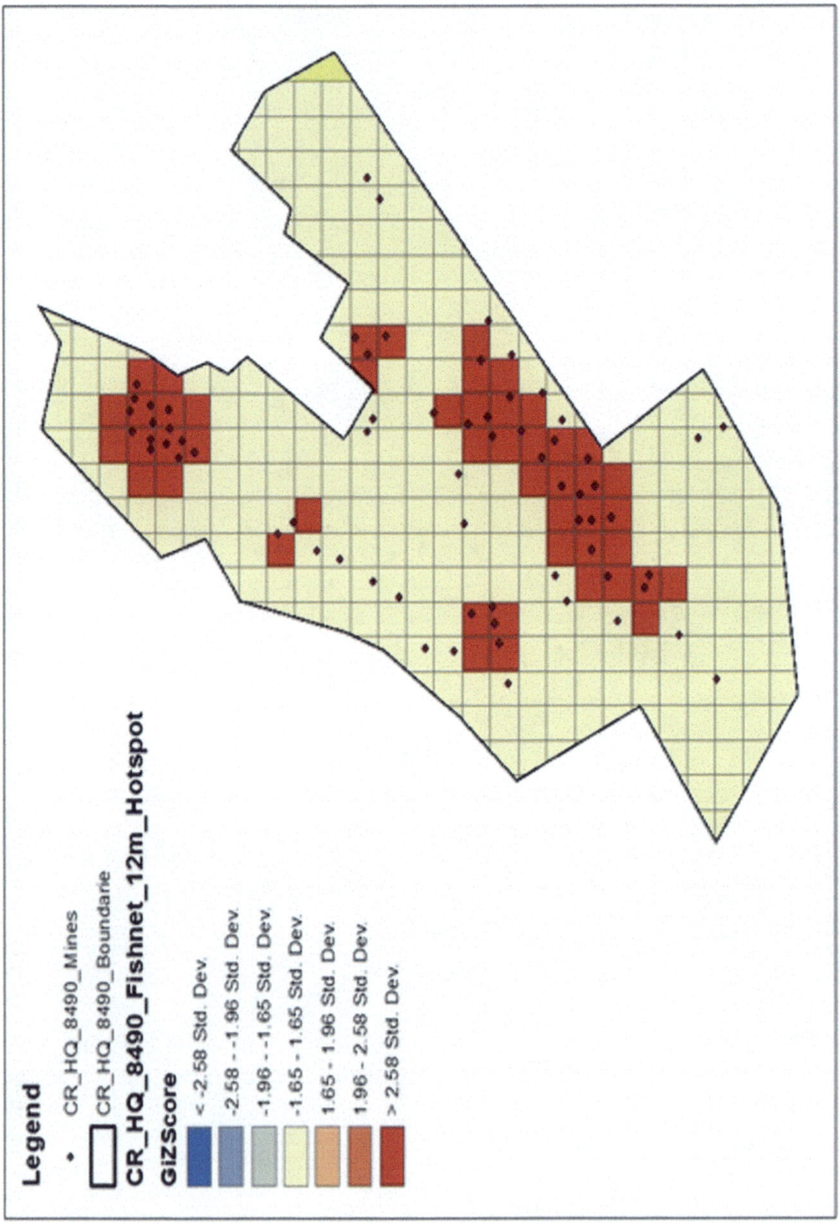

Fig. 7.8 Hotspot analysis with 12 m fishnet for CR-HQ-8490

organisations, whose policies and aims in funding HMA activities may vary considerably over time. It was with this variation in the focus of HMA activity funding that the spatial Multi-Criteria Decision Analysis (MCDA) tool shows its strength, as it provides the HMA GIS operator with the ability to prepare the base data, comprising the various criteria describing the HMA activity decision milieu, and then repeatedly run the multi-criteria analysis step with varying weights allocated to the criteria. The resulting output from the MCDA tool provides the decision makers with clearly prioritised hazard clearance tasks based on the current priorities of funding institutions.

7.6.1 Methodology

The Spatial Multi-Criteria Decision Analysis Tool development process focused on delivering a highly portable and lightweight toolset that would allow the user to perform MCDA on spatial features using several predefined decision criteria to prioritise land clearance operations. Heymans highlighted that the operational level is the primary focus for these decision support systems; this requires that the MCDA tool allows the user to operate on tactical-sized areas and up to national level in order to increase the usefulness of the MCDA Toolbox.

The weighted sum model was identified to be the best fitting scheme for the MCDA Tool, since the number of criteria were small and therefore did not require an Analytical Hierarchical Process (AHP). This decision was supported by the availability of discrete values on the same scale, for all the decision criteria. A weighted product model would have resulted in unnecessarily large result values, compared to the weighted sum model (Claassens 2016). The multi-criteria decision analysis methodology consists of the identification of several relevant and often conflicting criteria that will be used to decide on the identification of the most suitable candidates from the set of available candidates, followed by the allocation of a rank to each candidates criteria using a consistent scale range. This ensures that the criteria are ranked using values in the same order of magnitude. The criteria are then assigned a relative weight, which are then used to calculate a cumulative value for each candidate. These cumulative values are then split into a predefined set of groups to implement a ranking classification for the candidates.

The multi-criteria decision analysis result classification can be formulated as shown in Eqs. 7.1 and 7.2:

Where Sigma represents the hazard area cumulative value calculated for each hazard area, the value is obtained by multiplying the ranking of each criterion by the user-assigned weight of that criterion and then summing the criteria values.

Equation 7.1. Hazard area cumulative value calculation

$$\sum = (X_1Y_1) + (X_2Y_2) + \cdots + (X_nY_n) \tag{7.1}$$

Given the user-supplied values for Alpha (α) and Beta (β), denoting the boundaries between the high priority/medium priority rankings and the medium priority/low priority rankings, the hazard priority classification or Tau (T) of each hazard area is calculated by determining whether Sigma (Σ) is larger than or equal to Alpha, in which case it is classified as a high priority hazard clearance task, or smaller than Beta and thus is classified as a low clearance priority hazard area. Where Sigma falls between Alpha and Beta, the hazard area is classified as a medium priority classification.

Equation 7.2. Hazard area priority classification

$$T = \begin{cases} \sum \geq \alpha \\ \infty > \sum > \beta \\ \sum \leq \beta \end{cases} \tag{7.2}$$

7.6.2 Data Sources

The Spatial MCDA Tool focused on the usage of mine action data already at the disposal of the HMA actor that will be used in conjunction with freely available spatial datasets to increase the Return on Investment (ROI) in the decision support system. The mine action data consists of the reported hazard areas and mine accidents in the HMA actor's area of operations, while the remaining decision criteria consisted of the landcover type, slope and aspect values of the hazard area and the number of people, water bodies, infrastructure features, such as roads and railways, key features, such as markets or community centres, and points of interest are present within a user-defined radius from the hazard area's inside centroid.

With the exception of the mine action and LandScan population data, which is licensed to non-commercial organisations on a case-by-case basis (Oak Ridge National Laboratory 2012), all the other spatial data can be obtained freely from public sources, such as the ESA GlobCover dataset (European Space Agency 2010), MODIS Global Landcover (NASA 2003), and OpenStreetMap (2017).

The MCDA Toolset was developed with the goal of aiding the user in the preparation of the primary criteria data sets, but where the primary criteria may not be readily available, the GIS analyst is free to substitute the criterion with a data set that provides an indirect indicator of the required criterion, for example the census data may be used to represent the population distribution in lieu of high-resolution population data. It is as always important to measure that spatial resolution and temporal attribute of each dataset against the required outcome granularity to determine the fitness of purpose of each dataset.

After the MCDA results are derived, the output is shared between authorized mine action stakeholder staff in the collaborative cloud in the form of either a map service or downloadable files, both of which are then utilised in desktop GIS applications for further analysis and task planning. By hosting the results in the

Cloud, they are available to stakeholders at all times from a central location, reducing the lead time required to source the data through more laborious channels such as email or telephonic requests.

7.6.3 Toolbox Components

The Spatial MCDA Toolbox was developed as a Python toolbox for ArcGIS for Desktop, using the arcpy library and supported by freely available Python libraries required to implement the supplementary functionality of the tools in the toolbox. The tools were developed so that the lowest ArcGIS for Desktop license level would be sufficient to use the toolbox components, reducing the overall cost of the DSS and the level of support and training required to use the Spatial MCDA Toolbox.

The Spatial MCDA Toolbox caters for the less experienced GIS operator by providing simplified, automated tools that aid in preparing the data sets for the final MCDA step, where the prepared data set is kept intact and the MCDA calculations are applied on a replica of the prepared data set. This is to ensure data integrity and provide the GIS operator with a portable final data set that can be disseminated easily, via the Cloud, and functions independently of the source data sets.

The tools consist of three primary functional groups, i.e. preparation of the mine action data set, preparation of the additional criteria data sets and the final multi-criteria decision analysis. The additional dataset preparation tools are designed to focus on a specific dataset and may be adjusted with little effort to utilise other data sets. Once the GIS operator has stepped through the sequentially labelled data preparation tools, they will be in possession of a new dataset that combines the mine action and ancillary criteria datasets, all of which have been left unchanged, and can proceed with the multi-criteria analysis step. The multi-criteria step, as in the preceding phase, replicates the source data set and then applies the weights assigned by the user to calculate the cumulative value of each hazard area and, using the user-supplied values, classifies the hazard area into a high, medium or low priority task.

7.6.4 Contribution to Humanitarian Mine Action

The development of the MCDA Toolbox, according to aforementioned constraints and goals, provides a tool that can be widely used with ease, and at a low cost to the mine action stakeholder. The ability to repeat the multi-criteria analysis on the prepared dataset ad infinitum, without affecting the source data, makes it easy for the organisation to continuously adjust their hazard clearance priorities to keep up with changes in donor priorities or in the tactics of organizations responsible for the disposal of mines, ERW or IEDs. The ability to disseminate the results of the multi-criteria analysis using the Cloud improves timely and secure access to the data and enables organizations to respond faster to changes in survey and hazard clearance priorities. The logical evolution of the MCDA Toolbox, seen in context of the growth of the Cloud in the GIS milieu, is the development of a similar tool using only cloud-based resources that are accessible to Internet-capable devices across the multitude of operating systems and GIS client apps, including web browsers.

The ability of organisations to align their mine action activities to donor requirements or preferences enables the organisation to better deliver the outcomes required by donors, which in turn helps to countermand the decrease in funding for mine action activities to a large extent. By adjusting the criteria focus, mine action organisations also can align the potentially contradictory requirements of their donors and focus on those locations with overlapping or clustered hazard areas, thereby optimising the utilisation of labour resources and equipment and in turn improving the perceived performance of the mine action organisation.

7.7 Conclusion

This chapter discussed the use of the collaborative cloud with respect to decision support and mapping to prioritise areas for humanitarian demining. This is necessary since funding from donors reduces with each year necessitating the better utilization of available funds.

As mentioned in Sect. 7.4, the Cloud is a convenient way of accessing computing resources on demand, ranging from simple relational databases to high-end computing activities on a needs basis, which reduces the overall cost for computing power and systems for the user. Cloud computing currently has three service models namely Software as a Service, Platform as a Service and Infrastructure as a Service. Agrawal et al. (2011) argue that cloud computing can also serve as Database as a Service or as Storage as a Service. With regards to use in humanitarian demining, the Cloud can be a combination of Software as a Service, Infrastructure as a Service and Database as a Service.

Feedback from research conducted in the two case study areas, Western Sahara and Afghanistan, was positive and suggests that such a system on a global level, linked to strategic objectives, will provide value to decision makers. However, concerns were also raised about the data quality and national authorities' willingness to share data for comparison at a global level.

In the Western Sahara case study, GIS as a decision support system was found to be effective, since it provided relevant data to classify and prioritize hazards for clearance during the development of strategic and operational clearances plans at national levels. For example, by providing spatial information on minefields that were close to schools, enabled mine action programmes to prioritize these hazards for clearance.

The research conducted by Heymans (2015) and Claassens (2016) was to investigate the use of regression models and clustering tools using available historical clearance data from Afghanistan. Using available GIS-generated attribute data from the national mine action data, variables were selected to "predict" if mines would be found during clearance operations. The variables available did not yield any meaningful results from the regression analysis as there was no linear correlation between the variables.

The research was then extended to investigate potential clustering of landmines. Some of the historical clearance records had the location of landmines, found during clearance operations, recorded. Using the available data, various cluster analyses were performed. As in the case with the regression models, the research did not deliver any significant results.

From these case studies the spatial multi-criteria decision analysis tool was developed for use in the collaborative cloud. Further research is required on how to improve the various applications in the collaborative cloud as to make the applications even more user friendly. Furthermore, further research is necessary to allay certain data security fears when using the collaborative cloud.

References

Abadi DJ (2009) Data management in the cloud: limitations and opportunities. Bull Tech Committee Data Eng 32(1):3–12

Agrawal D, Das S, Abbadi AE (2011) Big data and cloud computing: current state and future opportunities. EDBT 2011, 22–24 Mar 2011, ACM 978-1-4503-0528-0/11/0003

Alencar P, Cowan D, McGarry F, Palmer MR (2014) Developing a collaborative cloud-based platform for waterhed analysis and management. s.l. In: IEEE international conference on collaborative computing: networking, applications and worksharing (CollaborateCom 2014)

Ali O, Soar J, Yong J (2015) Collaborative cloud computing adoption in Australian regional municipal government: an exploratory study. Calabria, IEEE

Anon (2016) Landmine monitor. Available at: http://www.the-monitor.org/en-gb/reports/2016/landmine-monitor-2016.aspx, visited Jan 2018

Armbrust M et al (2009) Above the clouds: a Berkeley view of cloud computing. University of California at Berkeley, California

Bradley T (2013) PC world. [Online] Available at: http://www.pcworld.com/article/244425/the_cloud_day_10_storing_data_in_the_cloud.html, visited Apr 2017

Claassens A (2016) Dissertation on the development of a spatial multi-criteria decision analysis tool for land clearance prioritisation in mine action. University of Leeds, Leeds

European Space Agency (2010) ESA GlobCover Portal. [Online] Available at: http://due.esrin.esa.int/page_globcover.php, visited May 2017

Hamdaqa M, Tahvildari L (2012) Cloud computing uncovered: a research landscape. In: Memon A (ed) Advanced in computers, vol 89. Academic Print, San Diego, pp 41–84

Hassan QF (2011) Cross talk: The Journal of Defense Software Engineering. [Online] Available at: http://www.crosstalkonline.org/issues/januaryfebruary-2011.html. Accessed 28 Apr 2017

Heymans JM (2015) Effectiveness of GIS as a decision support system for mine action. University of Leeds, Leeds

Hoang DB, Chen L (2010) Mobile cloud for assistive healthcare (MoCash). Hangzhou, IEEE

ICRC (2012) Anti-personnel mines. [Online] Available at: https://www.icrc.org/en/document/anti-personnel-mines, visited Jan 2018

Kshetri N (2013) Privacy and security issues in cloud computing: the role of institutions and institutional evolution. Telecommun Policy 37:372–386

Langholtz HJ (2014) Mine action and explosive hazard management: humanitarian impact, technical aspects and global initiatives. Peace Operations Training Institute, Williamsburg

Lehner B, Verdin K, Jarvis A (2008) New global hydrography derived from spaceborne elevation data. EOS Trans Am Geophys Union 89(10):93–94

Mell P, Grance T (2011) The NIST definition of cloud computing. U.S. Department of Commerce, Gaithersburg, MD

Michener WK (2015) Ten simple rules for creating a good data management plan. PLOS Comput Biol. https://doi.org/10.1371/journal.pcbi.1004525. 22 Oct 2015

Moniruzzaman ABM, Hossain SA (2013) NoSQL database: new era of databases for big data analytics—classification, characteristics and comparison. Int J Database Theory Appl 6 (4):2013

NASA (2003) MODIS global land cover. [Online] Available at: https://visibleearth.nasa.gov/view.php?id=61004, visited May 2017

NIST (2016) NIST. [Online] Available at: https://www.nist.gov/programs-projects/cloud-computing, visited Apr 2017

Oak Ridge National Laboratory (2012) LandScan data availability. [Online] Available at: http://web.ornl.gov/sci/landscan/landscan_data_avail.shtml, visited May 2017

OpenStreetMap (2017) OpenStreetMap downloading data. [Online] Available at: https://wiki.openstreetmap.org/wiki/Downloading_data, visited May 2017

Rong C, Nguyen ST, Jaatun MG (2013) Beyond lighting: a survey on challenges in cloud computing. Comput Electr Eng 39:47–54

Tutino M, Mehnen J (2013) Manufacturing paradigm shift towards better cloud computing in the military environment: a new model for collaboration in the operational information exchange networks. In: Li W, Mehnen J (eds) Cloud management. Springer, London, pp 243–256

UNICEF (n.d.) The state of the world's children. [Online] Available at: https://www.unicef.org/sowc96pk/hidekill.htm, visited 16 Apr 2017

UNOG (n.d.) Anti-personnel landmines convention. [Online] Available at: http://www.unog.ch/80256EE600585943/(httpPages)/CA826818C8330D2BC1257180004B1B2E, visited Apr 2017

UNOG (n.d.) State parties and signatories. [Online] Available at: http://www.unog.ch/80256EE600585943/(httpPages)/6E65F97C9D695724C12571C0003D09EF%3fOpenDocument, visited Apr 2017

Chapter 8
Modern Requirement Framework for Geospatial Information in Security and Government Services

Friedrich Teichmann

Abstract Appropriate requirement management for geospatial information is a very dedicate and sensitive task, because this phase majorly influences the success of the geospatial service delivery. During the requirement design the special needs of the service target group needs to be comprehended and translated into concrete map layouts, geospatial data content or topology and symbols, which should provide the necessary information for the end user. Especially the distinction between strategic (what?), operational (how?) and the lower tactical levels are useful differentiations during the requirement design, especially for hierarchical structured organizations. In addition, at the management level an optimal composition between time—quality—cost needs to be defined for the project. Main constraints for modern mapping activities are the short time-to-deliver requirements, completeness, accuracy and the possibility to easily share and comprehend the map. Rapid mapping supports a fast map production on one hand, geospatial intelligence collects relevant information for security actions on the other hand. This contribution focuses on the requirements of map production for security and government services, such as combined civil-military humanitarian relief missions.

Keywords Geospatial requirements · Thematic mapping · Geospatial service provider · Military humanitarian missions

8.1 Introduction

Identifying the concrete requirements for geospatial information products (most commonly physical maps, political maps, theme maps, charts, but also regional descriptions or country books) is perhaps the most rewarding task in the entire production process, because it is the key process to really fulfil the user/customer demand. On the other hand, it is the most challenging phase. It requires a good

F. Teichmann (✉)
IMG (Institut für Militärisches GeoWesen), Austrian Armed Forces, Vienna, Austria
e-mail: friedrich.teichmann@bmlvs.gv.at

J. Döllner et al. (eds.), *Service-Oriented Mapping*, Lecture Notes in Geoinformation and Cartography, https://doi.org/10.1007/978-3-319-72434-8_8

knowledge of both, the geospatial production potential and the scenario for which this product is used. Confronting the user or costumer with a requirements analysis, might result in a wide range of answers, from clear product definitions to "what do you have"? The more complicated the geospatial product tends to be, the more challenging this phase will turn into.

Marschall (2015) proposed in his book "*Prisoners of Geography, Ten Maps That Tell You Everything You Need To Know About Global Politics*", that maps are ideal means for explaining state of affairs and can also be used to convey complex information. This is in line with several previous authors, like Kaplan (2013) with "*The Revenge of Geography: What the Map Tells Us About Coming Conflicts and the Battle Against Fate*", Dockrill (1991) with "*Atlas of 20th Century World History*", Kidron and Smith (1983) with "*The War Atlas*", or Makower (1992) "*The map catalog: Every Kind of Map and Chart on Earth and Even Some above it*", who have used maps as the primary tool to express information. However, creating maps or geoinformation products, that are fully answer the costumers need and have a significant impact, are not easy to design. During the International Map Year 2015–2016[1] "*The World of Maps*" was produced under the guidance of the International Cartographic Association. In Chap. 4 "Map Design" following statement explains the challenge: "*When the objective of a map is specified, the target group of users, the way of working with the map and the volume of conveyed information are carefully formulated*". Following Loxton (1980) "Practical Map Production" one should carefully study the works of Tufte (1983, 1990) "*The Visual Display of Quantitative Information*" and "*Envisioning Information*" for extremely interesting analysis and suggestion to this matter.

This increasing challenge of map design or requirement engineering is perhaps the result for the widening gap between the possibilities of modern technology on the production- and the user side. Geospatial environments are characterized by data variety and the fast evolving framework for products, which are utilized in different use-cases by the end-user. Over the past decades, a company's specific portfolio (e.g. of geoinformation products) has been used extensively to bridge the gap between "what do you have" and "what does the user want". This product portfolio is usually the result of generalization and standard map compilation. It generally represents the most common and promising products of the business line. The portfolio list has proven its merits numerous times for common situations. However, with increasing complexity of multi dimensional data fusion, especially from distributed sources, the established product portfolio cannot represent all production options and is especially weak in answering timing and costs. In addition, moving more and more towards purely customized defined products, a predefined standard portfolio is ineffective. Therefore it is no longer appropriate for modern geoinformation management to predefine map content and manufacture excellent static theme maps.

[1]http://mapyear.org/the-world-of-maps-book/.

In order to be a successful geoinformation provider, additional attention towards the interaction with the end user is opportune (see "*The World of Maps*" (See Footnote 1)). There seems to be a need for a better understanding between the geospatial production lane and the end-user (costumer), hence a modern requirements development. Three key ingredients need to be balanced with a clear understanding of both sides: the costs, the time dimension as well as the product specifications (see "*The World of Maps*" (See Footnote 1)). The initial basis for a successful interpretation or arbitration between the geospatial producers and the users (costumers) has to use a common semantic (see Tufte 1983, 1990), hence definitions and meaning. In addition to a common language as well as the mentioned resources (costs) and the timing, the main discussion shall focus on "what should be depicted?" and "for what purpose?" in the geospatial product. This might also depend strongly on the availability of required georeferenced data. Both, the content and the data for a complex geospatial information product can be solved satisfyingly only by an iterative process between producer and costumer (e.g. based on year-long personnel experience like it has been shown in the UN exercise ModEX).

8.2 Current Work Environments

One of the currently most challenging, but also rewarding environments for delivering appropriate geospatial information is for combined civil-military humanitarian relief missions, because not only do the end user differ greatly (for example training, equipment or semantic), but also of the rapid reaction needed in this scenario, see Fig. 8.1.

Fig. 8.1 Participation of UN Exercice ModEX (http://www.bmi.gv.at/204_english/download/start.aspx, http://www.johanniter.de/die-johanniter/johanniter-unfall-hilfe/home/news/news-2016/eu-module-exercise-in-lower-austria/, https://www.youtube.com/watch?v=WjLWX2EXRg0) (MO DUL ISTAN) 2016. The field exercise was carried out in June 2016 in Lower Austria, included 9 different humanitarian relief teams for across Europe and was training "Urban Search and Rescue" operations.; both operational as well as tactical geospatial information was procured to support the exercise

trend	effects
Globalization	Global Area of Interest, Interoperability, Standards, Metadata
High Speed	Quick Decision Making and Fast Innovation Cycle
Modern Warfare	New Threads (asymmetric or hybrid warfare, cyber), special legal aspects
Budget Control	Cost Control and Innovative Solutions, Synergies
Digitalization	Analogue to Digital, Simulations, Virtual Realities, M2M (machine to machine), Big Data - Data Exploitation
Geo and Space	GNSS and PNT-Services, Earth Observation / Remote Sensing
Information Society	Internet, Social Media, Network Enabled Capabilities, Digitally Natives
High Complexity	Comprehensive Approach, Cooperation and Partnerships, Pooling&Sharing, Additional Actors in the Mission

Fig. 8.2 Overview of the key factors sharpening the current work environment

Before going into the detail of geospatial product specification, our current work environment for security and/or emergency activities needs to be addressed. These trends and effects are summarized in Fig. 8.2.

The globalization of our work environment of the last decades has various effects. This leads to an increasing global area of interest, not only for decision makers, but also for almost all other social-economic aspects. This again resulted in the need for enhanced coordination between the relevant stake holders and interoperability during execution, which further drove the path for more homogenization of the metadata or standard development (see *The World of Maps* (See Footnote 1) or Tufte 1983, 1990). The establishment of an information society based, among others factors, on Network Enabled Capabilities with high speed internet, distributed data centers, machine-to-machine technologies allowed new geospatial technologies to enter the mass market, such as Google Maps and Bing Maps including 3D or virtual realities. In order to support these new geospatial productions, "Big Data" and "Data Exploitation" needs to be fine-tuned (for example see various subject presentations during DGI 2017 "*Geospatial Intelligence for National Security*"[2]). The advents of high quality space applications, especially Satellite Communication, Satellite Navigation and Earth Observations (e.g. GALILEO and COPERNICUS for the EU[3]) have tremendous influence on the modern geospatial production lanes and allow the delivery of impressive products (example Fig. 8.3).

One other highly interesting aspect is the increasing speed within the work environment and therefore the geospatial production requirements. The fast

[2]https://dgi.wbresearch.com/.

[3]https://ec.europa.eu/growth/sectors/space_en.

Fig. 8.3 The power of Google Earth—Haiti earthquake before and after

changing world combined with the wide area of interest for decision makers or analysis results and information need that need to be covered as quickly as possible, time for delivery decreased from days or weeks down to hours and sometimes minutes! Similarly, the requested "*time to deliver*" for geospatial information is driven by a very quick decision making process and has resulted in a massive time span reduction for the producers. This fast production requirement is extremely challenging for the producers and has to be addressed by innovative methods, like its own production lane "*rapid mapping*" (based on year-long personnel experience). However, not only has the production increased speed, also the Innovation Cycle for the products themselves turns faster and faster and needs to be compensated for example with close cooperation with specific R&D (research and development) institutions (for example see Fig. 8.11).

Especially for governments and/or security organizations, the new work environment is characterized by a high complexity. It requires a "*Comprehensive Approach*" between state and non-state actors, resulting in extensive cooperation's and partnerships in order to achieve success. This multi-dimensional approach is needed to successfully confront the new threads like asymmetric or hybrid warfare, or especially cyber security issues. A pragmatic approach to deal with this complex work environment is to identify different target groups and use cases.

8.3 Hierarchy Levels

One highly successful approach to discuss the requirements for geospatial information follows hierarchy levels, which have been applied in the military domain for the last decades (Fig. 8.4). The military based hierarchical division below is very suitable to address for each level a specific target group, its purpose and specific requirements for prospective geospatial products.

Although the above presented division into strategic, operation, tactical is based on military doctrines, it can also be applied in analogy in the civil world. Just like a state or a government, business companies also follow strategic guidance's or carry out specific operations or tasks. In addition to these three hierarchy levels, geospatial support needs to pay special attention also to the weapon system or platform level, below the tactical. In order to support the necessary decision making process or provide relevant geo-related data for running the business, geospatial

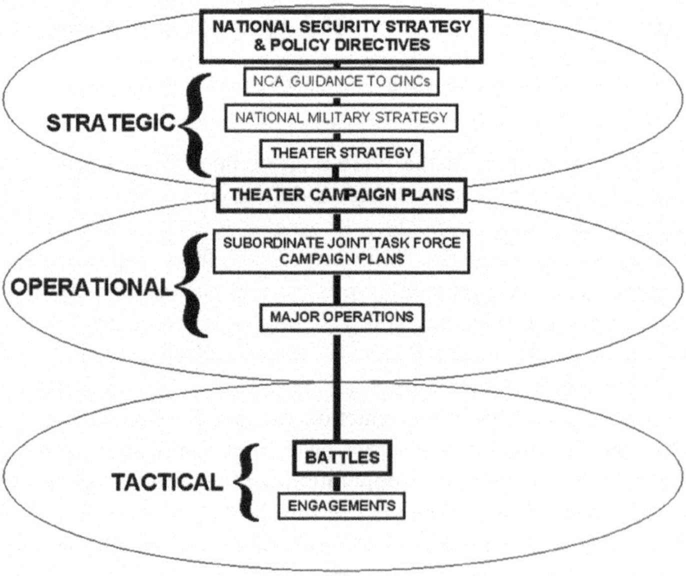

Fig. 8.4 Levels of guidance hierarchy, modified after US army field manuals (http://www.globalsecurity.org/military/library/policy/army/fm/9-6/9-6CHAP1.HTML)

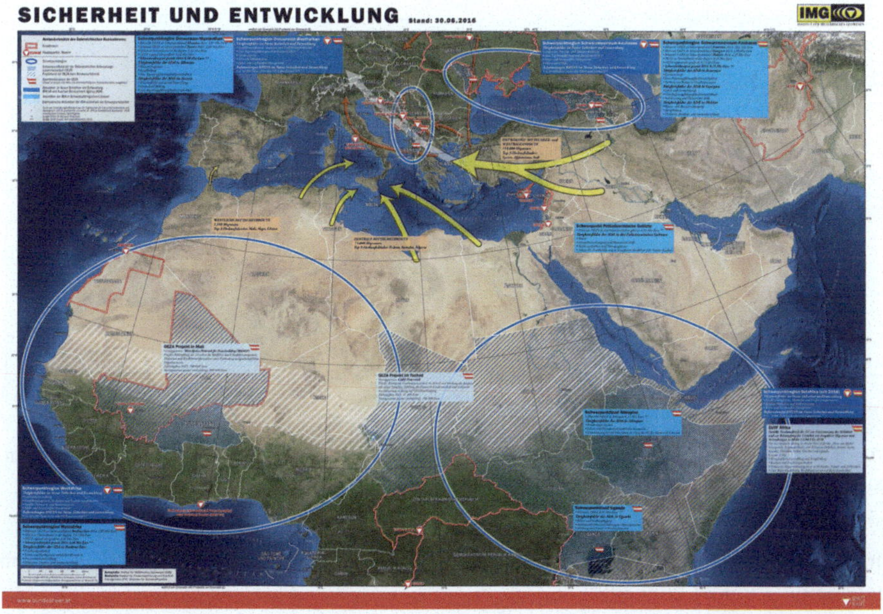

Fig. 8.5 Example for a strategic product, produced by the IMG (Institut für Militärisches GeoWesen), Austrian Military, 2016

Fig. 8.6 Example for an operational product, produced by the IMG (Institut für Militärisches GeoWesen), Austrian Military

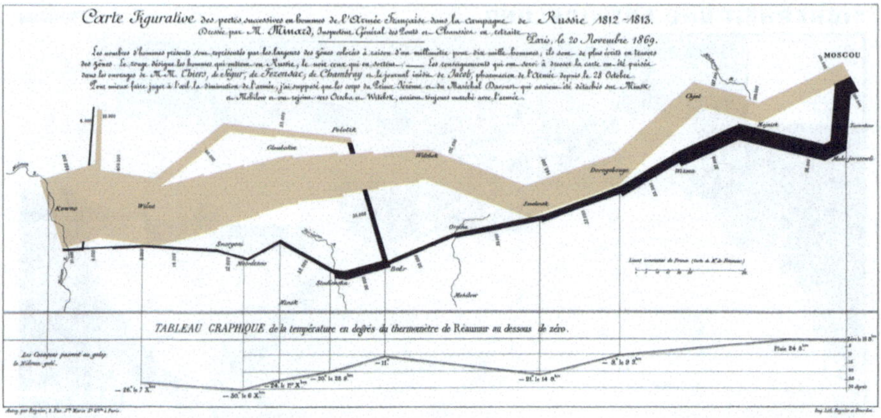

Fig. 8.7 Charles Minard's map of Napoleon's disastrous Russian campaign of 1812 (https://en.wikipedia.org/wiki/Charles_Joseph_Minard)

products are necessary for a wide range of operations (see national and coalition geospatial support concepts).

The following section will illustrate this idea with one map example each (Figs. 8.5, 8.6, 8.7, 8.8 and 8.9).

8.4 Requirements and Examples for Geospatial Information

As discussed in Sects. 8.1 and 8.2, in addition to a clear focus for each geospatial product, especially related to (1) target group (=whom), (2) purpose (=why) and data content (=what), the requirements for geospatial information services might also be dominated by following aspects: current mission or task framework, scenario and application, delivery network, interoperability and types/data sources.

Geospatial products are successful and task supportive, if the following seven primary criteria are fulfilled:

- **comprehensible** (semantic and readability of the geoproduct in line with the end user)
- **accuracy** (e.g. geodata precision correspond to the user needs)
- **timely** (e.g. map delivered on time)
- **relevant** (e.g. geoproduct answers the question)
- **sharable** (e.g. map can be shared with relevant partners or other users, depend on classification and releasability rules)
- **complete** (e.g. geoproduct covers all necessary topics
- **available** (e.g. proper delivery means to the end user in place)

Fig. 8.8 An example of a tactical product, produced by the IMG (Institut für Militärisches GeoWesen), Austrian Military

In addition to the seven criteria cited above, the delivery of geospatial information products includes also such complex topics as classification, information assurance, information security, applied metadata or the style sheets, legends and symbols based on interoperability requirements.

The above listed criteria need to be matched with the target group, the purpose and the data content for the respective geospatial product. Most commonly this is achieved using an iterative process, similar to a spiral development during software

Fig. 8.9 Example for a weapon system product, produced by the IMG (Produced during a field exercise in 2017 by Captain Bernhard Metzler for an advanced placement course and other experts from the IMG) (Institut für Militärisches GeoWesen), Austrian Military

design (Boehm's spiral model[4]) where the product draft versions slowly merge with the user expectations. This can be supported by increasing awareness on both sides such as the producer and costumer, common training or joint development teams.

The next four subsections will discuss, based on real example maps, the special user requirements and additional challenges.

8.4.1 Strategic Level

Strategy is a detailed plan for achieving success in situations such as war, politics, business, industry, or sport, or the skill of planning for such situations as explained in the Cambridge Dictionary[5] or in "*Patterns in Strategy Formation*" by Mintzberg (1978). Therefore, the strategic level, based on vision and policy, addresses and focuses on doctrine, in order to achieve the desired end state. The critical issue at this level is therefore to identify and understand the strategic goals (What?) and the questions related to these goals, which can be addressed with geo-related information. Therefore, the product needs to focus on the goals and address the end state (What should be achieved?) or the path towards it.

Geospatial products for the strategic level are primarily for the support of key decision makers on critical matters. For that reason, the target group is clearly defined and focuses on the decision makers as well as their advisory bodies. The

[4]http://iansommerville.com/software-engineering-book/web/spiral-model/.

[5]http://dictionary.cambridge.org/us/.

purpose of the geospatial product at strategic level is either to provide background information or display critical factors which have a geo-relationship like political entities or borders, global or regional developments as well as key environmental facts in order to assist the strategic decision making.

Based on the complexity of the topics, the necessary (geo-referenced) data usually has to be derived from various distributed sources, increasing the complexity for the final product such as reliability and classification. Owing to the fact that strategic geospatial products are supporting key decisions, they might also be used as a tool to depict political information for external use, in the sense of "*a picture is worth a thousand words*"[6] as illustrated in Fig. 8.5.

This map in Fig. 8.5 addresses the complex situation in the Mediterranean and Northern Africa region, including mission areas of the Austrian military, key regions and priority countries of the Foreign Ministry, including the Austrian Development Agency, as well as the primary migration route.

8.4.2 Operational Level

Operation is the way that parts of a machine or system work together, or the process of making parts of a machine or system work together as explained in the Cambridge Dictionary (See Footnote 6) or Rogers's (2006) "*Strategy, Operational Design, and Tactics*". Geospatial products in support for operations are therefore directed primary for internal use, in support of the key process owner. They are supposed to guide, assist and steer the operations which have been put in place in order to achieve the strategic goals. Unlike at strategic level, where the "What" was in the centre, geospatial products at operational level need to address the "How" this goal is achieved, e.g. with what combination of means/forces/tools or in what sequence.

The target audiences for geospatial products at operational level are therefore usually staff officers, the top and middle management (and their advisory teams) or the planning branches. The key challenge therefore is what geospatial data can be applied in order to support the significant questions regarding the operations. These spatial related questions can range for the physical situation within the area of interest to the lines of communications (road or train network) or environmental details like climate or GEOMETOC (Geo-, Meteo- or Oceanographic data) data, but also include a never-ending list of special theme maps, from particular areas of interest to human geography layers to threat zones. In line with the large amount of theme maps, there is the need to integrate various data sets (in order to produce all the special theme layers), both from internal and from external sources, in order to deliver geospatial support. The clear focus in this category on ongoing operations has to be reflected by the geospatial information support service.

[6]https://en.wikipedia.org/wiki/A_picture_is_worth_a_thousand_words.

The foundation for successful geospatial support is an excellent understanding of the operation plan ("OPlan"). How do the elements interact and what geo-factors are relevant for success? The mission-critical key elements and/or factors need to be displayed within the proper time in order to best support the operational on hand. It is worth to note, that geospatial data at operational level is not only applied for the primary process, but also for various support activities, such as logistics.

An example of such an operational product is given in Fig. 8.6 which displays the Ethnic and linguistic groups distribution for Irak-Syria and was published 2016 in the Atlas Syria & Irak (Vogl et al. 2016), which also included special theme maps like Topography and administrative divisions, Control of territory, Major Clashes, Religious and sectarian groups, Oil and gas fields, Camps for refugees and IDPs or Asylum applications of Syrian/Iraqi citizens and refugees and IDPs.

A scientific discussion about geospatial information for government, military and security application use would not be complete, if Minard's map of Napoleon 1812 Russian campaign would not be included. This ground breaking work still serves as an excellent example of how distributed data (e.g. temperature) can be displayed and cross-linked with other information (e.g. number of troops) both simply and beautifully in one map. The graphic, as shown in Fig. 8.7, is notable for its representation in two dimensions of six types of data: the number of Napoleon's troops; distance; temperature; the latitude and longitude; direction of travel; and location relative to specific dates (Tufte 1983).

8.4.3 Tactical Level

Tactic is an action or method that is planned and used to achieve a particular goal as explained in the Webster Dictionary[7] or Rogers' (2006) "Strategy, Operational Design, and Tactics". Although this level usually addresses a limited area of interest, the resolution and amount of details is normally very high. The target group is middle management, who is responsible to carry out one specific task. High-quality geospatial support needs to centre on the one central task. The mutual support of different elements or capabilities, usually from the same organization, is one of the characteristics of this level (see national and coalition command and control concepts).

An example of a tactical map is illustrated in Fig. 8.8 and shows the "Urban Search and Rescue Exercise" area of the Austrian Armed Forces in Lower Austria which was produced on a 1:1500 scale and uses a high resolution areal image as the basis, overlaid with a geographic grid coordinate system and the important points such as landing sites, blasting area, decontamination site or tunnels for detailed tactical training.

---

[7]https://www.merriam-webster.com/dictionary/.

Interesting aspects to consider at this level are the usability and readability of geospatial products delivered for all stakeholders working in that area. In addition, the actuality of the generalized data set is extremely important to support tactical engagements. Examples are all versions of high resolution maps, images or products, usually boosted with various physical information (e.g. about the terrain) or other relevant tactical information. The Haiti earthquake showing the before and after images is another example from for a tactical product that is especially enabled by modern space based technologies, in this case Digital Globe (see Fig. 8.3).

8.4.4 Weapon System Level

Finally, below the tactical level is the need to deliver geospatial support for special tools, systems, platforms or weapons. Due to the technological advances, most modern platform or systems require specially formatted geospatial data in order to be operational. These data sets need to be produced to each platform or system and is therefore highly specialized. The key aspect is the proper timing in the production lane (including quality assessment), because any delivery in proprietary formats is usually time consuming and needs to be trained.

Due to the ongoing technological advances at platform or/system level, it is necessary for the geospatial service support to stay in close contact with industry in order to fulfill the system requirements. On the other hand, precise geospatial data is usually the base for a successful system operation and therefore the geoservice provider is a modern system enabler.

The image shown in Fig. 8.9 is a 3D visualization of a housing complex (used as training area for urban operations) and can be rotated and tilted according to the user needs.

The four sections above should have illustrated the broad range of maps or geoinformation products needed for security or government needs. However, the user needs need to be aligned with the other factors potentially limiting the production, like cost and time.

8.5 Discussions and Summary

As discussed above, in order to provide useful geospatial products, they need to fulfill the criteria of being comprehensible, accurate, timely, relevant, sharable, complete and available. Similar to the NEC vision (Network Enabled Capabilities),[8] the three key domains people, technology and information also need to be

[8]http://www.nato.int/cps/en/natohq/topics_54644.htm.

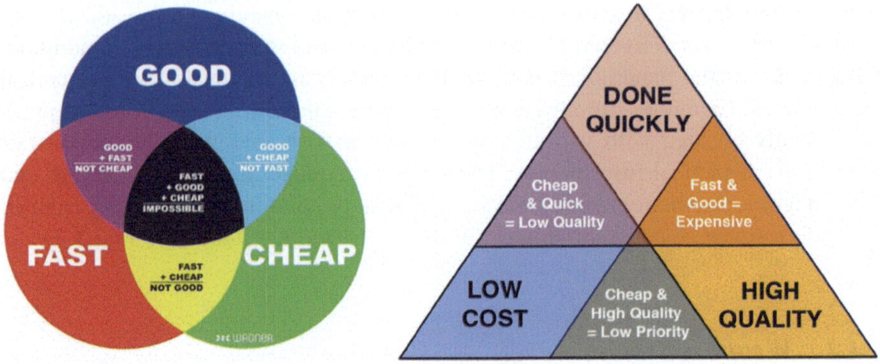

Fig. 8.10 The magic triangle, following project management: Left image by "Wagner" (http://www.wagnercompanies.com/good-fast-cheap-cant-three/), right side by "The Smarter Website Owner" (http://smarterwebsiteowner.com/the-project-management-triangle/)

synchronized during the production process. However, the key tradeoff that needs to be carried out during production or service delivery is time-quality-cost.

In analogy to project management principles, the magical triangle of time, quality and cost by Atkinson (1999) should govern our geospatial production or service delivery. The three factors time, quality and cost are not independent variables. Two of the three factors can be adjusted (high or low) by the producer or costumer, but the third factor is then a result of the first two. For example, if time is short and quality needs to be high, the resulting costs will be high (see Fig. 8.10). Or if costs need to be low but a fast delivery in needed, the resulting product will be usually be of poorer quality.

In Sect. 8.2 the fast decision cycle was discussed, which usually leads to tasks that need to be fulfilled quickly; this is usually matched with limited financial resources for the assignment. This would normally lead to low quality products, see Fig. 8.10. What are the possibilities a service producer has, to deliver a high quality product despite limited time and money? The first option could be described as contingency planning, where based on likely scenarios, initial preparation for production like style sheet, layout or formats can be predetermined and prefabricated before a task comes in officially. The second main option is the establishment of new and innovative production that is design to allow quick delivery. An example of "*rapid map making production*" is the 1:250,000 Map of Easter Turkey which includes a satellite image with a standard grid system, lines of communications and points of interest superimposed (see Fig. 8.11).

In addition to a substantial increase in the applications of space based technologies like in Fig. 8.11, future geospatial services will also deliver highly interesting new products in the fields of virtual reality, seamless analysis and visualization, combination of raster, vector and 3D data and the massive incorporate of information from other geosciences towards a comprehensive environmental

Fig. 8.11 Example for a quick delivery product of the IMG (Institut für Militärisches GeoWesen), Austrian Military

picture.[9] Furthermore, new developments will include, among others, time line animation, hyper linkage with other data (e.g. geo-tagged photos), semi-automated processing and multi-stereo analysis.

Modern work environment, technical innovations and user demands will require innovative developments for the provision of geospatial information, especially if time, quality and cost need to be balanced (see Fig. 8.10). An iterative, spiral shaped process between the end users and the producer will support this innovative requirement development. And the most common question “How long do you need to produce this for us?” can only be answered with “The more time we have, the better our product will be!”

Acknowledgements The excellent work of the team of the IMG (Institut Miltärisches Geowesen) of the Austrian Armed Forces as well as our national and international partners have supported this report.

[9]https://portal.dgiwg.org/files/?artifact_id=5443&format=doc.

References

Atkinson R (1999) Project management: cost, time and quality, two best guesses and a phenomenon, its time to accept other success criteria. Int J Project Manage 17(6):337–342

Dockrill ML (1991) Atlas of 20th century world history, Collins. ISBN 978-0004350615

Kaplan RD (2013) The revenge of geography: what the map tells us about coming conflicts and the battle against fate. Random House, USA. ISBN 978-1-4000-6983-5

Kidron M, Smith D (1983) The war atlas: armed conflict—armed peace. Heinemann, USA. ISBN 0435354965

Loxton J (1980) Practical map production. Wiley, New Jersey. ISBN 9780471277835

Makower J (1992) The map catalog: every kind of map and chart on earth and even some above it. Vintage Books, New York. ISBN 978-0679742579

Marschall T (2015) Prisoners of geography, ten maps that tell you everything you need to know about global politics. Elliot & Thompson, UK. ISBN 1783961414

Mintzberg H (1978) Patterns in strategy formation. Manage Sci 24(9):934–948

Rogers CJ (2006) Strategy, operational design, and tactics. In: Bradford JC (ed) International encyclopedia of military history. Routledge, UK. ISBN 978-0415936613

Tufte ER (1983) The visual display of quantitative information. Graphics Press, Cheshire. ISBN 9780961392109

Tufte ER (1990) Envisioning information. Graphics Press, Cheshire. ISBN 978-0961392116

Vogl M, Webinger P, Taucher W, Feichtinger W, Teichmann F, Wechselberger W (2016) Atlas Syria & Irak. Austrian Federal Ministry of the Interior & Austrian Federal Ministry of Defence and Sport, Vienna 2016. ISBN 978-3-9503643-7-8. http://www.ecoi.net/atlas_syria_iraq.pdf

Part II
Importance and Impact of the New Map Production Paradigm

infrastructures and process models such as SDMX, INSPIRE and the Generic Statistical Business Process Model (GSBPM).

Keywords Data integration · Statistical processes · Geocoding
Modern statistical production

9.1 Introduction

It is a challenging task for policymakers in Europe to capture what is happening in society and to design policies to improve the situation for people, economy and environment. Territorial policies on national and European level together with the UN 2030 Agenda for sustainable development puts pressure on the National Statistical Institutes, NSIs, in Europe to be able to produce relevant and timely statistics (Haldorson et al. 2016). The 2030 Agenda is aiming to leave no one behind, which from a geographical perspective means taking into account the urban-rural dimension as well as issues of accessibility.

Regional statistics on administrative geographies are not always good enough if policies need to target the development of urban and rural areas, for example deprived neighbourhoods. Disaster risk management or measures to improve accessibility to public transport, schools, health clinics or green areas are other examples when statistics need to be produced in a more flexible way. The European Statistical System (ESS 2018) has recognised this development by funding the GEOSTAT projects (GEOSTAT 2018), for example GEOSTAT 2 resulting in guidance for NSIs on the fundaments of a more point-based statistical infrastructure (Moström et al. 2017a, b).

The European Statistics Code of Practise (ESCP 2011) states that the mission of the European Statistical System is to "provide the European Union, the world and the public with independent high quality information on the economy and society on European, national and regional levels and make the information available to everyone for decision-making purposes, research and debate". Regional statistics are regularly produced for the countries in Europe, typically on NUTS regions (NUTS 2003) built on administrative geographies in all Member States. Eurostat provide harmonised regional statistics in their database, but for users looking for more geographic detail the situation is far from harmonised. A first step to improve this has been to introduce the use of statistical grids, degree of urbanisation etc. together with Local Administrative Units as new output geographies in a revised EU regulation, TERCET (2018).

Statistical grids (SG 2018) are also about to become one of the dissemination areas in the next European census round 2021 (EC 2018). Production of grid-based statistics calls for modernisation of the way countries do their censuses and lead the way to a more register-based statistical system (Wallgren and Wallgren 2007; UNECE 2018) as the ultimate goal. Increasing demand for statistics of high geospatial detail could also be met by increasing the sample size of traditional household surveys to allow for more geospatial detail. Taking into account that many NSIs face budget constraints and the need for timely statistics with yearly updates, only a register based approach is considered sustainable for Europe in the long run (Fig. 9.1).

The GEOSTAT projects have so far proven that it is possible to create a grid-based population map for Europe using various techniques (Backer et al. 2011), and that it is possible to geocode statistical data to coordinate locations as well as include geospatial processes in the Generic Statistical Business Process

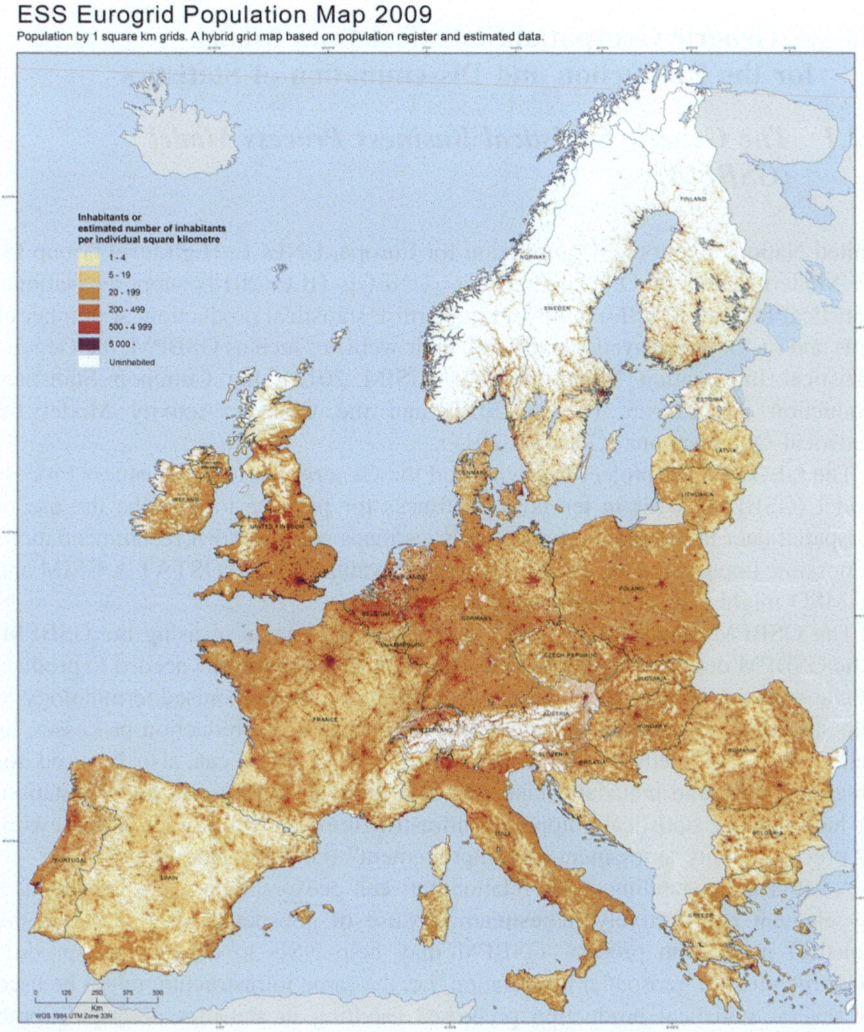

Fig. 9.1 Results from GEOSTAT 1: Population Map for Europe, total population 2009 on kilometre grids

Model (Moström et al. 2017a, b). There are a number of countries that already provide register based statistics, the Nordic countries are often used as examples of where the necessary fundaments are in place to allow for such production. Setting out to develop the European version of the Global Statistical Geospatial Framework for the ESS, ESS-SGF, in the GEOSTAT 3 project will provide guidance both on a strategic and a more technical level which steps to take to move towards a register based system. This register based system forms a fundament for an objective and more reliable statistical map production.

9.2 A Generic Geospatial-Statistical Data Model for the Production and Dissemination of Statistics

9.2.1 The Generic Statistical Business Process Model (GSBPM)

United Nations Economic Commission for Europe, UNECE, High-level Group for the Modernisation of Official Statistics (UNECE HLG 2018) supports National Statistical Institutes (NSIs) in how to modernise statistical production. A number of tools and methods are available through their website, such as GSBPM, the Generic Statistical Information Model (GSIM) (GSIM 2018), the Common Statistical Production Architecture (CSPA 2018) and the Generic Activity Model for Statistical Organisations (GAMSO 2018).

The GEOSTAT 2 project has evaluated the Generic Statistical Business Process Model (GSBPM 2018) in terms of its fitness for purpose to describe the use of geospatial data in the production of statistics/maps and provided recommendations on possible improvements. The CSPA will be evaluated in GEOSTAT 3, GSIM and GAMSO might also need evaluation later on.

The GSBPM documentation lists the following benefits of using the GSBPM: "The GSBPM describes and defines the set of business processes needed to produce official statistics. It provides a standard framework and harmonised terminology to help statistical organisations to modernise their statistical production processes, as well as to share methods and components. The GSBPM can also be used for integrating data and metadata standards, as a template for process documentation, for harmonizing statistical computing infrastructures, and to provide a framework for process quality assessment and improvement" (GSBPM Doc 2018).

A better understanding of the relation between geospatial data and statistics is a key element to effectively mainstream the use of geospatial information in the statistical production process. GSBPM may help NSIs to design cross-product infrastructures for geospatial information, i.e. the same infrastructure might be used for several statistical production processes resulting in statistics describing both society and economy.

The documentation of the GSBPM includes descriptions of how to use the model in the statistical production process. However, the geospatial dimension is completely absent. The GEOSTAT 2 project has evaluated the GSBPM in terms of its fitness for purpose, in order to describe the use of geospatial data in the production of statistics and provide recommendations on the possible improvement of the GSBPM to better address geospatial data management in the statistical production process. Chapter 6 in the GEOSTAT 2 report provides guidance on what could be considered as geospatial data management (Moström et al. 2017a, b).

9.2.2 Geospatial Information—Its Role in Statistical Production

In order to make recommendations, particularly on the implementation of a point-based geocoding infrastructure for statistics in the ESS, the full scope of geospatial information in statistical production first needs to be recognised. From a statistical production perspective, it is important to distinguish information needed as the **infrastructure** to geocode data from geospatial information needed to create **statistical content**. The different roles of geospatial information in statistical production can be described by means of a generic geospatial-statistical data model.

The GEOSTAT 2 project identified three different tiers of information constituting the geospatial-statistical data model:

- **Tier 1** consists of geospatial information used exclusively for the purpose of geocoding, geographically representing or disseminating statistical or administrative information. Examples of information in Tier 1 include address data, census enumeration districts, postal code areas, statistical grids or other statistical or administrative geographies.
- **Tier 2** consists of geospatial information which is used both to geocode, geographically represent or disseminate statistical or administrative information *and* to create statistical content. Typical information found in Tier 2 includes building information, cadastral parcels and transport networks but also new data sources such as traffic sensor information.
- **Tier 3** consists of geospatial information which is used only to produce statistical content. Despite its geospatial component, this category of information cannot be used directly to geocode statistical or administrative data. As such, information in Tier 3 can be regarded as *complementary* to, and *independent from,* information in Tiers 1 and 2. Some examples of data found in Tier 3 include DEMs, land use or land cover maps, topographic data, ortho-photo or satellite imagery, or other products derived from earth observation data. Typically, data in Tier 3 needs to be combined with data from tier 1 or 2 in order to be transformed into statistical information. The calculation of land area within a NUTS region can serve as an example: data on land mass and topographic maps data (Tier 3) is combined with a dataset containing NUTS regions (Tier 1) and a calculation is made on land (and water) area for each NUTS region.

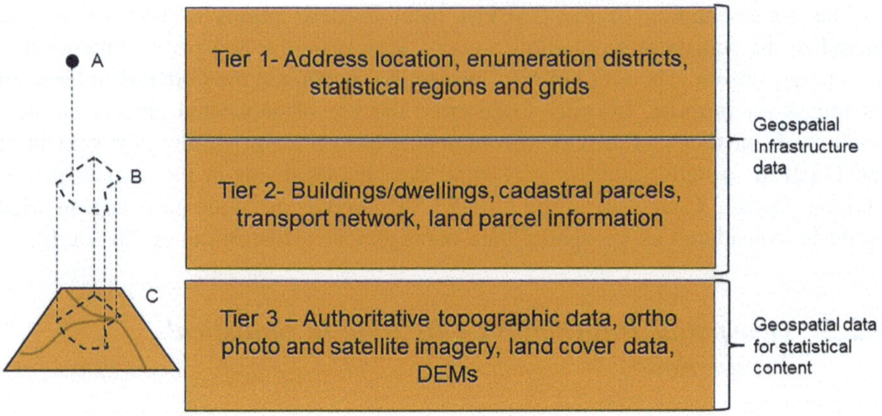

Fig. 9.2 Tiers of information in the generic geospatial-statistical data model for the production and dissemination of statistics. A workplace geocoded to an address location (A) can be linked to a cadastral parcel (B) from which land use can be assessed by combining the parcel with a land use map (C). The more consistent the system, the more opportunities for flexible linking of data

In conclusion Tier 1, 2 and 3, as illustrated in Fig. 9.2, in its totality represent the geospatial data that are required to geo-enable all relevant statistical information from enumerations, surveys, administrative and alternative data sources at the unit record and aggregate level throughout the statistical production process. Tier 1 and 2 are of a more fundamental nature and are therefore defined as the geospatial infrastructure data for statistics with the main purpose to geocode, spatially integrate, disseminate or represent statistics (e.g. on maps).

The categorisation of geospatial information outlined above is different from the other ways to categorise geospatial information, such as topographic data, core data, fundamental data or reference data. As an example, the working group on core data of UN-GGIM: Europe has put forward 14 INSPIRE themes as core data that comprise many of the themes of tier 1, 2, and 3 of this categorisation (UN-GGIM 2016).

Although both categorisations have their specific applications, it should be avoided that their applications create confusion and erect communication barriers. This is an area for further development.

9.3 A Point-Based Geocoding Infrastructure for Statistics

Building on the generic geospatial-statistical data model for statistics, as outlined above, this section introduces and elaborates on the foundation of a point-based geocoding infrastructure at the conceptual level. The aim is not only to facilitate a common understanding of the characteristics of this geocoding infrastructure, but also to outline the benefits, limitations and challenges associated with a point-based infrastructure specifically for statistics.

In a fundamental sense, a point-based geocoding infrastructure for statistics can be understood as a production setting where a record holding X, Y (and Z) coordinates of a location, along with a unique identifier (Id), can be linked to a record with statistical or administrative data which belongs to this point. This process is called "geocoding" of statistics or other data. The actual purpose of the point-based approach is to assign a single coordinate location to each unit record.

The term "point-based" should be understood in contrast to "area-based" which appears in traditional surveys and censuses where the population surveyed is assigned to a fixed output area, such as an enumeration district. It should be stressed that the proposed shift from an area-based to a point-based approach, as described here, only refers to the geocoding infrastructure itself and hence to the collection and processing of statistical information. The area-based approach is, and will continue to be, the primary method for the dissemination of statistics.

In Fig. 9.3, the conceptual differences between a point-based and an area-based geocoding infrastructure are illustrated. In both cases there is a record with statistical data consisting of four individuals, and a corresponding record containing location data. In the point-based approach, shown on the left, each individual in the statistical data record is linked to a unique dwelling location which has not been aggregated and is spatially represented by three different point locations. Two individuals have been assigned the same location as they are linked to the same dwelling. In the area-based approach on the right, all four individuals are linked to the same spatial object (Block A), as the area-based approach does not support spatial discrimination of their individual dwelling locations.

The underlying assumption of the GEOSTAT 2 project has been that a point-based geocoding infrastructure is far more flexible in terms of production and maintenance than a traditional area-based infrastructure with fixed output areas, such as enumeration districts or other small areas. The point-based infrastructure is

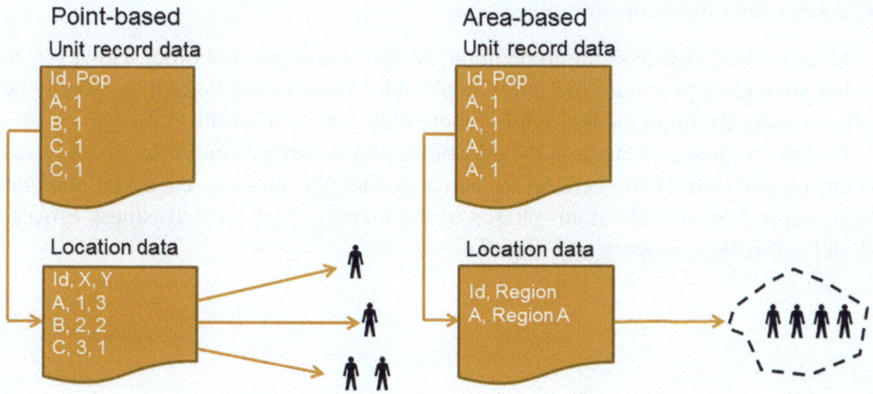

Fig. 9.3 The conceptual difference between point-based and area-based geocoding infrastructures

basically a system to integrate data in order to better exploit the geospatial dimension of statistics. As such, it does not presuppose a specific mode of data collection. The point-based approach can be implemented in the context of traditional Census data collection, as well as in the administrative data-based context. One of the key goals of the ESS Vision 2020 is to better exploit new data sources for statistics, which fits well with NSIs setting up a point-based infrastructure based on authoritative location data. Location can then be added to a number of statistical data sources, such as administrative data, through common identifiers.

9.4 Building and Maintaining Point-Based Geocoding Infrastructure for Statistical Production

There are a number of operational aspects of a point-based geocoding infrastructure that need to be tackled by NSIs in the process of setting up and maintaining such an infrastructure. A list of key tasks to effectively set up, maintain and use a point-based geocoding infrastructure include the following:

- Find out what the users need
- Promote geospatial statistics and the potential of geospatial information
- Recognise geospatial data sources
- Assess data processing capacity
- Specify geospatial statistics output
- Create a flexible statistical production set-up
- Build the geocoded survey frame
- Obtain and manage geospatial data
- Conduct geospatial data quality assessment
- Assess identifiers to enable correct data linkage
- Geocode data
- Define and prepare geospatial statistics products
- Assess data dissemination constraints

Ideally, these key tasks are to be implemented in a sequential order. However, in reality such ideal processes are rarely applicable. Some of the tasks may already be fully or partially implemented while others may not be relevant to all NSIs.

In order to demonstrate how the geocoding infrastructure can be set up and used as an integral part of the general statistical production process, each key task has been mapped against the main phases of the Generic Statistical Business Process Model (GSBPM), as shown in Fig. 9.4.

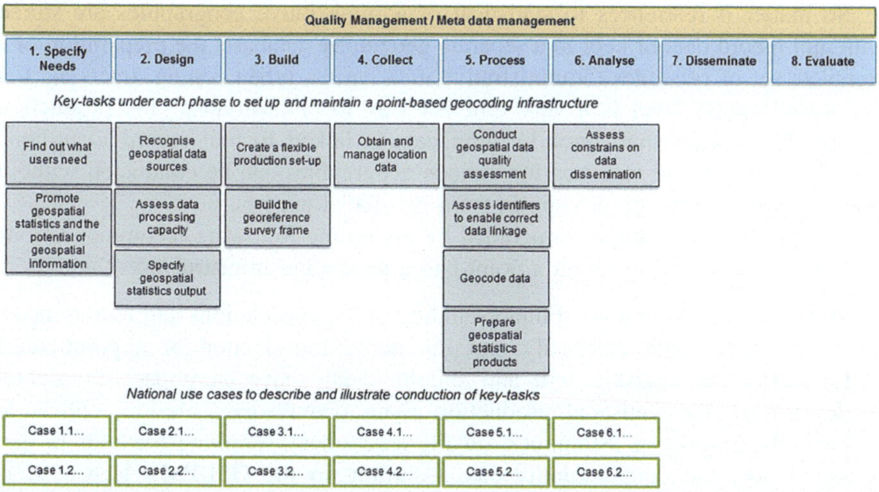

Fig. 9.4 Key tasks mapped to the phases of the GSBPM, case studies available in the GEOSTAT 2 report, Moström et al. (2017a, b)

The expectation is that this approach will help statisticians to better connect geospatial data management and geocoding to their professional reality. As the scope of GEOSTAT 2 was more on the geocoding infrastructure itself than the use of geocoded data and the dissemination of geospatial statistics, Phase 6 (Analyse) has not been fully pictured, and Phases 7 (Disseminate) and 8 (Evaluate) were not covered at all. The GEOSTAT 3 project running for 2017 and 2018 will cover these aspects, as map production is a powerful dissemination and visualisation method.

Each key task is briefly described and, wherever possible and suitable, generic recommendations are provided. For each key task a selection of the relevant use cases have been identified and described, based on the experience gathered by the countries forming part of the project consortium. The use cases are presented in Annex 1 (Moström et al. 2017a, b).

A flexible production environment, which supports a large variety of statistical output areas, is crucial for the efficiency and performance of the production process. A common approach among those NSIs that already have a point-based geocoding infrastructure in place is to set up a central "geography database" storing references, at the level of point-locations, to any relevant statistical or administrative geography. Using such a model, references to an unlimited number of geographies can be stored and kept separately from unit record data. The only reference that needs to be stored with unit record data is basically the unique identifier linking a unit record to its corresponding location in the geography database. This contributes to a smother spatial reference maintenance process: whenever the statistical or territorial geographies change, updates need to be applied only to the geography database, instead of updating multiple records of unit record data.

No matter if references to statistical or administrative geographies are stored with unit record data or kept in a separate geography database, the preparation of a complete set of geocodes (administrative units, grids, urban zoning, tracts etc.) at the point-location level is a very efficient way to "industrialise" the production setting. This means that, once location data is linked to unit record identifiers, tabulations to retrieve statistics for various geographies can be conducted without even using GIS software or support from geospatial experts.

The GEOSTAT 2 project concluded by giving the following recommendations on how to set up and maintain a point-based geocoding infrastructure:

- Make sure to obtain a good understanding of the expectations and requirements of the users, both externally and internally. Introduction of a point-based foundation for statistics will and should ideally have an impact on several domains of the statistical production chain (from survey design, collection methods, processing techniques, quality assessment, sampling methods to dissemination and confidentiality etc. according to the GSBPM). Hence, it is important to involve the organisation broadly to find out how and where a point-based infrastructure for statistics can improve business processes.
- Make a thorough assessment of location data sources that can be potentially used to set up a point-based geocoding infrastructure and decide on a reasonable level of quality needed. Fundamental and authoritative geospatial data from the National Spatial Data Infrastructure should be the first-hand option. Make sure that the provider or custodian has a consistent and systematic scheme of maintenance to keep the dataset updated and that data is properly described in line with metadata standards or any other official data specifications. It can be expected that the INSPIRE directive along with the work pursued by UN GGIM: Europe have triggered, or will trigger, a gradual improvement of access to geospatial data in many Member States.
- Build formal working relationships with external producers of geospatial information (e.g. NMCAs) as to safeguard long-term provision of data. Cooperation between institutions should ideally rely on (formal) agreements or legislation, but the agreements themselves are no guarantee for a good, flexible and solid cooperation (UNGGIM NIA 2018). Cooperation with producers or data custodians should entail establishment of feed-back routines for reporting and correction of errors found in geospatial information.
- Develop a strategy to manage the geospatial data streams in the best possible way. Such a strategy should address the human resources needed as well as the technical infrastructure allowing for efficient processing of geospatial data (software, data storage, computing capabilities etc.). The goal should be a flexible technical environment and to avoid duplication of data and smooth processes for maintenance.
- Develop routines on how to deal with the temporal aspects of data in order to obtain the best possible temporal cohesion between location data and statistical and administrative data to be geocoded (Moström et al. 2017b).

- Develop routines for a uniform approach to geocoding, most notably describing workarounds to handle erroneous data (erroneous ids, missing information etc.). Such routines may include address validation tools, homogenisation of address information or interpolation of address location points (Moström et al. 2017b).

The creation of a geocoding infrastructure for statistics and its integration into the statistical production process does not absolutely require a big-bang approach and a complete redesign of enterprise architectures, production processes and legislation inside and outside NSIs. Small and stepwise improvements are possible.

9.5 The Role of Maps in Statistical Production—Examples from Statistics Sweden

Increased possibilities to link location to statistical information in a flexible way need to be underpinned by increased capabilities in how to work with geospatial data in the NSIs. There are many GIS software tools which help in processing and understanding the data. Looking at your data together with various background information on a map, finding outliers, is one example. Another example is to extract information from one data layer and calculate its content using area boundaries. A third example is to visualise your statistical results on a map, or in some cases to let the statistical results determine the map features all together.

9.5.1 Maps as Editing Tool and for Quality Control

Use of administrative data and registers typically require consistency and quality controls. Even basic routines for checking consistency of location data and spatial accuracy (e.g. if a building coordinate falls inside a water body etc.) can be implemented in registers without involvement of visual interpretations. But some errors can be difficult to assess without the powerful implementation of geography offered by the visual interface of a GIS.

The map below shows an erroneous building coordinate. The drift of the coordinate could be easily detected through visual interpretation but hardly by machine based methods. However, visual interpretation is labour intensive and should ideally be applied as a complementary method only after execution of machine based rules for error detection (Fig. 9.5).

Fig. 9.5 A map showing an erroneous building coordinate. Integration of geospatial data and registers

9.5.2 Integration of Geospatial Data and Registers

Linking information on environment and economy together, sometimes require complex models involving numerous data sources. In a pilot study, Statistics Sweden wanted to produce basic land accounts where a number of land use categories were disaggregated according to established industry classification (NACE) (Fig. 9.6).

The basis for the method is the linked processing of data describing land use with information from registers on land ownership, industries and companies linked to spatially delimitated property units. This linked processing requires a geographical analysis on a low geographical level (Statistics Sweden 2016).

The first step was to define a proper spatial interface between land use and legal and economic status such as ownership and type of industry. In the project cadastral parcels (property units according to the cadastral map) were chosen as spatial interface. A cadastral parcel has a consistent delimitation in space, a well-defined ownership and by using unique keys links can be made between cadastral parcels and the Real Estate Register, and from there, further with economic data in the Business Register.

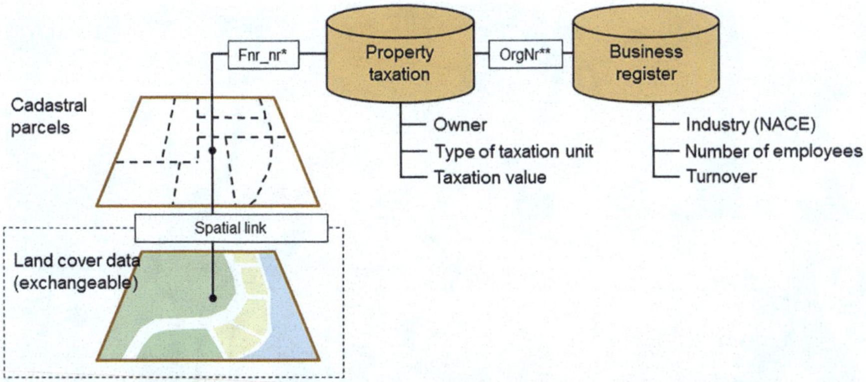

Fig. 9.6 Outline of the data model for land accounts. Explanation to the figure: *Fnr_nr is a unique code used to identify a cadastral parcel, which is also used as a key in the Real Estate Register. The code can be used to link the cadastral parcels in the cadastral map with data in the Real Estate Register and the Real Estate Taxation Register. **OrgNr is a company's unique corporate identification number, which is used as a key in the Business Register. The corporate identification number can be used to link data between the Real estate assessment and business statistics

The second step was to select geospatial data that describe the land use in a consistent manner. As the cadastral parcel was chosen to link ownership and economic data with the land, the data used must also have high geographic resolution.

The final step was to link the land use information with the cadastral parcels to aggregate the area of each type of land with the area of each individual cadastral parcel. Once this was accomplished, the cadastral parcel and the associated land type information could be linked to the landowner and sector or industry based on information in the Real Estate Assessment Register and the Business Register.

The disaggregation approach was also successfully tested on geospatial data describing above ground carbon storage in forests (by 25×25 m pixels) (Fig. 9.7).

9.5.3 *Visualisation of Statistical Results*

Statistical results can be presented as tables, graphs or maps. With increased territorial detail, leaving administrative geographies and producing statistics on geographies not so well known, maps are increasingly important. Maps provide the unique insight into the statistical information and show territorial patterns in the

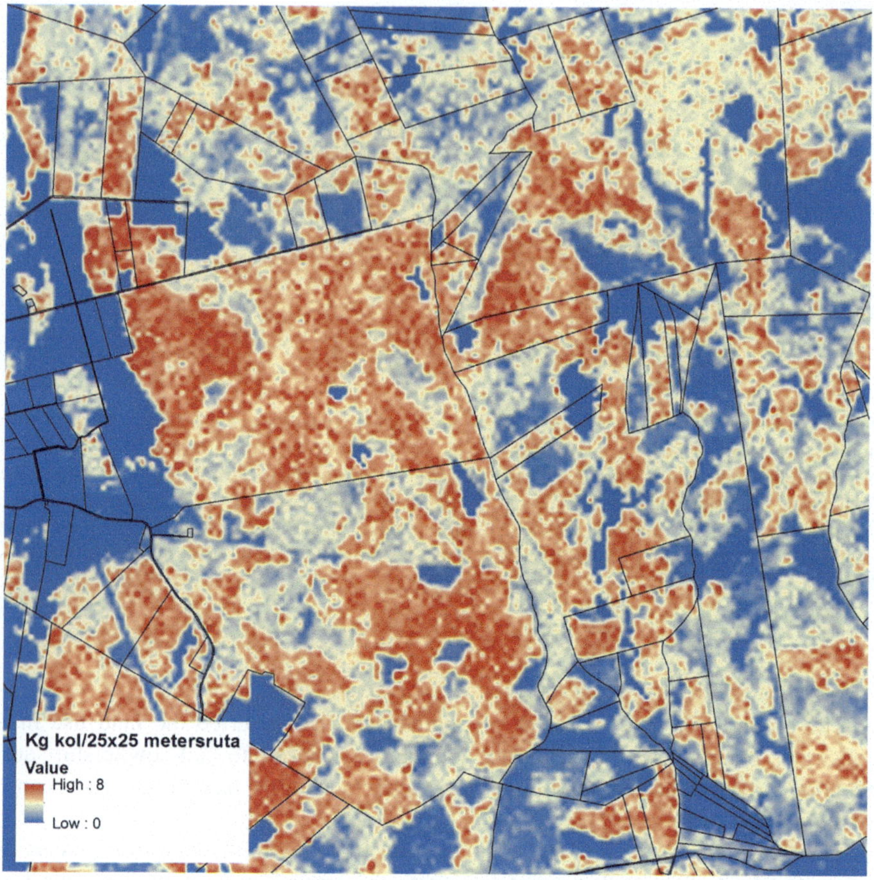

Fig. 9.7 A subset of the Forest Map 2005 where the content of above ground biomass per pixel has been translated into kg of carbon. The colour red indicates a high carbon content and the colour blue a low colour content. Cadastral parcel borders have been overlaid on the Forest Map. © Swedish University for Agricultural Sciences (Forest Map) and Lantmäteriet (boundaries of cadastral parcels)

data, such as where you can find clusters of the population with high and low income.

Statistics Sweden has explored some different ways of showing statistics on maps, sometimes even letting the statistics decide the map features all together! The map below, Fig. 9.8, is an example of a Dorling diagram, designed to illustrate the skewed distribution of population numbers in Swedish municipalities. Each circle represents a municipality.

The idea of the Dorling diagram is to let the bigger circles gently push the smaller circles aside but still keep the spatial relations between the objects so that it can be still interpreted as a map. The result is a piece of infographics combining the

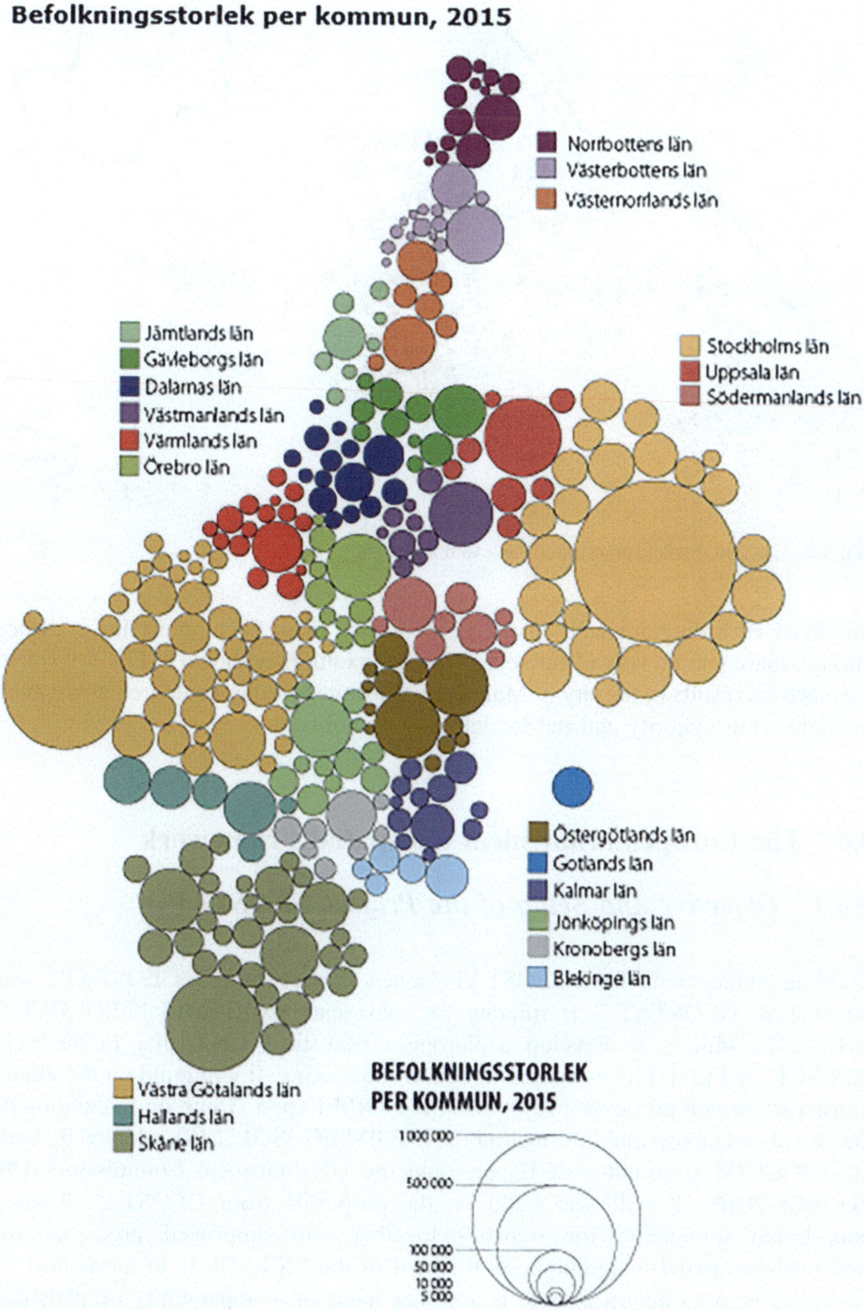

Fig. 9.8 Svanström Dorling map as example

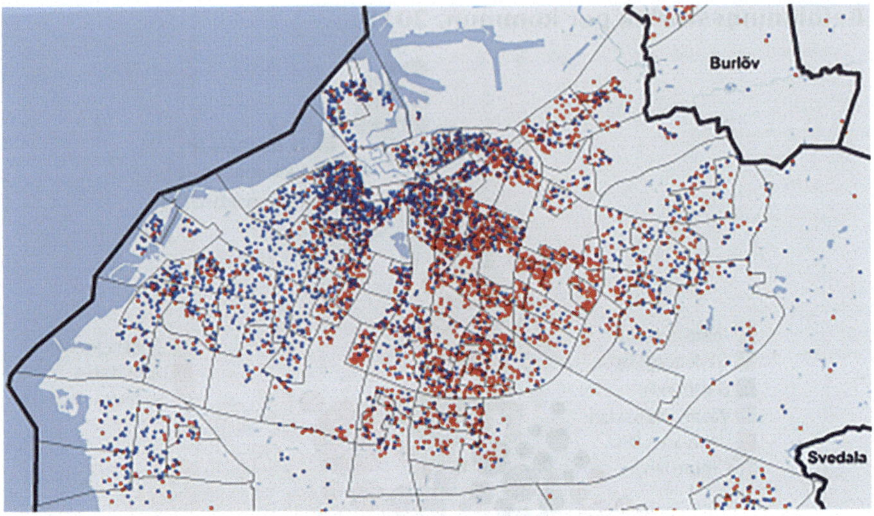

Fig. 9.9 Election results (point maps) as example

simplicity of a diagram with spatial element of a map. Communicating statistics through maps can be very effective and another example is shown in Fig. 9.9 where the electoral results in the city of Malmö are illustrated through coloured dots—blue for right wing majority and red for left wing majority.

9.6 The European Statistical Geospatial Framework

9.6.1 Objective and Setup of the Project GEOSTAT 3

Building on the results from GEOSTAT 2 a new project team for GEOSTAT 3 was established. GEOSTAT 3 is running for two years, 2017–2018 (GEOSTAT 3 2018). The aim is to develop a European Statistical Geospatial Framework, ESS-SGF, and to test its usefulness for SDG monitoring. It will build on the global framework principles developed by the UN-GGIM Expert Group on Integration of Statistical and Geospatial Information (UN-GGIM EG-ISGI 2018), adopted by both the UN-GGIM Committee of Experts and the UN Statistical Commission (UN EG-ISGI 2016). It will also build on the proposals from GEOSTAT 2 on a point-based foundation for statistics together with improved processes for statistical-geospatial integration. As the goal of the ESS-SGF is to harmonise its principles across countries, there is a strong focus on comparability of statistical outputs, geospatial data sources, common methodologies and interoperability of data sources and metadata.

The project team consists of members from eight NSIs, two National Mapping and Cadastral Agencies, NMCAs, and the secretariat of the European Forum for Geography and Statistics, EFGS. Drafting and implementing an ESS-SGF is a task that requires cooperation between NSIs and NMCAs, both on a strategic and a technical level. As in previous GEOSTAT projects also this one will benefit from engaging with the expert community of the European Forum for Geography and Statistics (EFGS 2018).

In the same way that the global framework has been built around the Australian framework the approach for an ESS-SGF is to review existing European and national initiatives and frameworks, in search for best practice to populate the principles of the ESS-SGF. From a strategic point of view, the review will consider institutional arrangements between NSIs and other data producing institutions along with legal, organisational and technical aspects of geospatial information management in statistics. UN-GGIM: Europe has done some significant work in this area in the Working Group B on Data Integration, which will be used (UN-GGIM: Europe 2018).

The GEOSTAT 2 project showed that the access to reliable geospatial data in Europe has improved substantially in recent years and establishing UN-GGIM: Europe has helped in increased cooperation between NSIs and NMCAs. While GEOSTAT 2 focused on three of the five principles from the NSIs point of view, see Fig. 9.10, the GEOSTAT 3 covers all five principles and look in more detail into for example mechanisms for data sharing between authorities responsible for administrative registers, such as population and tax registers. A challenging task, but crucial to really get the full potential from the ESS-SGF. The framework will include recommendations on human resources requirements and a roadmap for implementation will be prepared in consultation with UN-GGIM: Europe. The rationale for implementing the framework at European NSIs, supported by the NMCAs, is resource efficiency, relevant outputs and increased flexibility of statistics.

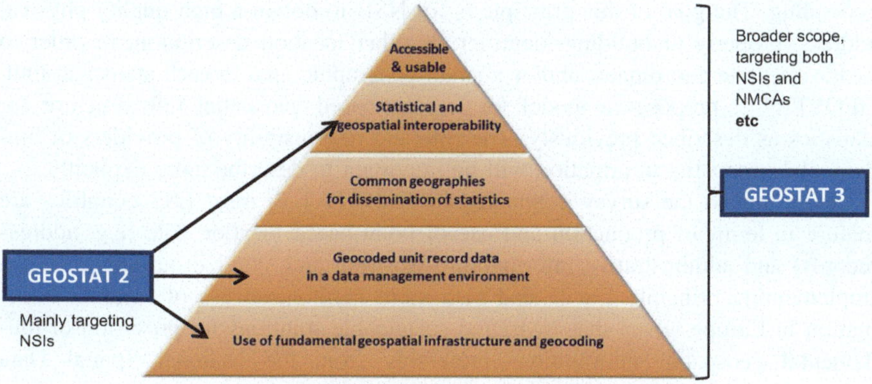

Fig. 9.10 Illustration of the link between the GEOSTAT projects and the SGF

The ESS-SGF will also relate to current best practise provided by UNECE (UNECE HLG 2018) in the Common Statistical Production Architecture (CSPA 2018) as data integration in general and integration with geospatial information in particular might need to be more visible in the CSPA. CSPA is a reference architecture for the statistical industry. The description of CSPA tells that "The scope of CSPA is statistical production across the processes defined by the GSBPM (i.e. it does not characterize a full enterprise architecture for a statistical organization). It is understood that statistical organizations may also have a more general Enterprise Architecture (for example an Enterprise Architecture used by all government agencies in a particular country). The value of the architecture is that it enables collaboration in developing and using Statistical Services, which will allow statistical organizations to create flexible business processes and systems for statistical production more easily".

The project will produce a proposal for the ESS-SGF during 2017. To test its usefulness the second half of the project during 2018 will select a number of 2030 Agenda indicators, preferably with a focus on accessibility. The idea is that by implementing the ESS-SGF the capacity to produce indicators relying on integration of statistical and geospatial information will be greatly improved.

The ESS-SGF is also important with regard to the 2021 Census round in Europe, as the successful production of a European kilometre grid in the GEOSTAT 1 project has led to grid statistics becoming a census output. Grids are thought of as an important complement to traditional census districts, especially if looking at cross-border issues or conducting accessibility studies where the equal size of grids are well suited as statistical input.

9.6.2 The Framework Principles Through a "European Lens"

Principle 1 says that NSIs should use a fundamental geospatial infrastructure and geocoding. The goal of this principle is for NSIs to obtain a high quality physical address, property or building identifier, or other location description, in order to assign accurate coordinates and/or a small geographic area to each statistical unit. GEOSTAT 2 proposes a model for a point-based geospatial infrastructure for statistics as described previously. The role and responsibility of providers of fundamental geospatial information will be important to describe more explicitly.

According to the survey conducted in GEOSTAT 2, most ESS countries are mature in terms of production and use of point-based location data (e.g. address records) and administrative information (Moström and Wardzińska-Sharif 2015). Implementing principle 1 will lead to a more harmonised use of location information in Europe, given that NSIs find sustainable solutions for provision of fundamental geospatial information—preferably from the National Spatial Data Infrastructures.

Principle 2 states that geocoded unit record data should be in a data management environment in order to ensure that all statistical data is consistently geospatially enabled. NSIs need to find smart setups for storage of their (point-based) infrastructure enabling integration with statistical micro data, producing statistical results in a flexible way. Many European NSIs have already developed, or are under way to develop, environments ("geography databases" or "keycode databases") for consistent linking of statistical information to location data.

This principle holds a special challenge considering the push for increased use of administrative data sources, typically kept by other public or private bodies. National administrations are different depending on legal aspects, institutional culture, and public sentiment regarding the use of data for statistical purposes. Producing European data on the basis of different national administrative systems is very challenging in terms of comparability.

Principle 3 stresses the importance of a common set of geographies to ensure that all statistical data is consistently geospatially enabled and that users can discover, access, integrate, analyse and visualise statistical information seamlessly for geographies of interest. In the EU context, a foundation for common geographies is already long since established through the NUTS and LAU frameworks along with the recent development of statistical grid systems. However compilation of data for pan-European statistical geographies is complex and the workflows needs to be improved. There are also reasons for global outlooks, namely the Discrete Global Grid Systems (DGGS) standard (DGGS 2018), developed under the auspices of OGC, to assess how this initiative may impact the future of the European grid system. The various practises in Member States will be addressed too, with regards to maintenance of national geographies.

Principle 4 deals with statistical and geospatial interoperability, including data standards and processes. Overcoming different "barriers" between data and metadata from different communities and providers will enhance the efficiency of discovery, access, and use of geospatially enabled data. The project will review standards such as SDMX and OGC (EGISGI 2015).

The guiding principle is to leave data at its source, which means that data, both geospatial and statistical, should be published in a way that allows for machine-to-machine transformation and integration of data. One of the key issues is to increase the interoperability between on one hand the Spatial Data Infrastructure regulated with binding, technical INSPIRE specifications – on the other hand the statistical SDMX standard (EGISGI 2015). To this end the project will review how the INSPIRE model and associated data formats relate to the SDMX standard and provide advise on how to adapt data formats to get increased interoperability.

The project will also make a concrete proposal on how to link geoservices and statistical data, namely through testing the INSPIRE models for statistical units and population distribution using OGC Table Joining Services (TJS), in order to obtain a higher degree of harmonisation.

One of the key elements of principle 4 is a consistent inclusion of geospatial processes into statistical business processes. The results from GEOSTAT 2 project show that the Generic Statistical Business Process Model, GSBPM, is well suited to also include geospatial elements.

Principle 5 gives advice on how to make your statistical output accessible and usable. It stresses the need to identify or, where required, develop policies, standards and guidelines that support the release, access, analysis and visualisation of geospatially enabled information. It aims to ensure that custodians release data in appropriate forms for users to discover, access, integrate, analyse and visualise statistical information seamlessly for geographies of interest.

There is a great variety of platforms to disseminate geospatial statistics among Member states, but a common feature is that they are typically built on INSPIRE services (INSPIRE 2018) (see also Chap. 3 of this book). The various solutions range from simple catalogue and file services to advanced portals with built-in capabilities for geospatial analysis. The project will identify some examples of good practice to encourage development of easy access and user friendly solutions for dissemination of geospatial statistics. The European Location Services will be evaluated for cross border dissemination of geospatial statistics (ELS 2018).

The trend towards more open data is strong (OKFN 2018) and there is a need for a good business case showing the possibilities of making interesting facts on a detailed regional level accessible to the society. Some NSIs release a broad range of geospatial statistics under open licenses, while others use various business models to provide only commissioned services with restricted licenses. One of the undertakings of the project will be to show-case the potential of geospatial statistics in an open data environment.

9.7 Conclusions

The 2030 Agenda (ASD 2018) aims at leaving no one behind, and from a geospatial perspective this would mean to implement policy measures aiming to improve the situation where it is needed the most. Having indicators only showing the situation on a country, county or municipality level is not enough. Implementing a Statistical Geospatial Framework in close cooperation between NSIs and NMCAs and aiming for increased register-based statistical production will help NSIs to stay relevant to policy makers and to produce valuable facts on all territorial levels.

The report "A world that counts" (UNDR 2018) talks about the data revolution that needs to take place for the 2030 Agenda to be achieved and measured:

> New institutions, new actors, new ideas and new partnerships are needed, and all have something to offer the data revolution. National statistical offices, the traditional guardians of public data for the public good, will remain central to the whole of government efforts to harness the data revolution for sustainable development. To fill this role, however, they will need to change, and more quickly than in the past, and continue to adapt, abandoning expensive and cumbersome production processes, incorporating new data sources, including administrative data from other government departments, and focusing on providing data that is human and machine-readable, compatible with geospatial information systems and available quickly enough to ensure that the data cycle matches the decision cycle.

The GEOSTAT 3 aims to help European NSIs taking part in the "geospatial data revolution" in close cooperation with NMCAs and other public authorities. The efforts of UN-GGIM: Europe has created a momentum where NSIs and NMCAs are working together, the challenge will be to create similar positive conditions for cooperation also with providers of administrative data at large. Even if the European NSIs are allowed by law to access administrative data for the production of official statistics, there still need to be formal agreements and supporting structures in place.

References

ASD, A/RES/70/1—Transforming our world: the 2030 agenda for sustainable development, https://sustainabledevelopment.un.org/post2015/transformingourworld, visited Jan 2018

Backer L, Holt Bloch VV et al (2011) GEOSTAT 1A—Representing census data in a European population grid, Final report

CSPA, Common Statistical Production Architecture, https://statswiki.unece.org/display/CSPA/Common+Statistical+Production+Architecture, visited Jan 2018

DGGS, Discrete Global Grid System, OGC. http://www.opengeospatial.org/projects/groups/dggsswg, visited Jan 2018

EC, European Census 2021, https://ec.europa.eu/info/law/better-regulation/initiatives/ares-2016-3268077_en, visited Jan 2018

EFGS, About EFGS. http://www.efgs.info/about-efgs/efgs-organisation/, visited Jan 2018

EGISGI (2015) UN-GGIM expert group on integration of statistical and geospatial information, background document, session 6: metadata interoperability based on DCAT, Lisbon 24 May 2015. http://ggim.un.org/docs/meetings/2nd%20UN-EG-ISGI/UN-GGIM%20EG%20Lisbon%20meeting%20Session%206%20background%20paper%20Metadata%20DCAT.pdf

ELS, European Location Services, http://www.eurogeographics.org/content/european-location-services, visited Jan 2018

ESCP European Statistics Code of Practise (2011) European statistics code of practice for the national and community statistical authorities. Revised edition 2011

ESS Euroean Statistical System, http://ec.europa.eu/eurostat/web/european-statistical-system/overview?locale=fr, visited Jan 2018

GAMSO, Generic Activity Model for Statistical Organisations, https://statswiki.unece.org/display/GAMSO/Generic+Activity+Model+for+Statistical+Organizations, visited Jan 2018

GEOSTAT 3, Project web page, http://www.efgs.info/geostat/geostat-3/, visited Jan 2018

GEOSTAT Project Descriptions, http://www.efgs.info/geostat/, visited Jan 2018

GSBPM Doc, GSBPM Documentation, Introduction, https://statswiki.unece.org/display/GSBPM/I._Introduction#I._Introduction_Toc375051192, visited Jan 2018

GSBPM, Generic Statistical Business Process Model. http://www1.unece.org/stat/platform/display/GSBPM/GSBPM+v5.0, visited Jan 2018

GSIM, Generic Statistical Information Model, https://statswiki.unece.org/display/gsim, visited Jan 2018

Haldorson M, Zaccheddu P, Fohgrub B, Petri E (2016) Geospatial information management in Europe—responding to the user needs. Stat J IAOS 32(4):481–487

INSPIRE, Infrastructure for spatial information in Europe. http://inspire.ec.europa.eu/, visited Jan 2018

Moström J (ed) et al (2017a) A point-based foundation for statistics. Final report from the GEOSTAT 2 project

Moström J (ed) et al (2017b). A point-based foundation for statistics. Final report from the GEOSTAT 2 project, Annex I—Chapter 5
Moström J, Wardzińska-Sharif A (2015) Spatialising statistics in the ESS. Results from the 2015 GEOSTAT 2 survey on geocoding practices in European NSIs
NUTS Regulation (2003) Regulation (EC) No 1059/2003 of the European Parliament and of the council of 26 May 2003, on the establishment of a common classification of territorial units for statistics (NUTS), OJ L 154, 21.6.2003
OKFN, The state of open government data, http://index.okfn.org, visited Jan 2018
SDMX, The official site for the SDMX community. A global initiative to improve statistical data and metadata eXchange. https://sdmx.org/, visited Jan 2018
SG, Statistical grids, INSPIRE data specification on statistical units, https://inspire.ec.europa.eu/Themes/125/2892, visited Jan 2018
Statistics Sweden (2016) Land accounts for ecosystem services. https://www.scb.se/contentassets/6d29a226fcd34a0fa192c97c8ec339dc/mi1301_2015a01_br_mi71br1701eng.pdf
TERCET, Territorial typologies of the European Union. http://ec.europa.eu/eurostat/web/nuts/tercet-territorial-typologies, visited Jan 2018
UN EG-ISGI Background Document (2016) United Nations expert group on the integration of statistical and geospatial information: background document on proposal for a global statistical geospatial framework. Advanced Draft as of 28 July 2016
UN-GGIM Europe WG A (2016) Core data scope. Working Group A—First Deliverable of Task 1.a
UN-GGIM EG-ISGI: Website of the UN-GGIM Expert Group on Integration of Statistical and Geospatial Information. http://ggim.un.org/UNGGIM-expert-group/, visited Jan 2018
UN-GGIM: Europe, Working group B, Data integration. http://un-ggim-europe.org/content/wg-b-data-integration, visited Jan 2018
UNDR, A world that counts, Mobilising the data revolution for sustainable development, http://www.undatarevolution.org/, visited Jan 2018
UNECE HLG, UNECE High-level Group for the Modernisation of Official Statistics, https://statswiki.unece.org/display/hlgbas/High-Level+Group+for+the+Modernisation+of+Official+Statistics, visited Jan 2018
UNECE statistiks wiki: https://statswiki.unece.org/display/adso/Towards+a+Register-based+Statistical+System, visited Jan 2018
UNGGIM NIA, UN-GGIM: National Institutional Arrangements: Instruments, Principles and Guidelines, http://ggim.un.org/docs/meetings/GGIM7/Agenda%207%20NIA%20Instruments,%20Principles%20and%20Guidelines.pdf, visited Jan 2018
Wallgren A, Wallgren B (2007) Register-based statistics: administrative data for statistical purposes (vol 553). Wiley, New Jersey

Chapter 10
Vizzuality: Designing Purposeful Maps to Achieve SDGs

Sergio Estella and Jamie Gibson

Abstract The amount of data readily available for people to access has been increasing rapidly over the last decades, due to the cumulative effects of the open data movement, the huge growth of cheap cloud storage, and broader access to the internet through smartphones. This poses a double-edged sword: we know more about our world than ever before, but making sense of all this data and communicating that to a general audience can be a big challenge. Vizzuality have been working with research organisations, universities, Non Government Organizations (NGOs) and governments across the world to develop interactive, understandable and usable data tools. Through a combination of maps, tables, charts and other data visualizations, they've helped these organisations communicate cutting-edge research on sustainable development to policy-makers, analysts, the media and the general public. From deforestation to government transparency and emissions from agriculture, this data can help these people make decisions to create a more sustainable world. This chapter will outline the techniques used by Vizzuality to understand what data needs to be communicated, how, and with whom, for a certain change to take place. The interdisciplinary approach, encompassing psychology, design, data science and software development, will be revealed through three case studies of Vizzuality's products.

Keywords Big data · Data visualization · Data design · Sustainable development goals

10.1 Introduction

Map-based data visualisation is a powerful means of communication. By using thematic maps to situate intangible or invisible data (pollution in the air, crime rate, etc.) within someone's lived experience, we can shape how they see the world.

S. Estella (✉) · J. Gibson
Vizzuality, Calle de Fuencarral, 123, 28010 Madrid, Spain
e-mail: sergio.estella@vizzuality.com

J. Döllner et al. (eds.), *Service-Oriented Mapping*, Lecture Notes in Geoinformation and Cartography, https://doi.org/10.1007/978-3-319-72434-8_10

Snow's map of Cholera cases in London,[1] for example, showed it was possible to create insight, and improve lives, by systematically displaying information with a spatial component on a map. Once they pinpointed the outbreak to one public well, using the map, effective action could be taken.

With the advent of personal computers and geospatial information applications, geospatial visualisations have moved into a digital realm. Advances in these technologies mean more people can view geospatial data, as well as a greater ability to interrogate, analyse and customise the data presented on the visualisation. The role of the modern map maker is to create tools that help people "choose a location to map and the features to include on the map" (Peterson 2007), rather than to create the one perfect view for a specific audience.

The emerging Information Society has also influenced the modern paradigm of map production. Over the last few decades, the creation, distribution and use of information has become one of the core means by which economies, society and personal identities are produced and reproduced (Van Dijk 2006). There are now billions of networked devices—from satellites to smartphones and sensors—collecting data about our world almost all the time. We're generating more data than ever before, and thanks to a range of satellite navigation systems and geo-tagging much of it has a spatial component. Digital geospatial visualisations are rising in prominence, helping people absorb new information by displaying it within the context of a map that is already familiar. In particular the web-based visualisation is a focus for map makers, as key data consumers spend an increased amount of their time on the internet (OFCOM 2017). The internet is also a means of cheap mass communication, allowing the visualisation to reach a larger potential audience than ever before (Meeker 2017).

This is the paradigm of map production that Vizzuality currently sits in: making versatile geospatial visualisations that present big datasets to a large and varied global audience. Visualisations have to be simple and understandable for any user, but also have the power and flexibility for more advanced or technical users to conduct deeper analyses and gain new insights.

At the same time, as maps move into the digital arena, the processes for making them are changing. On the one hand you have the Agile Method (Beck et al. 2001), an approach to software development that: places people and users first; prioritises collaboration with users and clients throughout a development process, and; proposes iteration, flexibility and adaptation. For thematic maps related to sustainable development topics, you also have the influence of the social theory of Participatory Development (Chambers 1997). As understood by authors like Chambers, Participatory Development challenges practitioners to deeply understand their stakeholders and allow them to influence and control the direction of development initiatives, in order to deliver better and more sustainable outcomes.

Both are affecting the way software, and some geospatial visualisations, are created. It is now widely recognised as a priority to align an organisation's

[1]For more information see https://en.wikipedia.org/wiki/1854_Broad_Street_cholera_outbreak.

knowledge with the needs of the people they aim to support or empower. By using techniques from different methodologies, map developers (such as our team at Vizzuality) have improved the outcomes of product solving tasks, in a way that uses fewer resources and responds nimbly to unpredictable changes in the scope of work.

This is all taking on increased importance in the era of the Sustainable Development Goals (SDGs) (United Nations General Assembly 2015). The 17 goals agreed by the United Nations in September 2015 lay out a framework for coordinating global action to improve the status of our planet and the people living on it. To take action successfully, the world needs better data to work out where to act, which actions to take, and how effective actions have been so far. Cartographers are increasingly building maps using data about Sustainable Development Goals to help achieve this vision.

In this chapter we will outline how the convergence of Information Society, a new paradigm of maps, and new ways of making geospatial visualisations, could help feed the world with the information needed to take action to meet these goals more effectively than ever before.

10.2 Information Society and the Sustainable Development Goals

10.2.1 Consequences of Information Society for Data Production, Management and Dissemination

"Information society" constitutes the latest revolution after the industrial revolution. We've moved from a period of machine-based expansion, to one of data and information-based expansion (Castells 1996). The term acknowledges the pivotal role of Information and Communication Technologies in the (re-)organisation and (re-)production of the political, cultural and economic realms (Sarrocco n.d.).

In recent years two predominant technologies—smartphone hardware and mobile application software—have been at the forefront of shaping information societies. They have provided people with information instantly almost everywhere in the world. You can see the economic revolution that has resulted in the creation of billion dollar valued companies in these two areas of technology (Apple, Google, WhatsApp, Facebook etc.), while the increasingly rapid creation and transmission of culture (through viral videos, memes and others) evidences the cultural influence (Levitin 2015; Smith 2014; Pew Research Centre 2015; Henkel 2013).

Predictions, like those presented in Fig. 10.1, for ownership and use of mobile technologies suggest a widening geographical reach (particularly into the Asia-Pacific region, the Middle East and Africa) and increased usage in the next few years. It is likely that Information Society will further change as a result, as more people become connected and use new services to complete new activities.

This trend could "overcome geographic constraints, improve communication, and limit asymmetrical information" (Tran et al. 2015), such that vital information can be provided to billions of people around the world and produce the kinds of behaviour that would create a sustainable economy.

This is just the dissemination element of Information Society. As information is the dominant force driving social, economic and cultural change, there are businesses dedicated to creating and storing data, like never before. The combination of data generated by all those smartphones plus data collected by satellites, sensors, social networks, and other instruments has led to exabytes (or billions of gigabytes) of data being collected every day (Wall 2014). The sheer size of these datasets can make processing difficult, and are often so large that it's hard for many people to navigate easily or effectively (Paroutis and Al Saleh 2009). There is a challenge here for all cartographers to find the best ways to display this information clearly, for anyone to actually use.

Following Castells' (1996) idea of Network Society, this data is increasingly transmitted and accessed in a networked approach using Application Programming Interfaces (API), a clearly defined method for communicating between software components. It's more efficient to store data in one place and use APIs to move it between multiple sites, than for all sites to duplicate the same data. Additionally, as information drives social, cultural and economic change in Information Society, there are incentives for data producers to make sure their data is used in as many applications and website as possible. APIs help data to propagate throughout the internet, and doing so confers power to the creator.

10.2.2 Information Design in the Context of the SDGs

On top of a range of corporate applications, overcoming the challenge of displaying and communicating big geospatial data could be key in reaching the sustainable development goals (United Nations General Assembly 2015). Data visualisation can lead to sustainable behaviours by helping "to inform and educate as well as encourage people to change their lifestyle" (Krätzig and Warren-Kretzschmar 2014), making huge global issues (like climate change) more understandable (Sheppard et al. 2011) or reinforcing a sense of responsibility to the group (Jonsson and Nilsson 2014). Many authors also talk about the 'locus of control' and the perception that a person can influence an issue with their actions; they are more likely to take action if they feel more control (Jonsson and Nilsson 2014; Bartiaux 2008; Kollmuss and Agyeman 2002). The United Nations Data Revolution Group actively acknowledges this too and proposes a number of avenues to leverage technologies at the forefront of Information Society for Sustainable Development (Independent Advisory Group on a Data Revolution for Sustainable Development 2014).

However it is not as simple as just providing information to change behaviours. Humans are not always rational actors, using perfect information without emotion to make the right decisions. Attitudes or intentions have to be considered (Bartiaux

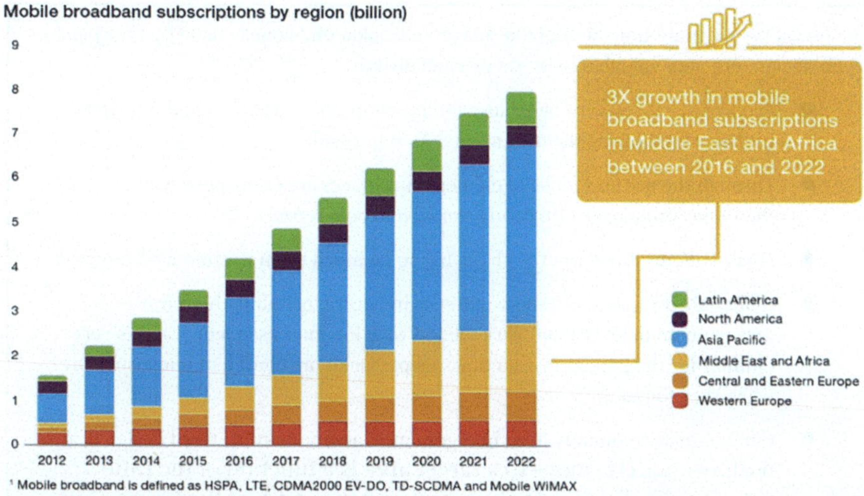

Fig. 10.1 Smartphone subscriptions per region, 2012–2022 (Ericsson 2017)

2008), as well as the frame within which information is provided (e.g. do a good thing as opposed to don't do this bad thing) (Van de Velde et al. 2010). Additionally, if that information isn't understandable or does not make sense in a local context, it may be easily disregarded (Sheppard et al. 2011). A number of authors have applied these arguments to explain why climate change actions have not been as rapid as would be expected: the way the problem is communicated is too global, too negative and abstracted from local realities, which reduces the amount of attention it gets and likely actions that result (Hulme 2008; Jasanoff 2010).

Innovation in the design of web cartography is starting to overcome these barriers, as map makers acknowledge that "a well-designed user interface is essential to explore the variety of atlas maps and media" (Schnürer et al. 2014). Where maps used to be an illustration or piece of context in literature, they're now the main focus. The combination of geo-information and interactivity in a geospatial visualisation often provides the dominant part of the story, with text in support.

Designers and developers are becoming increasingly conscious of the day-to-day work to be completed by the user on the application or tool, to avoid bad experiences that make them stop using applications. User-centred design, an approach to building products "that take into account the skills, knowledge and abilities of the users" (Haklay and Nivala 2010), is one technique that can help make sure interfaces are well designed. Specific design principles like progressive disclosure (Nielsen 2006), nudges (Thaler and Sunstein 2008), consistency and constraint (Preece et al. 2002), form follows function (Bradley 2010) and responsive templates (Marcotte 2010) also achieve this goal. We explain them in more detail in Fig. 10.2.

In the list below, there are some of the design principles traditionally used by designers in web design, are very valuable to tackle product design:

- *Progressive disclosure* helps users to maintain their attention and improves usability by showing the minimum data requested..
- Through the use and reuse of elements interfaces are easily perceived as part of the same system providing *consistency and coherence.*
- *Form Follows Function,* the shape has to be based upon the intended function.
- *Golden Ratio* appears in some patterns in nature including the spiral arrangement of leaves that very related with Fibonacci sequence, and is very familiar for the human eye, so that, people feel more confident unconsciously when they found it.
- *Fitts's Law* is frequently used in ergonomics and human-centered design is a predictive model to move to a target area is a function of the ratio between the distance to the target and the width of the target. (Fitts 1954)
- *Affordance:* using physical characteristics to influence its function.
- *Avoid cluttering,* meaning by this, to add only necessary elements of the interface, up to maximum five colours, ornaments and other unnecessary elements, users won't be disturbed and will focus in the main elements of the interface.
- *Basic Tasks Model,* a ranked set of visualisation characteristics that can be used to predict how easily viewers can identify patterns in the data being presented (Carswell 1992)

Fig. 10.2 Explanation of some common principles used in design of websites containing thematic maps

10.2.3 Building Purposeful Geospatial Visualisations

In the rest of this chapter we will demonstrate some applications of the concepts discussed so far using Vizzuality's experiences building geospatial visualisation tools. We explore the potential role of delivering big geospatial data to the whole world through geospatial visualisations, thanks to key trends in Information Society —increased production of and access to information. Combined with techniques for modern map making, we believe there are a number of opportunities for this data to be used to improve the state of the planet and the livelihoods of Earth's 7 billion (and counting) humans.

But in order to make the most of those opportunities, a number of principles need to be embraced, as we'll outline in the following. Geospatial visualisations will need to be more interactive, understandable and accessible, so they can adapt to the environment of the viewer. This is for two reasons. First, as datasets get bigger and more global, it's harder for the viewer to connect to it and feel the need to act in consequence. We need to allow users to bring the data within their locus of control: their immediate lived environment. Second is the fact that we're presenting this data to people for the first time, either because their new phone is their first connection to the internet, the data has only just been made publically available, or the data has just been created (derived from satellites etc.).

The three case studies outlined in the next section show different elements of this new paradigm, including community empowerment through visualisations, using machine learning to find patterns and getting end-users to participate in map design.

10.3 Connecting Organizations and People Through Data and Maps

In this section we present three case studies of geospatial visualisation for the Sustainable Development Goals. In each case, the visualisation was built to fill a gap in the amount of information available about a particular topic and the quality of its communication with people who could use it. The partnering organisations all have a mission to improve the state of the world and the sustainability of our economy. Using the principles of modern map design, and the opportunities of Information Society, these visualisations fill gaps in access and understanding, and ultimately empower people across the world to take actions consistent with the SDGs.

10.3.1 Global Forest Watch

Forests are a key focus for many efforts under the SDGs, especially Goals 12 (Responsible Production and Consumption), 13 (Climate Action) and 15 (Life on Land). They're fundamental for the lives and livelihoods of millions of people, limiting climate change, supporting biodiversity and improving water and air quality. There's also a set of political, economic and financial factors driving the decisions of which places we protect, the places we deforest, and the places we reforest (Kissinger et al. 2012).

- Forests provide humans with basic commodities such as food, medicines, wood, and fuel. It is estimated that more than 1 billion people depend on forests.
- Forests have an important role in the water cycle and they help reduce soil erosion, thereby acting as a buffer against drought and desertification.
- Forests are home to thousands of different species. About 80% of the world's documented species can be found in tropical rainforests. Deforestation equals habitat loss for those species and poses a threat to their survival.

- Forests absorb the carbon emissions that cause climate change. When they are cut down, this benefit is lost and the carbon stored into the trees is released back into the atmosphere. Deforestation is a major cause of carbon emissions.

Global Forest Watch (GFW) is a platform that helps monitor forest change globally and in near real-time, based on free, reliable and distributed data sources. In particular, it uses data derived from multispectral satellite imagery of the Landsat 5 thematic mapper, the Landsat 7 thematic mapper plus, and the Landsat 8 Operational Land Imager sensors (Hansen et al 2013). Detecting where forest has been lost is important, but the information becomes actionable if people can also know when it happened and some of the characteristics of that place. People can use the platform to put forest change data in the context of accurate data about drivers of deforestation—fires, land concessions or illegal logging roads—revealing the possible reasons why there has been loss or gain.

One of the first parts of the site to be built was the interactive map, as shown in Fig. 10.3. It still forms a central part of the site today: nearly 66% of all the time spent looking at GFW is focussed on the map pages. While a thematic map was an obvious choice to reveal the spatial distribution of forest change across the world, we decided to start with a global view because of the visual impact of seeing all the loss in the global context. By starting with the broadest overview possible, users could follow the story whichever way they wanted: narrowing down to specific areas as they saw fit.

Its design eschewed the standard ‘map layer menu’ commonly seen on interactive maps. Instead of a long list of nested groups that users must collapse or open to find information, the titles, labels and sequencing are intended to surface the most relevant data to people. They’re also grouped by potential uses—monitoring land use change, effects on biodiversity, effects on local communities—allowing for easier customisation by core audiences for their needs.

The role of the community that uses the geospatial visualisation is key to GFW’s success. The platform contains a few features which facilitate use of, or even ownership of, this data and this space like never before, helping to bring about changes in the way forests are managed. People can use the ‘stories’ feature to publish their own novel analyses, which means knowledge about the world’s forests can be generated by anyone with an internet connection. At the same time organisations can update the platform directly with their own data and information, to the benefit of the whole world. This is typical of maps within Information Society: they are now a place for people to come together around a shared concern, build a network and stay connected in pursuit of addressing that concern. In this way, a visualisation is elevated into a platform for knowledge exchange, generation and dissemination, and ultimately action. The designer’s job is to provide a flexible suite of tools the network can use to accumulate and grow.

As well as making the data intelligible to humans, the visualisations on the platform are also machine readable. We noted above how, in Information Society, it’s important to be able to propagate information across the web. While GFW has an API, allowing anyone to consume the raw data in their own applications, there is

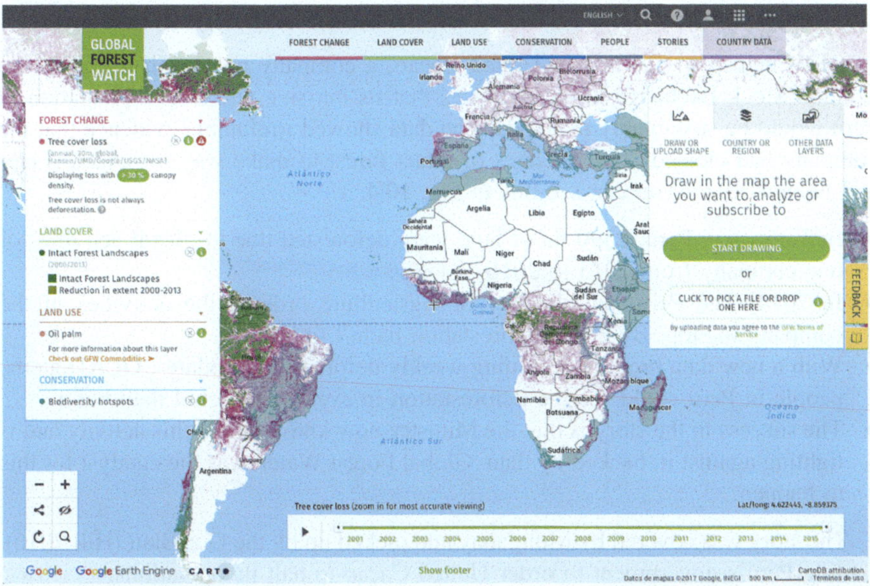

Fig. 10.3 The Global Forest Watch map, available at globalforestwatch.org/map

also a function allowing users to embed a whole customised visualisation that they've created on GFW in their own website. In a typical week, at least 15% of views of the GFW data are not on GFW.org itself. Again this acknowledges the role of community: visualisations are designed to empower others, not just to transmit knowledge to them.

Building on the success of the original application, Global Forest Watch added a complement of focussed applications, including GFW Fires, Commodities, Climate and Water. Each application visualises only a small portion of the dataset, in addition to a few other datasets, to serve a very specific purpose. In the case of GFW Climate, the focus is on the loss of carbon stored by forest ecosystems and carbon emitted when forests are cleared, by linking the data on forest change with global estimates of biomass. By stripping back the experience to fulfill the particular niche needs of a particular community, the visualisation can be much more effective.

Community needs are discovered through a participatory, consultative process. Discussions within the GFW Partnership have helped prioritise the key areas where interventions could be most effective, or are most needed. The team also regularly analyses the most viewed data layers on the map, and which places in the world are most looked at by people. Uncovering the areas of high demand can assist in the process of understanding if new, targeted applications could be useful.

GFW in Action

GFW helped to reveal that United Cacao, a company previously listed in the London Stock Exchange broke promises that its massive new cocoa plantation in Peru would be sustainably managed. The data showed irrefutable evidence that the company had cleared 2000 hectares of intact forest to make way for cocoa (Finer and Novoa 2016): you can see this in Fig. 10.4.

- In Tamiyacu, Peru, 2390 ha of land were deforested due to illegal activities of one company from 2013 to 2016.
- In 2014, The Peruvian Ministry of Agriculture ordered the activities of the company to stop. By 2015 deforestation hadn't finished.
- With a new data product providing weekly deforestation updates, GLAD alerts, people in Peru could see the deforestation in even more detail than before.
- The success in the story is that the Ministry now knows about this activity and is fighting against it, backed by data. Global Forest Watch was the catalyst for this to happen.

The story was covered by Mongabay and picked up by the Guardian (Hill 2016), leading Peru's government to order United Cacao to halt its operations.

By making data on forest clearing transparent and publicly available at a resolution accurate enough to be cited as evidence, GFW provided the foundation for environmental groups and the government to scrutinize United Cacao's land use practices.

10.3.2 Trase

In the last decade, a number of consumer goods companies have set out to achieve internal sustainability goals, including Unilever, Cargill and Starbucks. The complex system of international commodity trade makes it hard to find out the environmental impact of a particular product, which limits our ability to take appropriate actions, for example to reduce deforestation (Godar et al. 2016). Being able to link *who* is involved in a supply chain and *where* they produce their commodities would greatly enhance efforts by companies and other organisations to make their supply chains more sustainable. These discussions are also relevant for SDGs 12 (Responsible Production and Consumption), 13 (Climate Action) and 15 (Life on Land).

"Trase is an open-access platform…revealing the sustainability risks and opportunities that connect production landscapes with consumers worldwide" (Trase 2017). Vizzuality collaborated with the Stockholm Environmental Institute (SEI) and the Global Canopy Programme (GCP) to create the platform. Users can learn not just the volume being traded, but also the social and environmental impacts associated with the trade, and compare them with the vulnerabilities in the communities that produce them, and the commitments of the companies who trade them. You can see the main Trase tool in Fig. 10.5.

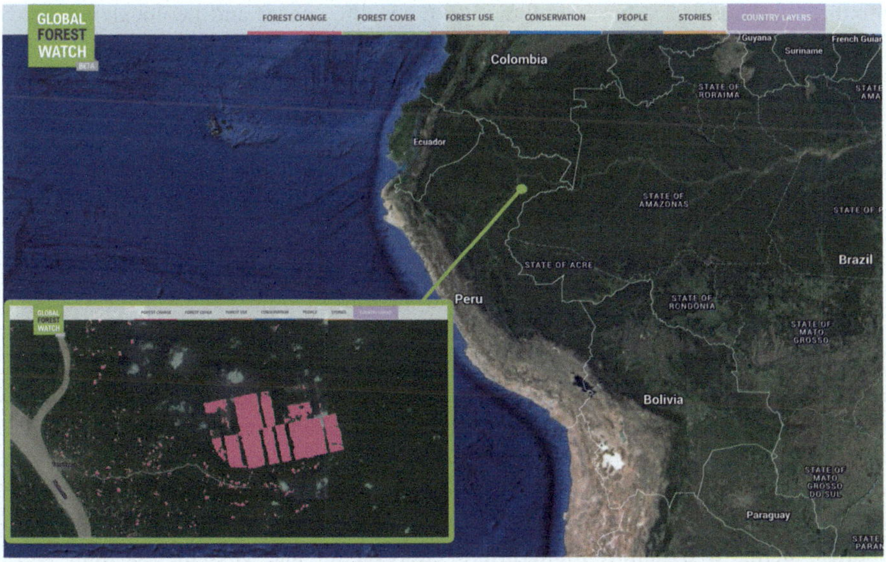

Fig. 10.4 Deforestation in Peru as visualised on Global Forest Watch, Payne and Mann 2015

Fig. 10.5 A screenshot of the map and sankey combination available on trase.earth

The data in Trase comes from shipment customs declarations, bills of lading, logistics data, taxation documents and self-declarations from companies, as well as environmental and social indicators in the countries of interest. As such there are a mixture of shapes (a value is assigned to a specific space) and geolocated records (this commodity was at this location at this time).

There is a core of geospatial data running through this dataset: one of the key impacts should be that a space is better managed, so that environmental and social impacts are reduced. As the analysis is often twofold—people want to look at the source of the commodity and how it travelled around the world—we used a combination of two different visualisation styles: a sankey diagram and a thematic map.

Usually the object of primary interest for users is a particular company or importer country: these are the actors who have most influence over the management of particular spaces, or who have made commitments to limit their impacts. The sankey diagram is the perfect choice for revealing those flows across borders. Beside the sankey there is a geospatial visualisation, ready to be expanded and explored once you've focussed your search down to the relevant municipalities. It shows spatial relationships between biomes, states and municipalities, which is something the sankey (and most other forms of visualisation) cannot do. While the sankey focuses on flows, the map allows for comparison and deep exploration of entities, using a dynamic multivariate choropleth. The two pieces thus work together in a symbiotic way, using the strengths of each to allow people to interrogate different dimensions of the geospatial data.

The interface design closely follows the idea of progressive disclosure—showing users only a few of the most important options (Nielsen 2006)—which has become commonplace in the new paradigm of map design. Trase has a huge and rich dataset: it's easy to get lost in it, or find it hard to work out where to start. By offering relatable items in the Sankey, from which to narrow a search, the geospatial visualisation becomes much more effective as a presentation tool.

Trase in Action

Trase can help people find out a number of things, based on their topic of focus. If, for example you were interested in the role of importers, you can select a particular importer on the sankey diagram: this will reveal the exporters, countries of import and the municipalities where the commodity was produced. The geospatial visualisation also updates automatically, to show impacts for only those municipalities that the selected importer sources from. When you expand the map, you can select a range of indicators to view on it to discover the risks related to that exporter's commodity sourcing. In the Fig. 10.6, you can see maximum soy deforestation and forest code deficit for the Brazilian municipalities that Bungee's soy comes from.

10.3.3 Global Fishing Watch

With Trase, the aim was to increase transparency in land-based supply chains for commodities, mostly grown in tropical areas. A similar problem exists at sea too.

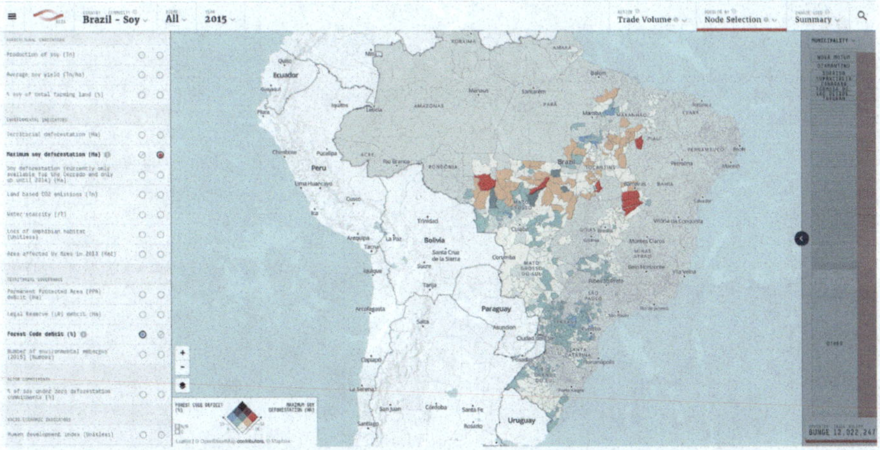

Fig. 10.6 Maximum soy deforestation and forest code deficit for Brazilian municipalities that Bungee sources soy products from

"Hundreds of millions of people depend on the ocean for their livelihoods and many more rely on the ocean for food" (Global Fishing Watch 2017), as well as a host of other benefits from healthy oceans like climate regulation, tourism and biodiversity (Costanza 1999). Pressures from overfishing and habitat destruction are putting that at risk, as summarised in SDG 14 (Life Below Water). In the absence of open, transparent data on commercial fishing fleets, unsustainable or illegal behaviour can continue unchecked.

Global Fishing Watch was launched in September 2016 to bridge this gap. The tool uses an interactive thematic map, as shown in Fig. 10.7, to visualise likely fishing events, as determined by a series of machine learning algorithms applied to data sent from the Automatic Identification System (AIS), a system designed to help large commercial vessels stay safe at sea by broadcasting their location, speed and direction (US Coast Guard 2017). These algorithms process the data to work out if a vessel is fishing or not—assigning it a "yes" score—what kind of vessel it is and what kind of species it's likely to be fishing (Cutlip 2016). The use of machine learning to extract patterns from large unstructured datasets is a key advance that is likely to rise further in prominence as Information Society creates more and more data.

The map makes something which is often unseen—what is happening at sea, away from the eyes of land-dwelling humans—more visible and more tangible. The abstract concept of 'fishing effort' is laid out bare as a series of points and tracks. The whole site makes use of WebGL, a piece of technology not typically used in thematic cartography (Escoffier 2017). This technology was particularly important to achieve a 'pulsing effect', to focus attention on certain places and certain times and allowing people to more easily distinguish the changing distribution and intensity of fishing effort between the seasons.

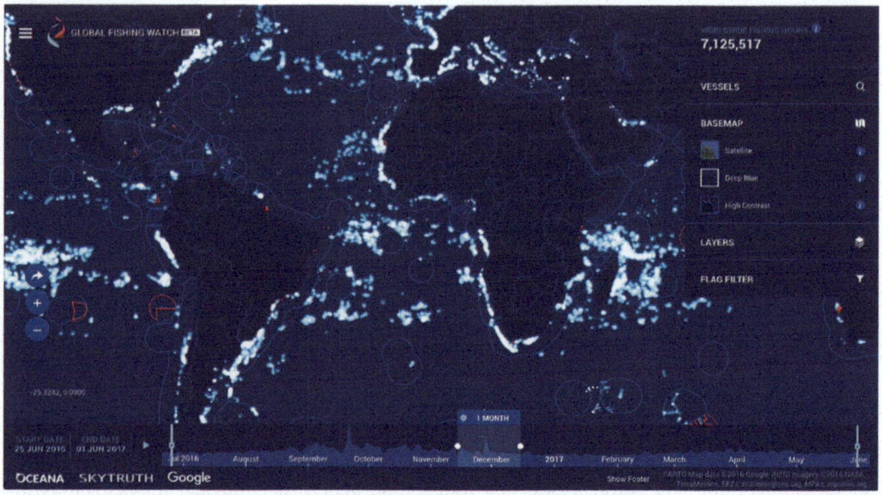

Fig. 10.7 The Global Fishing Watch map. globalfishingwatch.org/map

As with Global Forest Watch, there is an element of shared spaces that allow networks to analyse and discuss the data together. A certain 'state' (a combination of pinned vessels, filters and layers) can be shared as a workspace, so a group of individuals can work from a specific shared understanding of the issue. In this way Global Fishing Watch is a map tool allowing people to build their own bespoke geospatial visualisations.

The development process incorporated a participatory approach, using workshops and user testing to gather feedback and generate ideas for new features. The aim of each interaction was to generate new features to fit within a 'backlog' (Agile Alliance 2017), whether that was bugs to fix or new features to enhance the impact of the platform. Before a new feature was added to the backlog, we went through a series of exercises linking that feature to a persona, relevant job stories and the overall user journey (Government Digital Service 2017), to ensure things we added were useful and contributed to the purpose of the project. At the same time, these opportunities to contribute to the development direction of the tool builds ownership and increased use of the tool amongst potential users.

Global Fishing Watch in Action

Since its public launch the tool has been used by a range of different actors to reduce illegal, unreported and unsustainable fishing. In the case of the Revillagigedo Archipelago Protected Area "scientists used Global Fishing Watch to compare how much the fleet fishes inside the proposed Marine Protected Area with how much they fish throughout the rest of the Eastern Tropical Pacific" (Cutlip 2017). This evidence was compared with the revenues from diving and tourism in this area of rich biodiversity, to argue that the protected area should be expanded (to better protect that resource).

10.4 A View to the Future of Mapping

In these three examples, we've shown how the configuration of Information Society has led to the creation of new data and provided new opportunities to communicate that data to people. We've also outlined how we've tailored the design of geospatial visualisations to communicate data simply and effectively.

The modern paradigm of maps could be used effectively to meet a range of SDGs in a way that was not possible before. Maps have always been unique in their ability to display patterns of behaviour in a visual way. Using a host of modern digital technologies, and drawing from a more emotive design style, modern geospatial visualisations make the visual more visceral and arresting, shaping people's minds and, hopefully, closing the value-action gap which often stops people from acting in a way that's consistent with their beliefs (Kollmuss and Agyeman 2002). In the arena of SDGs—where organisations have often struggled to turn goodwill into impact—this could be extremely helpful.

In the next few years we would hope to see more thematic maps that represent the scale of the social and environmental challenges within the SDGs, as well as highlighting disparities among countries around the world. Considering the scale of ambition of the SDGs (17 Goals and 169 targets) and the amount of data to sort through, we will need to allow as many people as possible to be able to analyse the data. This will need interactive, customisable geospatial visualisation products that allow people to layer different datasets on top of each other and find new insights, as well as features that allow those insights to be shared widely within networks.

More and better geospatial visualisations can also help at a more operational level. Tools targeted at day-to-day decisions—was this fish caught illegally, is the soy produced by this company sustainable, where is the deforestation in this protected area—can help fill gaps in knowledge and lead to faster action to reduce unsustainable activity. This may need work on the data side too—to produce daily or hourly data—as well as design work to surface the most significant events within the data.

But in order to deliver more and better thematic maps there needs to be a strong base of open, accessible, compatible and well-documented databases. Initiatives to address any of these four issues are welcomed: from the basic act of making data openly available online to building APIs which allow for simple integration and exchange of data.

The more participatory approach—including potential users from the beginning of a project, testing with them, and allowing them to set the strategy of a product so they can use it to achieve the greatest impact—is also essential here. These user-centric elements of an agile software method are consistent with the idea of participatory development and, as we've seen in our work, can lead to increased and long-term use of the tools created.

In the next few years, Vizzuality will continue working in this vein, to make purposeful map-based visualisations that help people deliver the SDGs. We think there are a few areas of innovation that are particularly important for the world to focus on right now.

- Build open, publicly accessible APIs. By improving the discovery, exchange, querying and downloading of data, map makers can ensure people have the data they need at the right time, to deliver sustainable outcomes.
- Take that data to the place where the audience naturally lives. This goes beyond making sure content is embeddable and shareable on social media, to optimising data design and the content strategies behind that for the context of a viewer on Twitter, Facebook or whatever platform they use.
- Meet more users and continue discovering their contexts, ensuring each tool is useful, usable, and will successfully create and support a network of concerned actors.

References

Agile Alliance (2017) Backlog https://www.agilealliance.org/glossary/backlog/#q= ~ (filters ~ (postType ~ (~ 'page ~ 'post ~ 'aa_book ~ 'aa_event_session ~ 'aa_experience_report ~ 'aa_glossary ~ 'aa_research_paper ~ 'aa_video) ~ tags ~ (~ 'backlog)) ~ searchTerm ~ ' ~ sort ~ false ~ sortDirection ~ 'asc ~ page ~ 1). Accessed 12 June 2017

Bartiaux F (2008) Does environmental information overcome practice compartmentalisation and change consumers' behaviours? J Cleaner Prod 16. doi:1170e1180

Beck K, Beedle M, van Bennekum A, Cockburn A, Cunningham W, Fowler M, Grenning J, Highsmith J, Hunt A, Jeffires R, Kern J, Marick B, Martin RC, Mellor S, Schwaber K, Sutherland J, Thomas D (2001) Manifesto for agile software development. http://agilemanifesto.org/. Accessed 8 June 2017

Bradley S (2010) Form follows function? https://www.smashingmagazine.com/2010/03/does-form-follow-function/. Accessed 23 June 2017

Carswell CM (1992) Choosing specifiers: an evaluation of the basic tasks model of graphical perception. Hum Factors 34(5):535–554. https://doi.org/10.1177/001872089203400503

Castells M (1996) The rise of the network society: the information age: economy, society, and culture (vol 1, 2nd edn 2009). Blackwell Publishing, Oxford

Chambers R (1997) Whose reality counts: putting the first last. Intermediate Technology Publications, London

Costanza R (1999) The ecological, economic and social importance of the oceans. Ecol Econ 31 (2):199–213. https://doi.org/10.1016/S0921-8009(99)00079-8

Cutlip K (2016) Teaching machines to tell us about fishing. http://blog.globalfishingwatch.org/2016/09/teaching-machines-to-tell-us-about-fishing/. Accessed 26 June 2017

Cutlip K (2017) Where tourism and tuna overlap: how global fishing watch is informing policy discussions. http://globalfishingwatch.org/2017/04/where-tourism-and-tuna-overlap-how-global-fishing-watch-is-informing-policy-discussions.html. Accessed 23 June 2017

Ericsson AB (2017) Ericsson mobility report. https://www.ericsson.com/en/mobility-report/future-of-mobile-technology-usage-by-region. Accessed 20 June 2017

Escoffier E (2017) Saving the fish with rabbits: how we used WebGL and Pixi.js for temporal mapping. http://blog.globalfishingwatch.org/2016/09/teaching-machines-to-tell-us-about-fishing/. Accessed 23 June 2017

Finer M, Novoa S (2016) United cacao continues deforestation of primary forest in Tamshiyacu (Loreto, Peru). MAAP: #27. http://blog.globalforestwatch.org/gfw-community/partner-posts/

guest-post-maap-27-united-cacao-continues-deforestation-of-primary-forest-in-tamshiyacu-loreto-peru.html. Accessed 22 June 2017

Fitts PM (June 1954) The information capacity of the human motor system in controlling the amplitude of movement. J Exp Psychol 47(6):381–391. PMID 13174710. https://doi.org/10.1037/h0055392

Global Fishing Watch (2017) About the project. http://globalfishingwatch.org/the-project. Accessed 23 June 2017

Godar J, Suavet C, Gardner TA, Dawkins E, Meyfroidt P (2016) Balancing detail and scale in assessing transparency to improve the governance of agricultural commodity supply chains. Environ Res Lett 11(3). https://doi.org/10.1088/1748-9326/11/3/035015

Gov.UK (2017) User research. https://www.gov.uk/service-manual/user-research. Accessed 12 June 2017

Haklay M, Nivala AM (2010) User-centred design. In: Interacting with geospatial technologies, Wiley, Chichester. https://doi.org/10.1002/9780470689813.ch5

Hansen MC, Potapov PV, Moore R, Hancher M, Turubanova SA, Tyukavina A, Thau D, Stehman SV, Goetz SJ, Loveland TR, Kommareddy A, Egorov A, Chini L, Justice CO, Townshend JR. 2013. High-resolution global maps of 21st-century forest cover change. Science 342(15):850–53. https://doi.org/10.1126/science.1244693

Henkel LA (2013) Point-and-shoot memories: the influence of taking photos on memory for a museum tour. Psychol Sci 25(2):396–402. https://doi.org/10.1177/0956797613504438

Hill D (2016) Will London stock exchange bar firm over Amazon deforestation? https://www.theguardian.com/environment/andes-to-the-amazon/2016/may/25/london-stock-exchange-amazon-deforestation. Accessed 27 June 2017

Hulme M (2008) Geographical Work at the boundaries of climate change. Trans Inst Br Geogr 33 (2–3). https://doi.org/10.1111/j.1475-5661.2007.00289.x

Independent Advisory Group on a Data Revolution for Sustainable Development (2014) A world that counts: mobilising the data revolution for sustainable development. http://www.undatarevolution.org/wp-content/uploads/2014/12/A-World-That-Counts2.pdf. Accessed 12 Sept 2015

Jasanoff S (2010) A new climate for society. Theory, Cult Soc 27(233). https://doi.org/10.1177/0263276409361497

Jonsson AKE, Nilsson A (2014) Exploring the relationship between values and pro-environmental behaviour: the influence of locus of control. Environ Values 23(3). https://doi.org/10.3197/096327114x13947900181752

Kissinger G, Herold M, De Sy V (2012) Drivers of deforestation and forest degradation: a synthesis report for redd + policymakers. Lexeme Consulting, Vancouver, Canada

Krätzig S, Warren-Kretzschmar B (2014) Using interactive web tools in environmental planning to improve communication about sustainable development. Sustainability 6. https://doi.org/10.3390/su6010236

Kollmuss A, Agyeman J (2002) Mind the gap: why do people act environmentally and what are the barriers to pro-environmental behavior? Environ Educ Res 8(3):239–260. https://doi.org/10.1080/13504620220145401

Levitin DJ (2015) Why the modern world is bad for your brain. http://www.theguardian.com/science/2015/jan/18/modern-world-bad-for-brain-daniel-j-levitin-organized-mind-information-overload. Accessed 19 June 2017

Marcotte E (2010) Responsive web design. http://alistapart.com/article/responsive-web-design. Accessed 28 June 2017

Meeker M (2017) Internet trends 2017—code conference. https://www.ofcom.org.uk/__data/assets/pdf_file/0020/102755/adults-media-use-attitudes-2017.pdf. Accessed 23 June 2017

Nielsen J (2006) Progressive disclosure. http://www.nngroup.com/articles/progressive-disclosure/. Accessed 13 June 2017

OFCOM (2017) Adults' media use and attitudes. https://www.ofcom.org.uk/__data/assets/pdf_file/0020/102755/adults-media-use-attitudes-2017.pdf. Accessed 26 June 2017

Paroutis S, Al Saleh A (2009) Determinants of knowledge sharing using Web 2.0 technologies. J Knowl Manag 13(4). https://doi.org/10.1108/13673270910971824

Payne O, Mann SA (2015) Zooming. In: "Sustainable" cocoa producer destroys pristine forest in Peru. http://www.wri.org/blog/2015/06/zooming-%E2%80%9Csustainable%E2%80%9D-cocoa-producer-destroys-pristine-forest-peru. Accessed 23 Aug 2017

Peterson MP (2007). Cartography and the internet: implications for modern cartography. University of Nebraska-Omaha. http://maps.unomaha.edu/NACIS/paper.html. Accessed 8 Oct 2015

Pew Research Centre (2015) Cell phones in Africa: communication lifeline. http://www.pewglobal.org/2015/04/15/cell-phones-in-africa-communication-lifeline/. Accessed 20 June 2017

Preece J, Rogers Y, Sharp H (2002) Interaction design: beyond human-computer interaction. Wiley, New York, p 21

Sarrocco C (n.d.) Elements and principles of the information society. http://www.itu.int/osg/spuold/wsis-themes/Access/BackgroundPaper/IS%20Principles.pdf. Accessed 22 June 2017

Schnürer R, Sieber R, Çöltekin A (2014) The next generation of atlas user interfaces—a user study with "Digital Natives." In: Modern trends in cartography—Lecture Notes in geoinformation and cartography. https://doi.org/10.1007/978-3-319-07926-4_3

Sheppard SRJ, Shaw A, Flanders D, Burch S, Wiek A, Carmichael J, Robinson J, Cohen S (2011) Future visioning of local climate change: a framework for community engagement and planning with scenarios and visualisation. Futures 43. https://doi.org/10.1016/j.futures.2011.01.009

Smith D (2014) Poetry in motion: mobile site brings new audience to African writers. http://www.theguardian.com/world/2014/dec/29/africa-poetry-badilisha-mobile-phone-site. Accessed 3 June 2017

Thaler R, Sunstein CR (2008) Nudge: improving decisions about health, wealth, and happiness. Yale University Press, New Haven, USA

Tran MC, Labrique AB, Mehra S, Ali H, Shaikh S, Mitra M, Christian P, West K (2015) Analyzing the mobile "Digital Divide": changing determinants of household phone ownership over time in Rural Bangladesh. JMIR mHealth uHealth 3(1) https://doi.org/10.2196/mhealth.3663

Trase (2017) Trase flows. http://trase.earth/about.html. Accessed 27 June 2017

United Nations General Assembly (2015) Transforming our world: the 2030 agenda for sustainable development. Resolution A/RES/70/1

US Coast Guard (2017) Automatic identification system overview. https://www.navcen.uscg.gov/?pageName=AISmain. Accessed 14 June 2017

Van de Velde L, Verbeke W, Popp M, Van Huylenbroeck G (2010) The importance of message framing for providing information about sustainability and environmental aspects of energy. Energy Policy 38. https://doi.org/10.1016/j.enpol.2010.04.053

Van Dijk J (2006) The network society, 2nd edn. Sage, London

Wall M (2014) Big data: are you ready for blast-off? http://www.bbc.co.uk/news/business-26383058. Accessed 26 June 2017

Chapter 11
Geospatial Data Mining and Analytics for Real-Estate Applications

Gerhard Muggenhuber

Abstract Market information on housing is relevant for good decision making of households as well as for real estate economics and is also of systemic interest. The official statistics often provide only annual indices on national level based on transactions of previous year. The stakeholders of the real estate markets however request analysis of higher spatial resolution on local level based recent transactions. This paper focuses on methodology of automated data acquisition, analysis and visualization of data from the housing market. Data retrieved from observing real estate markets (to rent/to buy) are used for statistical modelling of value-descriptive parameters, for estimating and forecasting real estate market fundamentals, which also can reveal market risks. This paper elaborates methods of automated data acquisition based on web mining, reviews methods of econometric spatial modelling with impact from proximity in order to deduct efficiency parameters within different categories. The analysis and the resulting visualization at different granularity of spatial, temporal and typological effects provides relevant information for decision making.

Keywords Web mining · Real estate market · Econometric analysis
Spatial analysis · Visualization

11.1 Introduction

This paper focuses on gathering, analyzing and visualizing of data for the real estate sector in order to reduce uncertainties and to estimate risks in the phase of decision making. Information about the probability distribution of potential outcomes contributes to *risk mitigation* along with alternative approaches of transferring or diversifying risks as offered by insurances and by *real estate investment trusts* (REITs).

G. Muggenhuber (✉)
Federal Office of Metrology and Surveying, Vienna, Austria
e-mail: gerhard.muggenhuber@bev.gv.at

J. Döllner et al. (eds.), *Service-Oriented Mapping*, Lecture Notes in Geoinformation and Cartography, https://doi.org/10.1007/978-3-319-72434-8_11

Gathering, analyzing and visualizing data are steps within the *knowledge discovery process*, which consists of the components: data—infrastructure—analysis and discovery. It requires a process chain from acquisition, cleansing, segmentation, geocoding of data to econometric analysis of data for visualization of spatial, temporal and typological patterns with the goal to serve as decision support—see also data mining process models such as KDD,[1] CRISP-DM[2] and SEMMA.[3]

Related work and state of the art regarding the process chain of data mining according Fig. 11.1 is presented in Sect. 11.2, detailed steps and approaches such as for mass valuation will be discussed within Sects. 11.3–11.5.

11.2 Related Work

Data about the real estate transfer market including the house prices are of high relevance for systemic forecasting as well as economic decision making. In the optimal case data are up-to-date, of proper spatial resolution and directly accessible from partners involved in the transfer processes. Data can be gathered at different stages of the transfer process—see Shimizu et al. (2011) and Eurostat (2013). The private sector, governmental agencies (mass appraisal), statistical offices (Dechent and Ritzheim 2012), national banks (Schneider 2013) as well as EU (Eurostat 2013) are contributing with some datasets and analysis. Data acquisition via web mining is an alternative approach of market observation of data that is not yet available.

Within the time sequence of a property transfer there are several opportunities to observe value related aspects such as a*sk price—value for loan—transfer price—normative values for registration and taxation*. Thus the earliest opportunity to collect data is to collect *ask prices* of potential sellers. The resulting *thematic, spatial and temporal data* are of systemic relevance. Comparing data from different stages of the process chain provide relevant information for cost/benefit analysis as well as risk estimation. This paper focuses on the "*buy- to rent*" *ratio* as indicated in Fig. 11.2.

11.2.1 The ETL-Process

An ETL[4]-process chain is often based on multiple sources and can also include highly automated web mining processes. These data have to be cleaned, transformed, loaded and geocoded before econometric models can be applied on data

[1]KDD: Knowledge Discovery Databases process model (Fayyad et al. 1996: 41).

[2]CRISP-DM: CRoss Industry Standard Process for Data Mining.

[3]SEMMA: Sample, Explore, Modify, Model, Assess.

[4]ETL: Extract from data source—Transform into proper format—Load into target database.

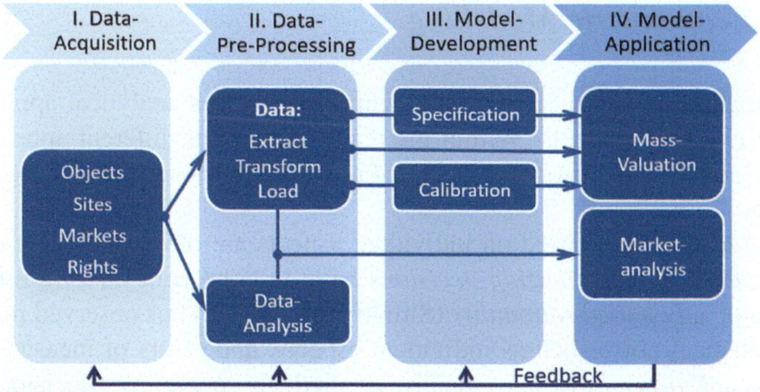

Fig. 11.1 Process chain of data mining for real estate market analysis and mass appraisal

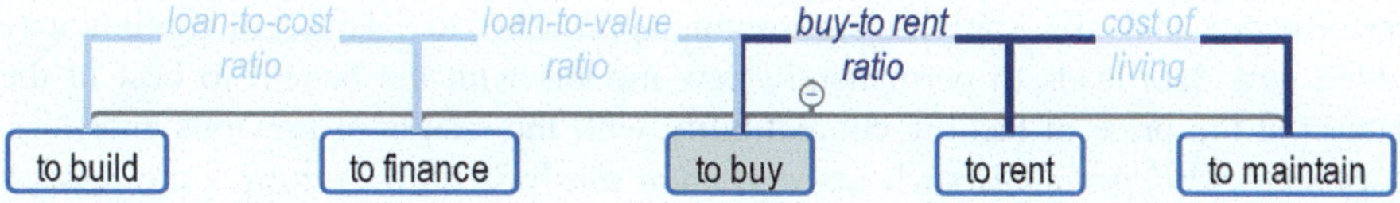

Fig. 11.2 Cost/benefit/risk analysis within the process chain: to build—to buy—to rent

and information can be visualized. The cleaning process includes preprocessing of textual data by tokenizing based on steaming and stop words. An alternative approach of transforming data with the purpose to integrate all sources into one single database is to keep the data in its natural form and define rule based extractions on the fly (Wood 2017).

11.2.2 The Need for Data Analysis

Residential property market parameters are of interest for households, investors and service providers as well as of interest from systemic perspective. Schneider (2013: 32–37) defines indicators to measure market fundamentals and clusters these indicators into three perspectives—see also Ambrose et al. (2013) and Campbell et al. (2009). National banks and statistical agencies regularly publish market indicators and indices on national and regional level—see the example of UK (Chandler and Disney 2014). Decisions are made under risks, uncertainties and ambiguity (Knight 1921). Decisions are still made under risk even when all information is available and the probability of outcomes is accurately ascertained. In practice however there is limited information about the situation as well as about the outcomes (Grossman and Stiglitz 1980). Especially data about situations at decision making level with regards to spatial, typological and temporal trends on local level are missing (Muggenhuber 2016: 41).

11.2.3 Econometric Modelling

Decision making is based on inference with more or less analytical approaches of estimating expected benefits. Within the real estate sector different approaches are applied (Kauko and D'Amato 2008). Figure 11.3. categorizes these different approaches.

Intuitive approaches based on individual aspects are often applied in daily life. Even *rational utility-maximizing decisions* have to cope with missing information which results in *bounded rationality* (Simon 1947: 179). Thus observed preferences include statistical noise such as individual aspects, and errors of measurements.

Rule based procedures are mainly applied for economic decision making regarding individual real estate objects with reference to market values.

Algorithmic approaches such as neural network or deep learning (Fig. 11.3) are applied for data driven modelling and forecasting such as for systemic market performance analysis. Applying algorithmic approaches to a dataset of continuously added new data leads to permanently new models with the benefit to best fit the dataset at the price of lacking compatibility with the results of previous analysis.

Stochastic approaches such as *regression analysis* or *geostatistics* are applied for modelling the economic value of real estate objects based on revealed preferences. The stochastic approach is preferable, when the same model should be applied even when new data are added to the dataset to be analyzed. The complexity of the model depends on the purpose (Bivand et al. 2017). Applications, like mass valuation for taxation purposes, often apply linear regression models with spatial classifications, which minimize deviations based on least square. The simple interpretation of these models is at the cost of artificial break lines and the need to revise class boundaries from time to time. Semi-parametric regression models like the *structured additive regression models*, as applied by Brunauer et al. (2013), optimize the nonlinear interactions from the data based on splines. *Hierarchical trend models* as applied by Francke (2010) allow to model even more complex space-time correlations.

11.2.4 Visualization of Results

The results of thematic, spatial, or temporal analysis are usually visualized in the classical static form on paper. Java-driven applications in web-browsers allow dynamic presentation of results with interactive user interfaces. This approach empowers the user to have a dynamic view on data and on variations of the model parameters. Such an interaction between users and data is based on web browsers. The user can choose his views on data based on his interest and knowledge based on parameters to choose without technical support. Parkkinen (2014) provides an overview of alternative visualizations in R and Ruefer (2016) visualizes real estate data for decision making based on R-shiny-applications.

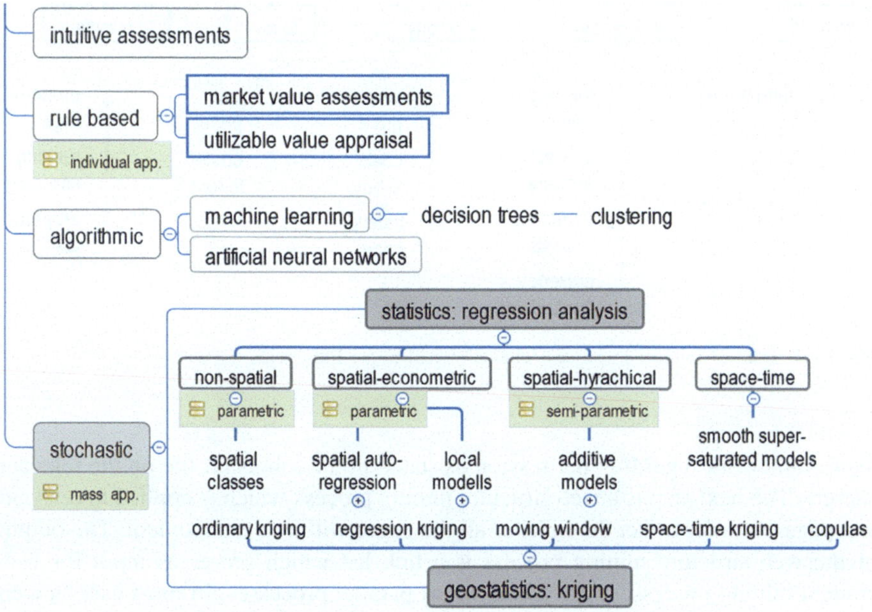

Fig. 11.3 Methodologies for assessment of value and modelling of real estate market

11.3 Data Acquisition

Within the real estate sector the data of interest from the process chains are: to build, to finance, to buy, to rent and to maintain (see Fig. 11.4). The opportunities of institutional cooperation for data capturing during the process chains of property transfer are by far not taken for granted and aspects like maintenance costs are still lacking standardization. Under optimal circumstances process-integrated data gathering is implemented, as recently introduced in Slovenia. Often alternative methods for data capturing have to be applied. The *ask prices* offers the first opportunity within the transfer process (see Fig. 11.4), which is of special interest in this paper. This approach provides data about the transfer markets '*to sell*' as well as '*to let*'. Especially rental market data are seldom available. They are essential for estimating the *buy-to-rent ratio* which is one component of the return on investment (RoI). RoI consists of a bond-like component, "the rental income", and an equity-like component, namely "fluctuation of value over time".

11.3.1 Web Mining, Cleaning, Storing

Web mining as part of a highly automated information retrieval is applied within several steps. The first step is to identify potential webpages and clarifying their

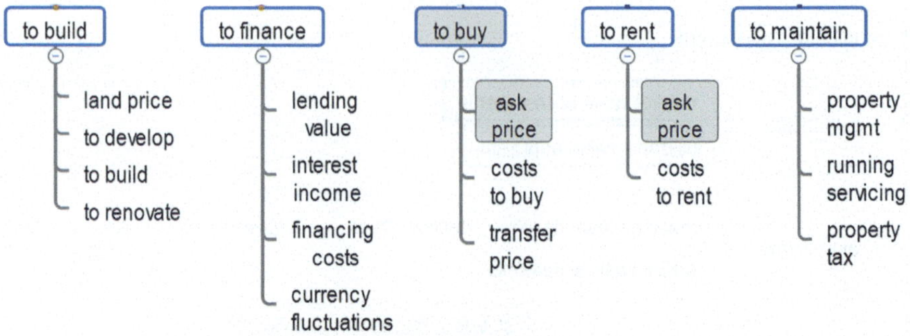

Fig. 11.4 Data of interest in process chains of real estate life cycles

legal framework regarding permission as stated in the robots.txt file on the relevant servers. The next step is a web structure mining process which is applied to relevant webpages in order to get the links of all the pages with relevant content. The output of the web structure mining process is a link list which serves as input for web content mining processes which can run as parallel processes. In most cases a web content mining process has to handle data in nested JavaScript Object Notation (JSON). The final outcome are records of semi-structured information as text and pictures which are stored in a database. There is a wide range of services available, ranging from commercial software tools to freeware for web mining on the market.[5] Data can be retrieved from the web with R-packages such as *rvest* for static webpages and *Rselenium* for dynamically generated data with *Javascript* and *Ajax*. R-packages such as *DOM* and *jsonlite* support interaction with web data on object level of *DOM* (*Document Object Model*) *and JSON*.

Preparing the data for analysis includes cleaning and scrubbing of data by removing noise and unrelated data in the source data. It also includes the processing of repeating data and imputation of missing variables. Additional steps are geocoding of source data and linking of data by converting information from different sources into a unified storage (Zhang and Yin 2008).

11.3.2 Cleaning and Normalization of Data

The semi-structured descriptions of real estate objects in tables as well as in free text formats has to be broken up by tokenizing based on content related stop words and steaming (see R-packages '*tokenizers*', '*stringr*' and Angeli et al. 2015). The result is a fully structured and dataset for each object offered for transfer.

[5]http://www.kdnuggets.com/software/web-content-mining.html.

11.3.3 Geocoding and Imputation

Geocoded address data are provided by governmental mapping agencies such as BEV in Austria (www.bev.gv.at) and by private sector service providers. Geocoding is also supported by R-packages such as *ggmap*. Using the Google maps API limits to 2500 queries a day and has also other limitations.

Imputation of missing variables is a challenging process since the imputed data should not interfere with the results of a following analysis of variance (*ANOVA*). The R-package *VIM*, *offers* tools for imputation by regression and diagnostics of variables to be processed. More complex R-packages such as *missForest* and *Hmisc* can even outperform *VIM*. As a result datasets of imputed data show the same variance as the datasets with fully observed variables.

11.4 Real Estate Market Parameters

The information retrieved from past events at the transfer market provides efficiency parameters for impending decision making. These parameters can be categorized into efficiency of transaction, efficiency of location and economic efficiency and serve as input for modelling the real estate market.

11.4.1 Efficiency of Transaction

Efficiency of transaction with main characteristics such as *density*, *bandwidth* and *spread* of market activities, but also *time on market* (*ToM*) and the negotiation margin (the gap) can be calculated by observing the prices at the different stages of the transaction. The final price is expected to be lower than the asking price, but can also be higher in certain circumstances of competition. Significant spatial typological and temporal variations of ToM are of specific interest. The temporal variations of ToM provide an early indicator for turning points in market trends (Anglin et al. 2003). Thus essential parameters for efficiency of transactions are:

- *Density, bandwidth and spread* of market activities.
- *Time-on-Market* (*ToM*) of properties as well as of actors is of interest for market analysis.
- *Negotiation margin is the* gap between the initial ask prices and final transfer prices. It can be calculated for objects to buy as well as to rent.
- *Costs and time to process a transfer* are benchmarked on national level by the annual "*Doing Business*" reports of the World Bank.

11.4.2 Efficiency of Location—Modelling Proximity

Efficiency of location can be measured by spatial typological or temporal *proximity* to (dis-)comfort based on deviations of observed user preferences. *Proximity* can be measured by parameters of distance and similarity. Parameters like spatial distance, density of data, and by time or cost of travel (Fig. 11.5) are of interest for measuring accessibility and availability of infrastructure to work, to shop and to recreate. Practically the following parameter are of interest for modelling:

a. *Distance to (dis-)comfort*: view, noise, air quality, risks.
b. *Availability of/accessibility to* work, education, infrastructure.
c. *Spatial deviation* of temporal price variations.

The potentials of a location can be expressed by absolute or relative proximity measures, which can be compared to other locations and evaluated by revealed user preferences (Palmquist 2005; Ekeland and Galichon 2012).

Location premiums for housing objects are well known at different scales in terms of regions, city or township, suburbs, and even within a building. The potential of a location for specific topics like the potential access to the labor market is an example for an absolute proximity measure and answers: "How many workplaces are accessible within a travel time of one hour?"

11.4.3 Economic Efficiency

Economic efficiency can be measured by parameters such as:

- *Investment efficiency* is measured by *loan to cost-*, *loan to value-* and *buy to rent ratio*. In this paper only the last measure will be discussed for estimating the RoI.
- *Energy efficiency*[6] shows some correlation with the market value within the ownership sector, but much less for the rental sector and thus reveals the "*landlord-tenant dilemma*" (Kholodilin et al. 2016).
- *Area efficiency* expressed as $\sum$floor space/area of plot is of special interest for office buildings and shopping malls.

[6]EU-Directives 2002/91/EG and 2010/31/EU.

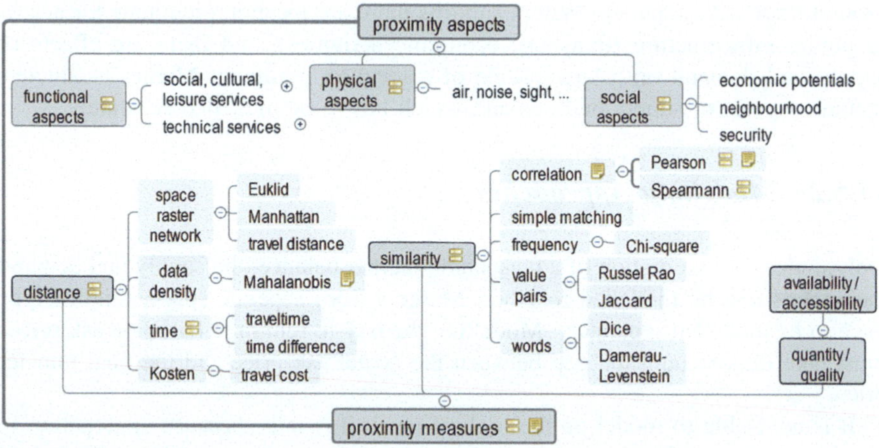

Fig. 11.5 Variables and measures for proximity

11.4.4 Information Efficiency—Modelling Transparency

Information efficiency is the degree to which market prices reflect the real values and serve an indicator for the market efficiency[7] and *market transparency* of real estate markets.

11.5 Analysis, Modelling and Visualization of Data

The purpose of visualization is insight, not pictures (Bostock 2017).

The results of data analysis depends on sound data acquisition processes at defined quality, parameters and models applied and granularity aspects.

11.5.1 Parameters for Modelling of Economic Value

Modelling the economic value of housing objects requires parameters about objects, locations, markets and legal facts.

Object parameters describe quantity, quality, usability and economic efficiency of an object. Location parameters describe (dis-)amenities, which can be observed at different scales. Within the building there are parameters such as the top floor or

[7]Benchmarks such as *Global Real Estate Transparency Index* and *Global Real Estate Bubble Index*. Information provided by stakeholders of the transfer market might be unbiased.

ground floor with a garden. Within suburbs there are location premiums for access to public infrastructure (transport, economic activities), and there are effects of supply and demand on a larger scale of a location premium. Market parameters include aspects of supply and demand which has to be modelled separately.

11.5.2 Modelling Approaches

Only under the assumption of a reasonable market balance, the supply and demand parameters can be modelled together. Markets from separate space, time or type have to be modelled separately. Modelling the buy-to rent ratio based on ask prices must take into account the gap between the actual ask price and the final transfer price.

It is advisable to model on the highest level of details, because aggregation in any dimension results in a stronger correlation between the variables (Goodchild 2011: 8) and would lead to models which cannot be compared. An inference of the economic value based on a few parameters has some advantages because of simplicity and interpretability and also in order to avoid the 'cure of dimensionality'.

11.5.3 Spatial, Temporal and Typological Effects

Sound decision making requires information with different granularity. On the time scale it can unveil short term and long term trends. Spatial effects like changing prices/m^2 are of interest on regional as well as local level. Also the typological dimension give insights when typological characteristics are viewed based on different aggregations. In the optimal environment the user can interactively select parameters such as spatial resolution, typological subunits and timespan of interest.

11.5.4 Spatial, Temporal and Typological Aggregation

Figure 11.6 shows such typological and spatial comparison for prices/m^2 in Euro for objects '*to let*' with objects '*to sell*' for new apartments (dashed blue line), apartments (blue), family houses (green) and rooftop apartments (red). The spatial comparisons have been applied with different spatial resolutions.

11.5.5 From Static to Interactive Visualization

Interactive visualization on the web combines the traditionally separated processes of visualization and communication.

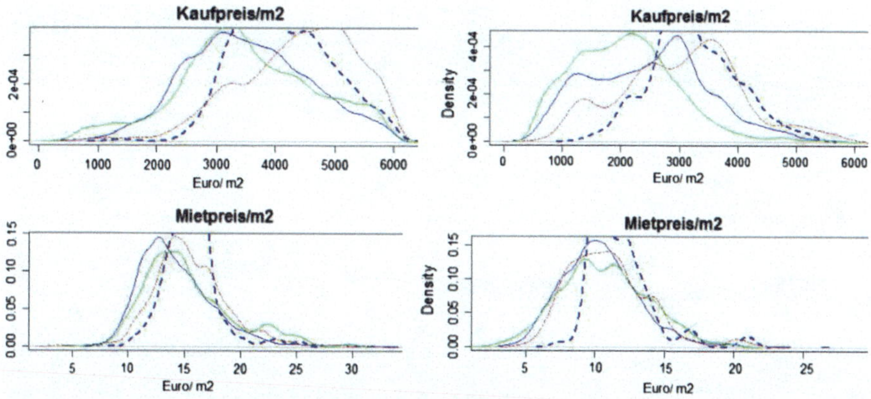

Fig. 11.6 Ask prices to let (bottom)/to sell (top) in Vienna (left)/Lower Austria (right)

Insights from static visualization are limited to the questions and background knowledge of the analyst preparing the topics in paper form. A professional data analyzer interprets the patterns differently from the experienced user in the phase of decision making. Thus static visualization limits the outcomes to the viewpoint of the data analyst.

The interactive visualization however is a combination of the skills of a data analyst and the experience of the user, who decides about the viewpoints on data based on sliders and buttons provided by the analyst—see examples like Parkkinen (2014), Ruefer (2016) and also Haseeb (2017) based on data from Zillow Research.

11.6 Case Study

11.6.1 Data Mining

Geospatial web mining and data analytics has been applied on the housing market of the region Vienna and Lower Austria. Within the framework of a Ph.D.-project at the TU-Vienna for a test period of 18 month about 90.000 observations of ask prices have been scraped from websites such as www.willhaben.at/iad/immobilien. The software tools applied included those R-packages mentioned before. In addition also *easy web extract* (http://webextract.net), *QGIS* (https://www.qgis.org) and *PostgreSQL* (https://www.postgresql.org) has been applied.

After extracting the web-extracting, geocoding and categorizing the data from the web the spatial and typological distribution of harvested data can be visualized (Fig. 11.7). The main typological categories are family house, apartments and rooftop-apartments. The year of construction allows subdivision of old and new objects.

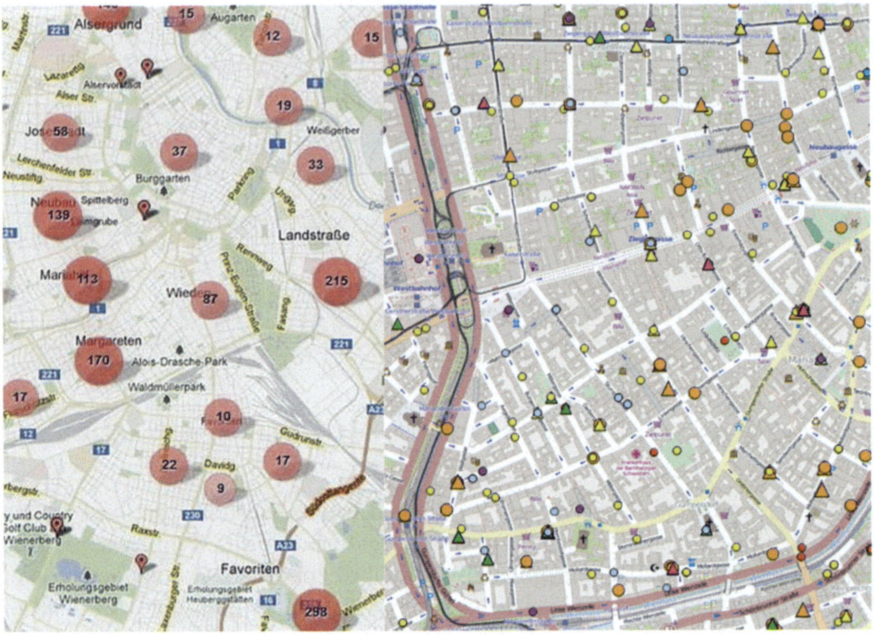

Fig. 11.7 Spatial distribution of ask prices in thousands of Euro based on web scraping

11.6.2 Validity of Sample Versus Population of Housing Objects

In order to prove validity of statistical findings these observed samples transferred housing units has to be compared with the population of all housing units. This comparison of sample with population has been applied for the main categories on regional level in order to prove significance and spatial variations of observed data. The temporal distribution with potential seasonal variations of the dataset has NOT been checked. Instead, an equal distribution within the time period of 18 month has been assumed.

Figure 11.8 shows the comparison of the population of all apartments in Vienna districts (1–23) with the observed samples. These deviations of the transfer rates between districts can be explained on the once side by apartments of social housing, which are not transferred on the open market (e.g. peak for the 1oth district). On the other hand apartments in noble districts seems to have a higher transfer rate.

Based on the assumption of a valid sample, the main typological categories of apartments (old/new/rooftop/penthouse) and 1–2 family houses can be analyzed and compared at different spatial resolutions, i.e. county, regional and local level.

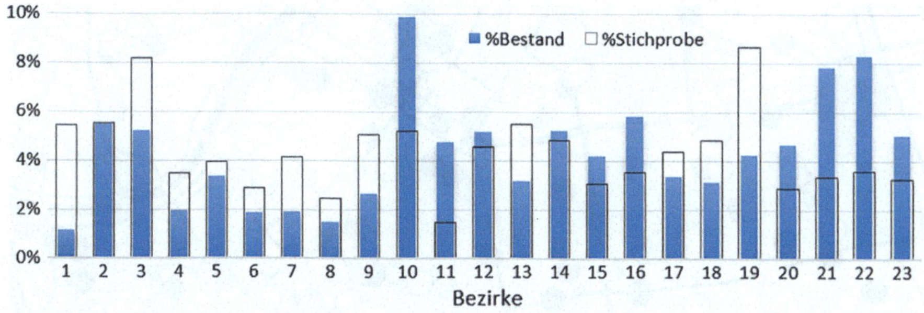

Fig. 11.8 Sample versus population of apartments in Vienna districts

11.6.3 Bi-Variant Data Analysis

The correlations between parameters can be viewed in a correlation matrix as digits or as point clouds (Fig. 11.9). The diagonal diagrams show the distribution of observations for the selected parameter.

Multidimensional data analysis comparing data from Vienna with data from Lower Austria proved significant variations (Fig. 11.6): Living space is more expensive in Vienna than in Lower Austria. The typological comparisons prevail that living at the rooftop is the most expensive compared with other categories of housing.

The bandwidths of objects in dependence with their characteristics unveil validity of potential inference. There are for example not enough observations of objects with less than 30 m^2.

These analysis may also unveil spatial and typological deviations of costs for living space. One aspect of all these analysis is the *RoI* as a figure of special economic interest.

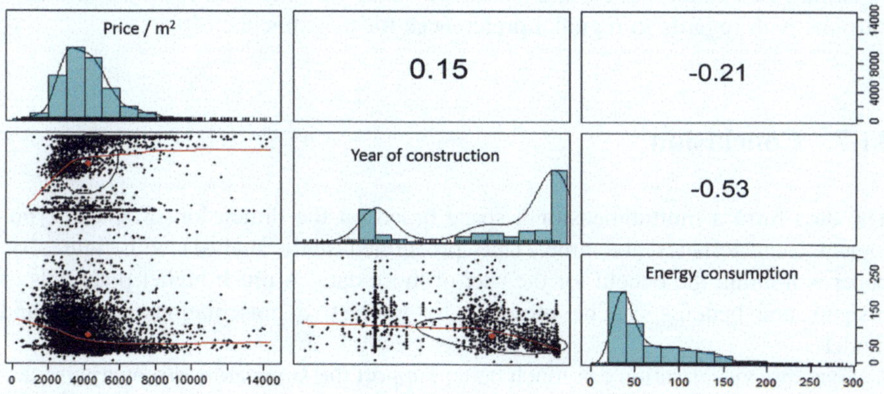

Fig. 11.9 Extract from a correlation matrix

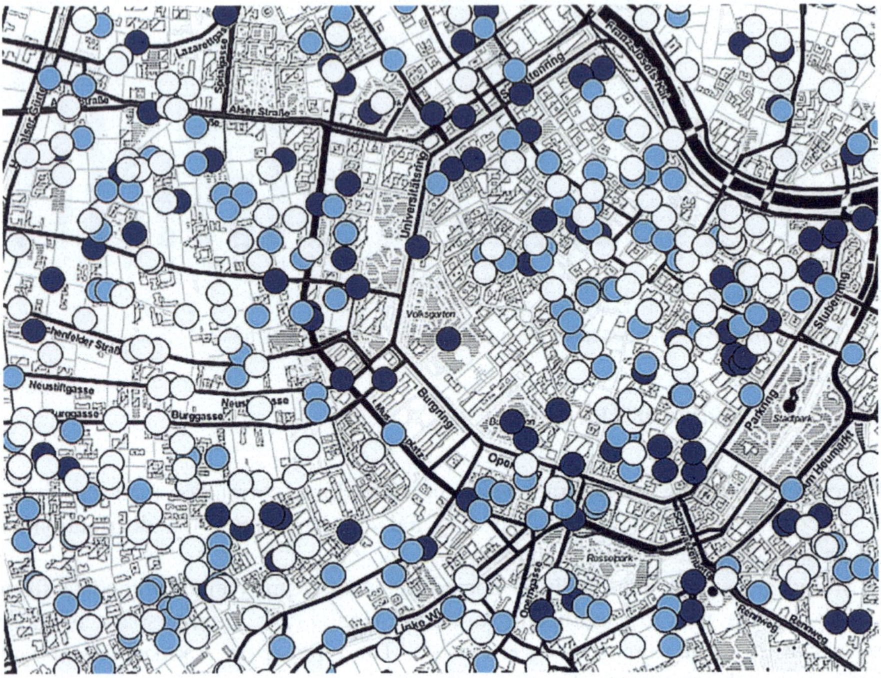

Fig. 11.10 Spatial distribution of 'buy-to-rent ratio'

For each observation we have got spatial, temporal and typological parameters such as location, time stamp and characteristics of the object observed. With the tools of spatial regression analysis, the prices *to buy* and *to let* can be estimated for each location.

Figure 11.10 shows the spatial variation of the *buy-to-rent* ratio for apartments in the center of Vienna based three categories [low (grey)—average (blue)—supreme (dark blue)]. The pattern however does not allow to make simple conclusions with regards to regional preferences for investments.

11.7 Conclusion

The data form a multidimensional space based on the dimension of space, typological characteristics and time. Thus predefined static analysis with figures on paper is limiting the benefit for the user of those data. A much higher potentials of insights and benefits can be provided by interactive presentations of data and models.

Dynamic visualization can much better support the *knowledge discovery process* within the real estate sector because it enables user of data to combine their business knowledge with the potential views on data and statistical inference based on the toolbox provided by data analyst.

Market observation based on web mining is an alternative tool for data acquisition and can unveil data on rental values, which are usually not public available on the market.

References

Ambrose BW, Eichholtz P, Lindenthal T (2013) House prices and fundamentals: 355 years of evidence. J Money Credit Banking 45:2–3, 477–491. http://nilskok.typepad.com/EPRI/AEL_JMCB_2011.pdf

Angeli G, Johnson Premkumar M, Manning CD (2015) Leveraging linguistic structure for open domain information extraction. In: Proceedings of the association of computational linguistics (ACL)

Anglin PM, Rutherford RC, Springer TM (2003) The trade-off between the selling price of residential properties and time on-the-market: the impact of price setting. J Real Estate Finan Econ 26(1):95–111

Bivand R, Sha Z, Osland L, Thorsen IS (2017) A comparison of estimation methods for multilevel models of spatially structured data. Spat Stat21:440–259

Bostock M (2017) A better way to code: introducing d3.express: the integrated discovery environment. https://medium.com/@mbostock/a-better-way-to-code-2b1d2876a3a0

Brunauer WA, Lang S, Feilmayr W (2013) Hybrid multilevel STAR models for hedonic house prices. Jahrbuch für Regionalwissenschaft 33(2):151–172

Campbell SD, Davis MA, Gallina J, Martina RF (2009) What moves housing markets: a variance decomposition of the rent-price ratio. J Urban Econ 66(2):90–102

Chandler D, Disney R (2014) Measuring house prices: a comparison of different indices. IFS briefing note, BN146. http://www.ifs.org.uk/bns/bn146.pdf

Dechent J, Ritzheim S (2012) Preisindizes für Wohnimmobilien: Einführung eines Online-Erhebungsverfahrens. Wirtschaft und Statistik, Okt 2012:891–897

Ekeland I, Galichon A (2012) The housing problem and revealed preference theory: duality and an application. Econ Theor 54(3):425–441. https://doi.org/10.1007/s00199-012-0719-x

Eurostat (2013) Handbook on residential property prices indices (RPPIs). Methodologies & working papers 2013. Publications Office of the EU, Luxembourg

Fayyad U, Piatetsky-Shapiro G, Smyth P (1996) From data mining to knowledge discovery. AI Magazine, 3. www.aaai.org/ojs/index.php/aimagazine/article/view/1230

Francke MK (2010) A state-space model for residential real estate valuation. AENORM 18:23–26. http://www.aenorm.nl/files/nlaenorm2012/file/article_pdfs/4abfm_A%20State-Space%20model%20for%20Residential%20Real%20Estate%20Valuation.pdf

Goodchild MF (2011) Scale in GIS: an overview. Geomorphology 130:1–2, 5–9. http://www.colorado.edu/geography/leyk/GIS1/Readings/Goodchild_2011.pdf

Grossman SJ, Stiglitz JE (1980) On the impossibility of informationally efficient markets. Am Econ Rev 70(3):393–408. http://people.hss.caltech.edu/~pbs/expfinance/Readings/GrossmanStiglitz.pdf

Haseeb DM (2017) Real estate investment: buy to sell or buy to rent? NYC data science academy blog. http://blog.nycdatascience.com/student-works/zillow/

Kauko T, D'Amato M (eds) (2008) Mass appraisal methods: an international perspective for property valuers. Wiley-Blackwell, Chichester

Kholodilin KA, Mense A, Michelsen C (2016) The market value of energy efficiency in buildings and the mode of tenure. Urban Stud

Knight FH (1921) Risk, uncertainty and profit. The Riverside Press, Cambridge. https://ia600303.us.archive.org/11/items/riskuncertaintyp00knigrich/riskuncertaintyp00knigrich.pdf

Muggenhuber G (2016) Immobilienmarktbeobachtung via web-mining von Angebotsdaten. Dissertation. TU-Wien. http://repositum.tuwien.ac.at/obvutwhs/download/pdf/1709414?originalFilename=true

Palmquist RB (2005) Chapter 16: property value models. In: Mäler KG, Vincent JR (Hrsg.) Handbook of environmental economics, vol 2: valuing environmental changes 2. Handbooks in economics 20. Elsevier, Amsterdam, 763–819

Parkkinen J (2014) Interactive visualizations with R—a minireview: a minireview of R packages ggvis, rCharts, plotly and googleVis for interactive visualizations. http://ouzor.github.io/blog/2014/11/21/interactive-visualizations.html

Ruefer S (2016) Basic shiny application. https://www.steffenruefer.com/2016/11/basic-shiny-application/

Schneider M (2013) Are recent increases of residential property prices in Vienna and Austria justified by fundamentals? Monetary Policy Econ Q4/13:29–46

Shimizu C, Nishimura KG, Watanabe T (2011) House prices at different stages of the buying/selling process. IRES working paper series, No. 69

Simon HA (1947) Administrative behaviour: a study of decision-making processes in administrative organization. The Macmillan Co, New York

Wood AM (2017) Data in its natural format: integrate and access any form of data using a multi-model database. https://www.oreilly.com/ideas/data-in-its-natural-format?imm_mid=0f40b1&cmp=em-data-na-na-newsltr_20170705

Zhang X, Yin X (2008) Design of an information intelligent system based on web data mining. In: Zhou J (ed) ICCSIT '08: 2008 International conference on computer science and information technology. Piscataway, NJ, 88–91

Chapter 12
SDI Evolution and Map Production

Serena Coetzee

Abstract Cartography and spatial data infrastructures (SDIs) are mutually dependent on each other: SDIs provide geographic information for maps; and maps help to understand geographic information provided by an SDI. SDIs emerged in the 1990s when paper maps were replaced by map production from digital geographic information. They evolved from government-funded initiatives within the geographic information science (GISc) community to SDIs with a variety of data producers and an audience much beyond the relatively small GISc community. SDIs continue to evolve in response to new technologies and developments. This chapter traces SDI evolution over the last three decades and considers how SDI technical developments have impacted map production and how they are likely to impact map production in the future.

Keywords Spatial data infrastructure (SDI) · Map · Map production
SDI development · Cartography · Standards

12.1 Introduction

Cartography is generally defined as the art, science and technology of making and using maps (ICA 2003). Map production starts with map planning, followed by data collection (e.g. through fieldwork or through discovery in a geoportal), data preparation (e.g. geo-referencing and harmonizing data), data processing (e.g. statistical or spatial analysis), map design and finally, rendering or printing of maps on media. Maps are produced from geographic information or spatial data, which is often provided by a spatial data infrastructure (SDI).

A spatial data infrastructure (SDI) is an evolving concept about facilitating and coordinating the exchange and sharing of spatial data and services between

S. Coetzee (✉)
Centre for Geoinformation Science, Department of Geography, Geoinformatics & Meteorology, University of Pretoria, Pretoria, South Africa
e-mail: serena.coetzee@up.ac.za

J. Döllner et al. (eds.), *Service-Oriented Mapping*, Lecture Notes in Geoinformation and Cartography, https://doi.org/10.1007/978-3-319-72434-8_12

stakeholders from different levels in the spatial data community (Hjelmager et al. 2008). Drawing on the definitions for 'infrastructure' in Dictionary.com (2017) and Wiktionary (2017), a spatial data infrastructure (SDI) can be defined as the facilities, services, systems and installations to provide a country, city or area with spatial data and services that are required for the functioning of society.

The geographic information products and services provided by an SDI are relevant for map production, because they determine the maps that can be produced. For example, the level of detail available in geographic information determines the scale at which maps can be produced. Other determining factors include the quality of the geographic information, licensing (terms and conditions for using the data), size of the dataset, encodings (formats), data models and services through the data can accessed.

The range of SDI products and services has changed significantly in the last few decades, from isolated digital datasets to static online maps and today's integrated location-based services. In the 1990s, SDIs emerged when paper maps and corresponding cartographic production arrangements were being replaced by digital geographic information (Masser and Campbell 1991). Early SDIs were strictly top-down government funded initiatives within the geographic information science (GISc) community. Since then, SDIs have evolved to include private sector producers and a wider audience beyond the GISc community. Today, SDIs continue to evolve in response to crowd sourcing, mobile technologies and the challenges presented by big data (Coetzee et al. 2013).

This paper reviews SDI developments and how they have influenced map production. SDI evolution is organized into four periods, roughly corresponding to major technological developments that impacted SDI products and services: Digital SDIs (impacted by personal computers); Read-Only SDIs (impacted by the advent of the internet); Read-Write SDIs (impacted by Web 2.0 developments, such as user-generated content), and Ubiquitous SDIs (impacted by sensors and the cloud). In Sects. 12.2–12.5, the major developments for each SDI period are described and the impacts of these on map production are discussed. Section 12.6 concludes.

12.2 Digital SDIs

National SDIs emerged when paper maps and corresponding map production arrangements were being replaced by digital geographic information (Harvey et al. 2012). In early SDIs, national mapping agencies were the drivers in creating digital geographic information products, mainly for use by the public sector (Rajabifard et al. 2006). In the 1990s, 'New Public Management' (NPM) strategies were prominent in many western countries. NPM stresses running governments like a business in order to save on costs and improve efficiency and effectiveness. This led to the disaggregation of large bureaucracies into small specialized and competitive organizations, and inevitably resulted in the fragmentation of governmental information. SDIs aimed to counter this by promoting effective and efficient management of information through sharing and by avoiding duplication of data collection (Sjoukema et al 2017).

The rise of affordable personal computers meant that data capturing was not restricted anymore to a small number of users with access to a few powerful but expensive machines. Desktop products for capturing, managing, analysing and visualizing geographic information were developed to take advantage of this opportunity. Initially, much focus was on digitizing data from existing paper maps, first into files, later into database management systems (DBMS), mostly through the user interfaces of desktop GIS products. A DBMS provides functionalities, such as a data dictionary, access control, multi-user access, transactions, backup and recovery, which made it possible to coordinate teams working on a dataset. DBMS significantly improved the way in which geographic information could be managed and maintained.

Before the advent of the internet, SDI products were distributed on disks. As a result, there was a time lapse from release in the production environment to the time when the disk arrived at the user. After arrival of the disks, data had to be loaded into the map production environment before map production could start. New versions of data were typically released annually, sometimes quarterly, but seldom more frequently than that. Naturally, map production cycles followed the release frequencies of datasets.

SDI products were typically distributed as files in vendor dependent formats. As a result, it was easier to use that same vendor's software for map production. Using a different vendor's software meant importing the data and any incompatibilities between the data models of the two vendors could result in loss of data. To overcome some of these challenges, first standards for metadata and for data exchange emerged during this period.

Before the advent of digital data, the map was the database (Longley et al. 2005). One advantage of digital datasets was that they could be shared and used for analysis and map production by many different organizations. The proliferation of mapping software along with the ability to infinitely reproduce copies of data opened possibilities for sharing geographic information way beyond the capabilities of traditional map production (Coetzee et al. 2013).

Affordable personal computers meant that many more users could now produce maps. Figure 12.1 shows the linear nature of map production with digital SDI products and personal computers. While the software on personal computers could automate map production, early map production software had many technical limitations, often resulting in sub-optimal map design.

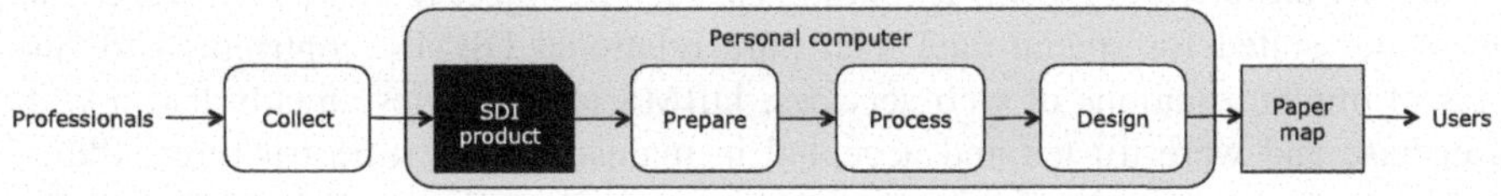

Fig. 12.1 Map production on personal computers in a Digital SDI

Despite the data changing from analogue to digital data, many paper maps were still produced in this period. While users could view electronic maps on their personal computers, society at large still relied on paper maps.

12.3 Read-Only SDIs

The advent of the internet made it possible to share SDI data online. Online data catalogues and clearinghouses emerged, changing the way in which data was discovered and distributed. They also helped to raise the awareness of spatial data. Online sharing of data eliminated the cumbersome process of distributing data via disks. This shortened the time from data producer to data user, but data collection was still a lengthy process. Unfortunately, due to bandwidth limitations, downloading data was often just not practically feasible and many datasets were still distributed on disks.

In early SDIs the focus was on creating digital data. Now that digital data was available, the focus shifted from the data in itself to the use and application of data, including the introduction of web services for providing data access (Rajabifard et al. 2006). SDI objectives changed from data creation towards an infrastructure for providing access to SDI products.

To facilitate the development of this infrastructure, standards were required. ISO/TC 211, *Geographic information/Geomatics*, and the Open Geospatial Consortium were established in 1994 and published the first international geographic information standards in subsequent years. Amongst others, standards were developed for a common feature model, the geography markup language (GML) and web services for viewing static maps over the internet. The common feature model and GML led to improved data exchange (export/import) so that cartographers had more freedom to use the software of their preference for map production.

Online maps led to a further increase in the number of people who accessed and used maps. Figure 12.2 illustrates online map production with read-only SDI data products. Each internet user was now a potential mapmaker because they could create their own map design by changing the colours and labels on a map. However, similar to the early desktop mapping products, technical limitations of those early online maps led to many sub-optimal map designs. Furthermore, tried and tested map design principles for paper maps had to be adjusted for display on new electronic media, such as computer screens, and for less real estate on handheld devices and mobile phones.

Further advances in DBMS technologies, such as object-relational databases that are better suited for spatial data than pure relational DBMS, contributed to successful implementations of web services. DBMS technologies already had a wide user base and were trusted and accepted in mainstream IT infrastructure. Adding spatial data was a simple addition to the existing infrastructure, so that web services

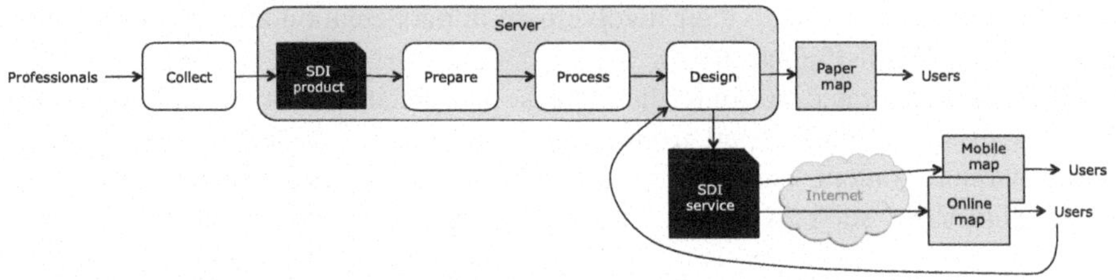

Fig. 12.2 Map production in a Read-Only SDI

could now access spatial data in an object-relational DBMS via a map server. Cartographers could perceive this as a loss of control over the data.

12.4 Read-Write SDIs

Further technological advances made it possible to edit and update information over the internet, commonly referred to as Web 2.0. User-generated content, i.e. content created by users and published on an online platform, is a prominent feature in Web 2.0. Volunteered geographic information, i.e. geographic information provided by the man-on-the-street, emerged in this period. It is the geospatial equivalent of user-generated content. OpenStreetMap (2017), an editable map of the world, is a classical example of volunteered geographic information. Standards, such as the Web Feature Service (WFS) and Filter Encoding, were published during this period and reflect the read-write nature of SDIs.

Traditionally, cartographers relied on data from authoritative sources who verified the quality of data before publication and who could be held accountable for the quality. In a Web 2.0 environment, data producers are not necessarily professionals in respected organizations. Suddenly, every internet user was a potential spatial data contributor. This has serious challenges for the quality of volunteered geographic information (Cooper et al. 2011), and one has to be aware of these when producing a map. Based on a comparison of authoritative and volunteered geographic information, Siebritz et al. (2012) concluded that volunteered geographic information is unlikely to be of a quality as good as the authoritative data from a national mapping agency, but that it could be useful in other ways, for example, to identify gaps or inaccuracies in authoritative data (Coetzee et al. 2013).

In a Web 2.0 environment, users play a much more active role. They are not only viewers of maps, but can also contribute data to be used in the map, as illustrated in Fig. 12.3. Widespread use of crowd sourcing and mobile technologies are changing the perception that authoritative data is a prerequisite for good decision support, because citizens can now participate in decision-making by contributing information (Coetzee et al. 2013). Scientific literature commented on a shift from top-down SDI governance to bottom-up SDIs driven by the users who were giving direction to SDI developments due to their involvement in data creation and map production (Coetzee and Wolff-Piggott 2015).

Web 2.0 accelerated not only the increase in spatial data volumes but also the diversity of spatial data. Apart from human- and man-made features portrayed in user-generated content, such as OpenStreetMap, spatial data generated by social media platforms, such as twitter and facebook, was now available. This new diversity created challenges (e.g. heterogeneous data models) and opportunities (e.g. additional kinds of data) for map production.

SDIs were influenced by ideas about governance that changed around the 2000s. Most western countries moved away from a market perspective towards a network perspective on governance, i.e. not competition, but networks and relations are important for coordination. At the same time, 'trust' and 'transparency' gained importance, leading to the 'open government' concept in which freedom of information (including geographic information) is deemed important for ensuring accountability, trust and public participation, the facilitators of the democratic process (Sjoukema et al. 2017).

As a result, SDI datasets were now made available as 'open spatial data', which resulted in increased usage of spatial datasets, also for map production, but created challenges for business models for maintaining data. One such example is Poland, where one of the biggest benefits of the national SDI was improved functioning of the government by facilitating access to spatial data. However, departure from the existing business model based on the sales of maps and data created some concerns (Harvey et al. 2012).

12.5 Ubiquitous SDIs

Today, spatial data creation and usage are ubiquitous. Sensors everywhere around us are producing location-based data at unprecedented speeds. Geostreaming refers to the continuous generating of location-based data from multiple sources, such as sensors, leading to new challenges for cartographers: large volumes of heterogeneous spatial data have to be visualized and mapped at break-neck speed before loosing their relevance.

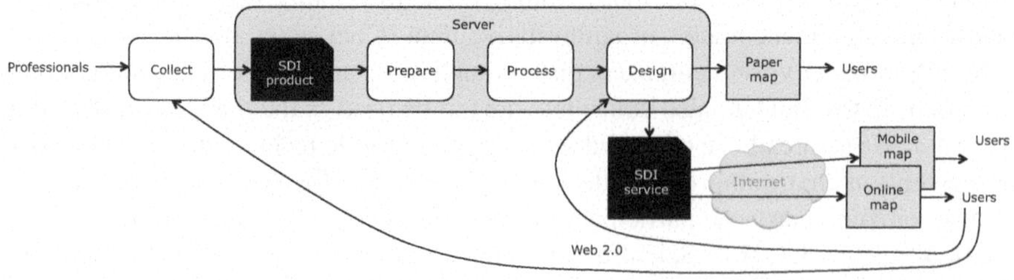

Fig. 12.3 Map production in a Read-Write SDI

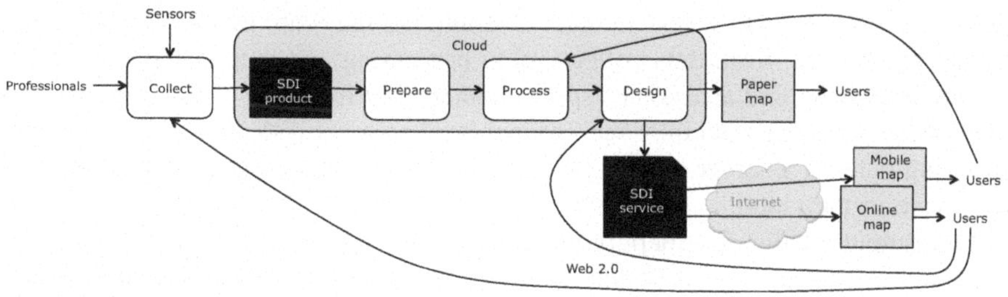

Fig. 12.4 Ubiquitous SDI

Cloud platforms are maturing and are transforming the way in which spatial data is processed, analysed and visualized. Instead of downloading spatial data, map production services are performed on SDI data products stored in the cloud (Fig. 12.4). This is especially useful for large datasets and processing intensive SDI data products, such as satellite imagery. Intelligent on-demand mapping or chaining of web services to produce maps are now possible (Asche et al. 2013; Tristan et al. 2013, Rautenbach et al. 2013). This has led to many more map production tools at the disposure of the cartographer.

Spatial data is integrated into our everyday activities, from navigating through traffic on a smartphone, to finding the closest restaurant and ordering an Uber taxi. Instead of downloading large volumes of spatial data once or twice a year, millions of small chunks of spatial data are transmitted between devices every day. Instead of large effort production of (static) map series, maps are continuously produced from small chunks of data updates.

The general user does not even know if and when spatial data is used in these services. Rajabifard (2007) refers to a spatially enabled society. By definition, an infrastructure should strive to be invisible. Users are interested in useful services provided by the infrastructure, not abstractions of the services or physical components that make the services possible (Blanchette 2012). In a spatially enabled society, the services provided by the SDI are visible to cartographers and used by them, but the SDI itself is becoming blurred.

12.6 Discussion and Conclusion

Over the years, SDI objectives shifted emphasis, from a focus on data sharing and coordination to one on supporting policy, from a top-down approach to a bottom-up approach, from centralised to distributed and service-orientated approaches (Williamson et al. 2006; Van Loenen et al. 2009).

These SDI developments influenced map production in many different ways. The way in which data is collected and discovered for map production changed from a dedicated search for spatial datasets to using map production services that 'automatically' provide access to data. Overall, end users are involved much more in map production, from data collection, to processing and map design.

The processing and analysis of spatial data moved from personal computer, to a thick client accessing data on a server via a local area network, to a thin web client accessing data on a server via the internet. Today, the processing and analysis can take place in the cloud.

The output media for map production changed from paper to all kinds of electronic media, such as computer screens and smartphone displays, each with a different set of parameters for map design.

The changing landscape of cartography is reflected in the top 15 most frequently used words in abstracts of peer-reviewed publications in the three journals affiliated with the International Cartographic Association, namely Cartographica, The Cartographic Journal and Cartography and Geographic Information Science (CaGIS). Figure 12.5 shows that 'digital', 'GIS' and 'analysis' appeared between 1994 and 2002, i.e. when the first SDIs were being established. Also, the use of 'map', 'maps' and 'cartographic' decreased during the time when the use of 'data' increased. When read-write SDIs emerged, the frequency of 'accuracy' first appeared in the top-15, possibly pointing to the aforementioned challenges related to the quality of volunteered geographic information. The inclusion of 'algorithm' from 2003 onwards may suggest the increased need for algorithms to process, analyse and map big spatial data.

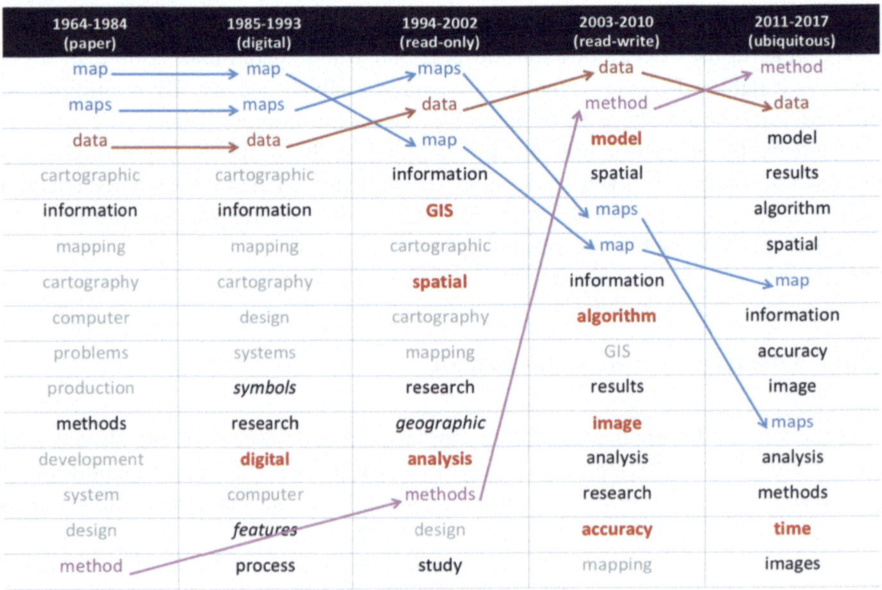

Fig. 12.5 The 15 most frequently used words in abstracts of peer-reviewed cartographic publications

As we go into the future, new data sources are emerging, such as point clouds for three-dimensional visualizations of buildings and landscapes collected by unmanned aerial vehicles. The volumes and the speed at which data is collected call for new ways to process and visualize the data. At the same time, the ever-increasing numbers of users are increasingly spatially enabled. They want to contribute to the data, determine the processing and analysis, and design their own maps and visualizations. The demand for tools to facilitate map production will continue to grow.

References

Asche H, Stankute S, Mueller M, Pietruska F (2013) Towards developing an integrated quality map production environment in commercial cartography. In: Proceedings of computational science and its applications—ICCSA 2013. Ho Chi Minh City, Vietnam, 24–27 June 2013

Blanchette JF (2012) Computing as if infrastructure mattered. Commun ACM 55(10):32–34

Coetzee S, Harvey F, Iwaniak A, Cooper AK (2013) Sharing and coordinating SDIs in the age of crowdsourcing and mobile technologies. In: 26th international cartographic conference, Dresden, Germany

Coetzee S, Wolff-Piggott B (2015) A review of SDI literature: searching for signs of inverse infrastructures. In: Robbi Sluter C, Madureira Cruz CB, Leal de Menezes PM (eds) Cartography—maps connecting the world. Springer, Berlin, pp 113–127, ISBN 978-3-319-17737-3

Cooper AK, Coetzee S, Kaczmarek I, Kourie DG, Iwaniak A, Kubik T (2011) Challenges for quality in volunteered geographic information. In: AfricaGEO 2011, Cape Town, South Africa, 31 May–2 June 2011, 13 pages

Dictionary.com (2017) Infrastructure. In: Dictionary.com Unabridged. Source location: Random House, Inc. http://www.dictionary.com/browse/infrastructure. Accessed 15 Sept 2017

Harvey F, Iwaniak A, Coetzee S, Cooper AK (2012) SDI past, present and future: a review and status assessment. In: Rajabifard A, Crompvoets J, Kalantari M, Kok B (eds) Spatially enabling government, industry and citizens: research and development perspectives. GSDI Association, pp 23–38

Hjelmager J, Moellering H, Cooper AK et al (2008) An initial formal model for spatial data infrastructures. Int J Geogr Inf Sci 22(11-12):1295–1309

International Cartographic Association (ICA) (2003) A strategic plan for the international cartographic association 2003–2011. http://icaci.org/files/documents/reference_docs/ICA_Strategic_Plan_2003-2011.pdf. Accessed 4 Apr 2017

Longley PA, Goodchild MA, Maguire DJ, Rhind DW (2005) Geographic information systems and science, 2nd edn. Wiley, Chisester

Masser I, Campbell H (1991) Conditions for the effective utilisation of computers in urban planning in developing countries. Comput Environ Urban Syst 15(1–2):55–67

OpenStreetMap (2017) OpenStreetMap Website. www.openstreetmap.org. Accessed 13 Sept 2017

Rajabifard A, Binns A, Masser I, Williamson I (2006) The role of sub-national government and the private sector in future spatial data infrastructures. Int J Geogr Inf Sci 20(7):727–741

Rajabifard A (2007) Towards a spatially enabled society. University of Melbourne, Parkville

Rautenbach V, Coetzee S, Iwaniak A (2013) Orchestrating OGC web services to produce thematic maps in a spatial information infrastructure. Comput Environ Urban Syst 37:107–120

Siebritz L, Sithole G, Zlatanova S (2012) Assessment of the homogeneity of volunteered geographic information in South Africa. In: XXII ISPRS congress, ISPRS archives XXXIX-B4, pp 553–558. https://doi.org/10.5194/isprsarchives-xxxix-b4-553-2012

Sjoukema J-W, Bregt A, Crompvoets J (2017) Evolving spatial data infrastructures and the role of adaptive governance. ISPRS Int J Geo-Infor 6:254
Tristan AC, Garza EJT, Calderon OAA, Perez JJ, Tagle MAG, Nemiga XA (2013) Spatial technologies to evaluate vectorial samples quality in maps production. Inv Geográficas, Boletín del Inst Geografía, UNAM 80:111–112
van Loenen B, Besemer J, Zevenbergen J (2009) Spatial data infrastructure convergence. In: van Loenen B, Besemer J, Zevenbergen J (eds) SDI convergence. Research, emerging trends, and critical assessment. Nederlandse Commissie voor Geodesie, Delft, The Netherlands
Wiktionary contributors (2017) Infrastructure. In: Wiktionary, The Free Dictionary. https://en.wiktionary.org/w/index.php?title=infrastructure&oldid=47478939. Accessed 15 Sept 2017
Williamson I, Rajabifard A, Binns A (2006) Challenges and issues for SDI development. Int J Spat Data Infrastruct Res 1:24–35

Chapter 13
Handling Continuous Streams for Meteorological Mapping

Chris Schubert and Harald Bamberger

Abstract Providing Weather and Climate Data Service need complex operating IT systems and infrastructures as well as 24/7 operation running software systems. Meteorological data streams used in both, in the field of sciences and for operational tasks, e.g. in national meteorological weather services. Application for weather services and products have a strong cross domain impact, hereby interoperability, standard conformal data use are essential for everyday tasks. Three use cases will be shown: (1) Approach to proper data management, apply dynamic data citation, (2) services for numerical weather prediction and (3) trajectories.

Keywords Dynamic data citation · Data management · Data life cycle
Persistent identifier · Query store · Interoperability · OGC service
Numerical weather prediction · Trajectories · GIS

13.1 Introduction

The Austrian weather service, the "Zentralanstalt für Meteorologie und Geodynamik" (ZAMG) has to deal with a variety of heterogeneous data formats, e.g. station data, model grids in different raster formats, vector based information and text based- or narrative information. At least each dataset has one additional dimension, such as one or more time dimensions and elevation. In general these two or more dimensions can be highly dynamic and make use of relational dependencies. Another characteristic of meteorological data is the high update frequency, which can range from days to minutes.

C. Schubert (✉)
CCCA Data Centre, Vienna, Austria
e-mail: chris.schubert@ccca.ac.at

H. Bamberger
ZAMG, Department of Software Application Development and Data Management, Vienna, Austria

J. Döllner et al. (eds.), *Service-Oriented Mapping*, Lecture Notes in Geoinformation and Cartography, https://doi.org/10.1007/978-3-319-72434-8_13

Due to the demands of customers, any data provision has to be in near real time. Therefore efficient and stable algorithms are needed for the production chain. For any usage in customer applications, complex meteorological data have to be adopted in widely used format and service standards e.g. OGC (see Chap. 15) or ISO (see Chap. 16).

In this chapter we report how to deal with aforementioned requirements and how to orchestrate software applications, data services, interfaces and IT components. Three use cases will be elaborated: (1) an approach to proper data management, which applies dynamic data citation, (2) services for numerical weather prediction and (3) trajectories.

- The use case **Dynamic Data Citation** highlights a technical pilot implementation that explains how to cite datasets and their derivatives in a dynamic process chain.
- The example **Numerical Weather Prediction** describes issues for raster information in a service-oriented environment. This example gives some insight to the meteorological community, especially how to make use of best practice guidelines in order to adopt OGC WMS standards for meteorological use.
- The use case **Trajectories** describes how to store and visualize trajectory input data. Basically these form of data are transport paths of air masses that are processed via OGC WMS Services. Limitations of this methodology will be shown.

13.2 Dynamic Data Citation

13.2.1 Overview and Requirements

Datasets change during time. Errors become corrected, outdated data are updated and others are replaced. In order to re-use data in a reproducible manner or to cite a dataset for a study, researchers need to identify the exact version, extension and quality of a used data set. If only parts of a dataset were used, the correct citing process becomes particular complex. Citing the original data and its correct version can be challenging. The goal of dynamic data citation is an exact, precise and comprehensive description of creation processes aiming at data subsets.

The current pragmatic procedures for the processing of large files, such as geospatial datasets, is mainly characterized by downloading the data onto a local workstation and editorials according to the customer- or application needs.

Generally GIS tools are used to select an individual 'area of interest' or to intersect distinguished layers. Such kind of intersected datasets are then stored again for further processing steps. In many cases those integrated data will be published on a data portal or directly via the cloud, ftp, etc. All metadata-information concerning the relation to the original datasets as well as its different versions may be lost and therefore has to be described from scratch. This manual procedure is time consuming and error-prone. Describing all the processes, their arguments and values that were used for the intersection procedure, will be highly imprecise. The CCCA Data Centre as a department of ZAMG wants to overcome those troublesome processes. Methods for the correct citing of aggregated data or its subsets, especially related with the complex data structures for the climate modeling are in the focus of research activities.

In the scientific meteorological community the citing of datasets in an appropriate manner is an agreed scientific practice and well established. To deal with the original source of data is fundamental for any verification of a data integration outcome or research results. Reproducibility is a core element against data manipulation and falsification.

Currently most of such citations base on static and unchanging data sets. If only parts of a dataset are needed a more or less dynamic citation is needed. The approach comprises an exact identification of those data-parts that are used in a study. As a consequence the updating and changing of datasets or parts thereof will lead to affected research studies that reused this dataset.

The CCCA Data Centre developed a web based tool for dynamic data citation. The main motivation was to contribute components for (1) a proper data life cycle management, (2) set up a versioning by time stamps, (3) assign all related derivations of data with a persistent identifier (PIDs), (4) identify resources by storing queries and make those queries available again, which will help to reduce storage redundancy and (5) automatically create a landing page for data subsets which provides a citation text as well as inherited metadata in a dynamic manner.

13.2.2 Institutional Framework

The CCCA Data Centre develops a structured and web-based service that describes the relations of originals and derivatives. This service provides an automated citation text based on the versioning. This application is called "Dynamic Data Citation".

In the cooperation with the research project e-infrastructures Austria and the Data Citation working group within the framework of the RDA Research Data Alliance,[1] a concept for a pilot implementation has been developed. Following institutional partners were involved: Vienna University of Technology (Institute of Software Technology and Interactive Systems), University of Vienna (Computer Center), Austrian Institute of Technology (Dep. Digital Safety & Security) and the Wegener Center in Graz.

The applied approach on the "dynamic data citation" based on recommendations, which were developed by the RDA Working Group on Technical

[1]https://www.rd-alliance.org/groups/data-citation-wg.html.

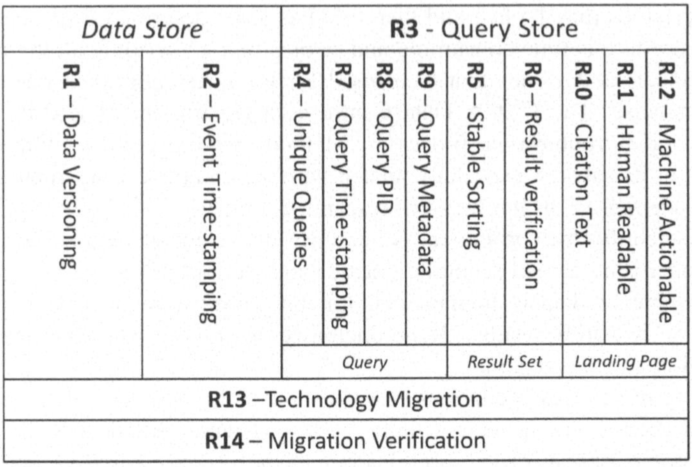

Fig. 13.1 14 Recommendations of RDA Guidelines, making data citable (modified after Rauber et al.)

Implementation (Rauber et al. 2015, 2016). Technical implementation concepts are to ensure versions with a time stamp, the allocation of persistent identifiers (PID) in the data set and automatically generated citation text: The main recommendations are shown in Figs. 13.1 and 13.2.

The big challenge was to avoid redundancy for generated sub data records. Instead only queries including arguments, such as the required time range or geographical region, shall be saved. This method saves a lot of disk space. In addition available queries are assigned with a permanent identifier and will therefore be always executable. This method can be applied to different versions of a data record or even used as a template for other data sets.

The main motivation for this development is the support of software-driven data management by contributing components. Data life cycles will be established with the existence of provenance information and data sets should be provided in an excellent, as well automatically manner. The components identify, which parts of the data are used. Whenever data change, the system identifies which queries (studies) are affected.

This use case implementation of the Dynamic Data Citation at the CCCA Data Centre focused on the CF[2] standard and used compliant NetCDF data of high resolution climate scenarios for Austria until December 2100, called OeKS15. NetCDF is an open standard and machine-independent data format for structured multidimensional data. It includes attributes, dimensions, and variables. For example, modeled temperature records on a daily basis are available as 1 × 1 km

[2]https://cfconventions.org/.

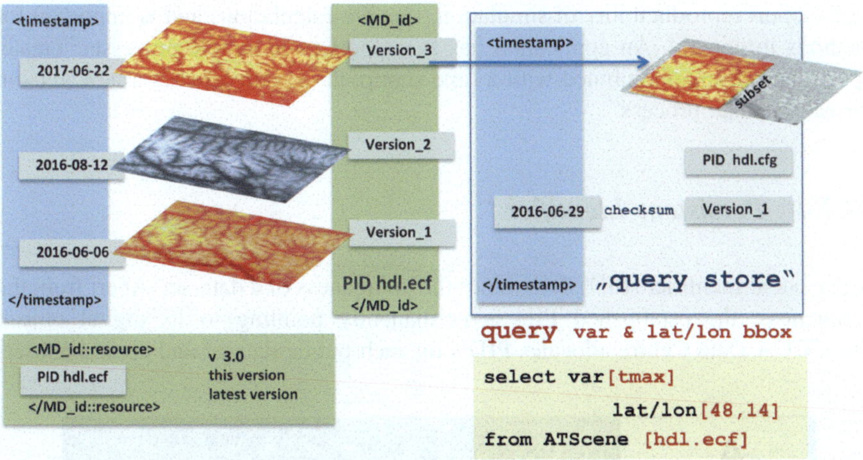

Fig. 13.2 Schematic draft of subset creation process (adopted from Baumann 2014). Imported data gets a time-stamp, a PID and metadata elements, like a version. The query store provides arguments for identifying data subsets and valuable meta-information

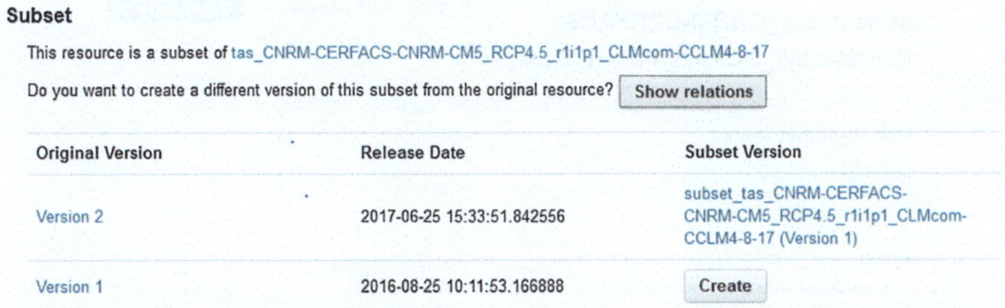

Fig. 13.3 Combined view on implemented versioning. On the left the aforementioned relation, from which a version subset was created, is listed. In addition the user could create a subset based on oldest versions with the same arguments

rasterized, geo-referenced data in multiple single files, which include different "Representative Concentration Pathways" (RCP's),[3] ensembles and model[4] runs.

The entire data package is openly accessible. For Austria it includes over 900 files with a size up to 16 GB per file.

13.2.3 Data Versioning

Versioning ensures that previous states of records are maintained and made retrievable. When being able to refer (Fig. 13.3) to previous versions of a data-set,

[3]https://www.aimes.ucar.edu/docs/IPCC.meetingreport.final.pdf.

[4]https://cmip-pcmdi.llnl.gov/cmip5/docs/cmip5_data_reference_syntax_v0-25_clean.pdf.

will support reproducibility of simulation, specific calculations and is important for methods in general. An application has to support three simple processes: create, update and delete. Combined with a time stamp, the system creates a new version for each update process.

13.2.4 Persistent Identifier

A Persistent Identifier (PID) determines the uniqueness of a data-set. Apart from the uniqueness, the established PID is permanently pointing to its digital object. The CCCA Data Centre allocates PID's on each public dataset and creates subsets

Fig. 13.4 Landing page on a new created subset, inclusive citeable text, versioning and their relations (CCCA, Schubert)

by using the HDL.NET® Registry. If the same filter and primary data-set are used for sub setting, an identical sub-set is available. In order to avoid this phenomenon, each identical subset gets only one PID.

Following item is used as PID at the Data Centre: "hdl.handle.net/20.500.11756/88d350e9".

The *hdl* PID resolves to a landing page, where detailed metadata are provided. These metadata are inherited from the primary data-set and amended with additional information. This information covers the contact information, the subset creator (who created this subset), a subset timestamp (when the sub-setting was done), a list of used filter arguments, such as changing the geographical extend or time range, as well as the link from the subset to the primary version.

13.2.5 Dynamic Citation Text

The web application generates automated citation texts. It includes predefined text snippets, such as title, author, publishing date, version and data repository. For any described subset, filter arguments based on queries are used and provided as text information (Fig. 13.4). This way of generating citation texts shall lower any barriers for data sharing and reuse.

13.2.6 Architecture

The following components are used for the CCCA Portal: (i) ckan web server, (ii) the application server for access, data management and usage as query store, (iii) Handle.NET® Registry Server for the PID allocation and (iv) the Unidata Thredds Data Server (TDS) retrieves the subset from the data (Fig. 13.5).

The application server records the requests (subset request) via the web server, generates arguments for the required time range and/or the geographical extent, links the requests with a persistent identifier and manages all relations for versioning.

13.2.7 Application

All the requests are stored in a Query Store in conjunction with the PID. The Thredds server retrieves the subset directly from the data store, where the original NetCDF data are saved, with the defined arguments and passes it to the web server in the chosen export format. Additionally it provides a view service (OGC Web Map Service) for web based visualization.

The user receives a dynamic generated citation text, which contains the original author, label of the data set, versions, selected parameters applied for intersection

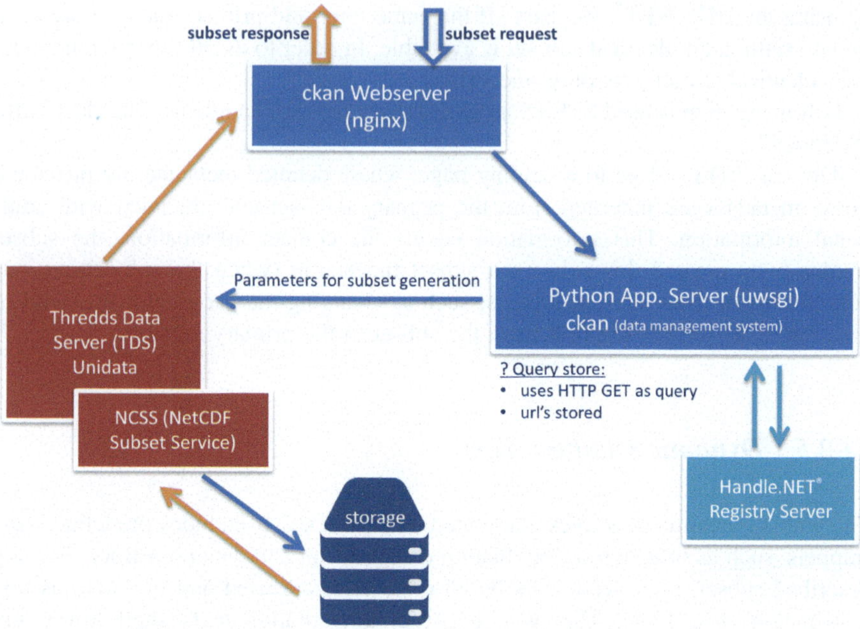

Fig. 13.5 Architectural structure of components for the Data Citation (CCCA, Schubert)

and the persistent identifier. The given result can be used for correct citing in studies, publications, etc. It is well accepted and comprehensible to the whole research community.

This implementation of Data Citation by the CCCA data centre accounts for an enormous strengthening of the potential and attractiveness for researchers using this data portal. This service has been regarded as a reference implementation for the ISO 690 "Information and documentation—guidelines for bibliographic references and citations to information resources".

13.3 Numerical Weather Prediction

There is quite a variety of meteorological models around. There are global models as the Global Forecast System (GFS), which is produced by the National Centers for Environmental Prediction (NCEP), or the one produced by the European Center for Midranged Weather Forecast (ECMWF). And there are Local Area Models (LAM), such as ALADIN/ALARO/AROME from MeteoFrance, COSMO from Deutscher Wetterdienst (DWD) and the COSMO Consortium, the Unified Model from the British MetOffice or the Weather Research and Forecasting (WRF), which is a collaboration of some US governmental agencies.

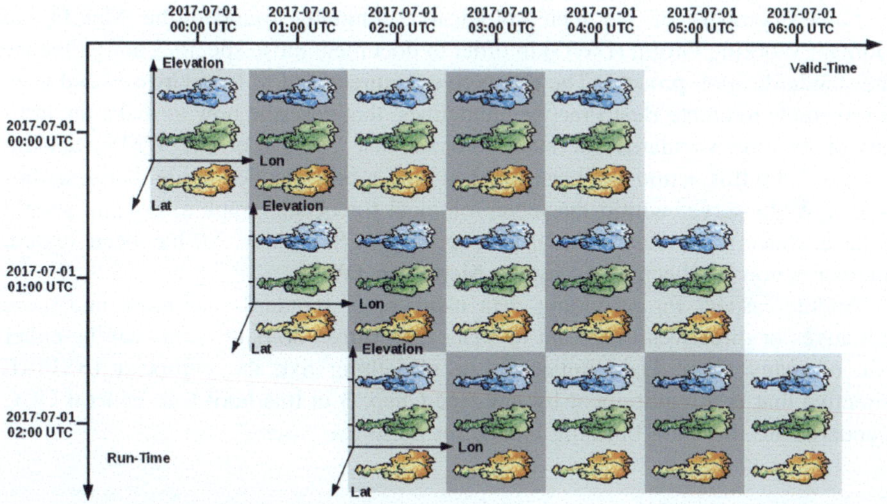

Fig. 13.6 Example of 5-dimensional Weather Data

The data structure of Numerical Weather Prediction (NWP) data is generally a datacube with at least 4 dimensions (see Chap. 14 in this book for datacubes). Normally there is, in addition to latitude and longitude, at minimum one time dimension and elevation. If there is more than one parameter (e.g. temperature, precipitation, wind speed, wind direction, etc.) in the datacube, this adds an additional dimension, see Fig. 13.6.

The Open Geospatial Consortium (OGC) has specified many standards for geospatial content and services (see Chap. 15 in this book for OGC standards). There are standards for metadata, data-formats and services. A Catalog Service for the Web (CSW) specifies access to metadata catalog information. Prominent OGC standard examples for data-formats are the Geography Markup Language (GML) or netCDF, which is quite common in the meteorological and climatological domain. Services as the Web Map Service (WMS), Web Feature Service (WFS), Web Coverage Service (WCS), Sensor Observation Service (SOS) and many more complete the standards portfolio of the OGC.

The aforementioned standards are widely used and many supporting libraries and software components support implementations. It is obvious that the acceptance of standards is high in the meteorological domain. There is a strong interest to use standards for meteorological data. Basically meteorological data is geospatial, but has some specifics: It does not always fit into the provided framework of OGC standards.

Generally there is a strong interest by customers and users to overlay meteorological data and their domain specific data. Therefore interoperability is a main topic.

Some members of the meteorological community founded the Met Ocean Domain Working Group (DWG) in order to document those specifics and influence the standardisation process. The distinctive features need to be addressed and used for example to create Best Practice Guidelines, that describe how to make the most out of existing standards for meteorological use cases. For the WMS standard version 1.3.0 this action has been finished with a best practice guide that describes how a WMS server within the meteorological (or oceanographic) domain should behave. Since the standardisation process for WCS Version 2.0 has been started, another action has been taken by the MetOcean DWG.

Another aspect for providing data using open standards are more and more initiatives or directives that want to share data across domains. This can be either one of many Open Data initiatives or something like the european INSPIRE directive that is a requirement by law (see Chap. 3 in this book). In general OGC standards are the main building blocks for these use cases.

13.3.1 Visualisation of N-Dimensional NWP-Data as OGC WMS Services

ZAMG is using the software product Geoserver (http://www.geoserver.org, [GeoSolMD]) to visualize the output NetCDF of a WRF-chem NWP Model (e.g. forecast of Ozone O3). Geoserver offers the plugin Image Mosaic for this task, which was intended to "assemble a set of overlapping geospatially rectified images into a continuous image" (GSDocIM 2017). The principle of the Image Mosaic plugin has been extended to be used with images covering the same geospatial area, but overlapping along other dimensions (e.g. TIME or ELEVATION).

The NetCDF file contains multiple parameters, such as the mentioned "Ozone O3". Every parameter can be mapped to a WMS layer. The header information of the NetCDF is structured as shown in Fig. 13.7.

As storage for the Image Mosaic index information, a RDBMS (e.g. PostgreSQL/PostGIS) can be used. It has to be configured in the file datastore. properties in the data directory, as shown in Fig. 13.8. Then an indexer can be configured that will index new NetCDF files in the data directory and create the appropriate entries in the database.

Furthermore the indexer has to be configured with the help of an XML file. The indexer file and a shared auxiliary file is shown in Fig. 13.9.

The next step is the configuration of the data store for the Image Mosaic itself. Through the Geoserver Web-GUI most of the values can be set. The Web-GUI is used as source for the WMS Service. The WMS Capabilities document contains the layer for the Ozon O3, see Fig. 13.10.

Whenever a higher complexity of the input data is observed, the Web GUI of Geoserver will not fit for all required configurations. A higher complexity means that the input data exceeds 2D GIS data and handling this data is getting more

dimensions

```
netcdf wrfout_d02c {
    dimensions:
          Time = UNLIMITED ; // (1 currently)
          DateStrLen = 19 ;
          west_east = 255 ;
          …
          dust_erosion_dimension = 3 ;
          emissions_zdim = 1 ;
          one = 1 ;
          vprm_vgcls = 8 ;
          termite_vgcls = 14 ;
```

variables

```
variables:
    floato3(Time, bottom_top, south_north, west_east);
          o3:FieldType = 104 ;
          o3:MemoryOrder = "XYZ" ;
          o3:description = "O3 mixing ratio" ;
          o3:units = "1e-6" ;
          o3:stagger = "" ;
          o3:coordinates = "west_east south_north" ;
          o3:long_name = "O3 mixing ratio [ppmv]" ;
          o3:grid_mapping = "WRF_LCC" ;
```

projections

```
int WRF_LCC ;
          WRF_LCC:grid_mapping_name = "lambert_conformal_conic"
          WRF_LCC:standard_parallel = 30.f, 60.f ;
          WRF_LCC:longitude_of_central_meridian = 12.f ;
          WRF_LCC:latitude_of_projection_origin = 50.f ;
          WRF_LCC:semi_major_axis = 6370000.f ;
          WRF_LCC:inverse_flattening = 0.f ;
          WRF_LCC:false_easting = 0.f ;
          WRF_LCC:false_northing = 0.f ;
          WRF_LCC:spatial_ref = "PROJCS[\
"Lambert_Conformal_Conic\",GEOGCS[\"GCS_unnamedellipse\
",DATUM[\"D_unknown\",SPHEROID[\"Unknown\
",6370000,0]],PRIMEM[\"Greenwich\",0],UNIT[\"Degree\
",0.017453292519943295]],..."false_northing\",0],UNIT[\"Meter\
",1]],";
```

attributes

```
 // global attributes:
          :TITLE = " OUTPUT FROM * PROGRAM:WRF/CHEM V3.4.1
 MODEL";
          :START_DATE = "2015-06-22_00:00:00";
          :SIMULATION_START_DATE = "2015-06-22_00:00:00" ;
          :WEST-EAST_GRID_DIMENSION = 256;
          :SOUTH-NORTH_GRID_DIMENSION = 223;
          :BOTTOM-TOP_GRID_DIMENSION = 48;
           …
          :Conventions = "CF-1.6" ;
```

Fig. 13.7 Structured header information of NetCDF files for NWP-data

datastore properties

```
user=dbusername                                     max\ connections=10
port=5432                                           min\ connections=1
passwd=dbpasswd                                     validate\ connections=true
url=jdbc\:postgresql\://localhost/wrf_chem          Loose\ bbox=true
host=localhost                                      Expose\ primary\ key=false
database=wrf_chem                                   Max\ open\ prepared\ statements=50
driver=org.postgresql.Driver                        preparedStatements=false
schema=public                                       Estimated\ extends=false
Estimated\ extends=false                            Connection\ timeout=20
SPI=org.geotools.data.postgis.PostgisNGDataStoreFactory
fetch\ size=1000
......
```

Fig. 13.8 Example of data store properties

complex. In this case some manual configuration has to be done directly on the file system.

13.4 Trajectories

Trajectories are basically transport paths of air masses. There are two types of trajectories: forward- and backward trajectories. Forward trajectories indicate where the air masses from a certain 3D-point in time are transported to. Backward trajectories describe where the air masses on a certain 3D-point in time come from.

The data structure of trajectories is a list of 3D-points in time with several parameters as attributes. The calculation of trajectories uses the data from a numerical weather forecast.

Figure 13.11 shows a prototype implementation of a web based viewer [MetOcBP], which uses OGC WMS for serving the trajectories. In the right column elements of the graphical user interface (GUI) are shown. These elements allow to select values for the multiple dimensions (TIME, ELEVATION, RUN and LOCATION).

Following parameters are used to structure the input data for a trajectory: runtime (Secs), spatial position, absolute altitude, altitude above ground, ECMWF

Indexer File

```xml
<Indexer>
  <domains>
    <domain name="Time">
      <attributes>Time</attributes>
    </domain>
    <domain name="soil_layers_stag">
      <attributes>soil_layers_stag</attributes>
    </domain>
    <domain name="bottom_top_stag">
      <attributes>bottom_top_stag</attributes>
    </domain>
    <domain name="bottom_top">
      <attributes>bottom_top</attributes>
    </domain>
    <domain name="one">
      <attributes>one</attributes>
    </domain>
    <domain name="runtime">
      <attributes> ref="runtimeCollector">runtime</attributes>
    </domain>
  </domains>
  <coverages>
      <schema name="o3">
<attributes>the_geom:Polygon,imageindex:Integer,location:String,Time:Date,bottom_
top:runtime:Date</attributes>
      </schema>
      <name>o3</name>
      <domains>
        <domain ref="Time" />
        <domain ref="bottom_top" />
        <domain ref="runtime" />
      </domains>
  </coverages>
  <collectors>
    <collector name="runtimeCollector">
      <value>MODIFY_TIME</value>
      <spi>RuntimeExtractorSPI</spi>
      <mapped>runtime</mapped>
    </collector>
  </collectors>
  <parameters>
    <parameter name="AuxiliaryFile" value="_auxiliary.xml" />
    <parameter name="AbsolutePath" value="true" />
  </parameters>
</Indexer>
```

Fig. 13.9 An example of an Indexer file were all included parameters are described

vertical coordinate information (Longitude, Latitude, Z, Z-ORO, ETA), air pressure (PRESS), potential vorticity (PV), potential temperature (THETA) and specific humidity (Q). An Example looks like Fig. 13.12.

For the use as source in the WMS, the input data is read to a PostgreSQL/PostGIS Database with the data model as shown in Fig. 13.13.

WMS Capabilities for the Ozon Layer


```xml
<Layer queryable="1" opaque="0">
  <Name>o3</Name>
  <Title>o3</Title>
  <Abstract/>
  <KeywordList>
    <Keyword>WCS</Keyword>
    <Keyword>ImageMosaic</Keyword>
    <Keyword>o3</Keyword>
  </KeywordList>
  <CRS>EPSG:971807</CRS>
  <CRS>CRS:84</CRS>
  <EX_GeographicBoundingBox>
    <westBoundLongitude>4.28...</westBoundLongitude>
    <eastBoundLongitude>19.02...</eastBoundLongitude>
    <southBoundLatitude>41.88...</southBoundLatitude>
    <northBoundLatitude>50.35...</northBoundLatitude>
  </EX_GeographicBoundingBox>
  <BoundingBox CRS="CRS:84" minx="4.28..." miny="41.88..." maxx="19.02..."
              maxy="50.35..."/>
  <BoundingBox CRS="EPSG:971807" minx="-534000.0" miny="-870000.0"
              maxx="486000.0" maxy="18000.0"/>
  <Dimension name="time" default="2015-06-25T00:00:00Z" units="ISO8601"
           >2015-06-25T00:00:00.000Z</Dimension>
  <Dimension name="BOTTOM_TOP" default="0.003124695736914873"
units="">0.003124695736914873,...,0.9965000152587891</Dimension>
  <Style>
    <Name>O3</Name>
    <Title>O3</Title>
    <LegendURL width="143" height="424">
      <Format>image/png</Format>
      <OnlineResource xmlns:xlink="http://www.w3.org/1999/xlink"
xlink:type="simple"
  </Style>
</Layer>
```


Fig. 13.10 The WMS GetCapabilities output for the Ozon Layer

13.4.1 Visualization of Trajectories as OGC WMS Services

The aforementioned trajectories data are stored in a PostgreSQL/PostGIS database (DB). Since this DB is widely supported in the open-source GIS software community, it can be used as data-source for various mapserver products, such as UMN Mapserver (http://www.mapserver.org) or Geoserver. In the given example a custom PHP implementation using UMN Mapservers Mapscript extension has been used to provide trajectories as WMS service.

The use of the Mapscript extension gives the flexibility to modify the configuration file of Mapserver, which uses a mapfile, on each request. In this way, it can determine the current available values for every dimension of each layer from the database and output them on-the-fly in the getCapabilities XML-document, which describes the possible content and functionality of a WMS.

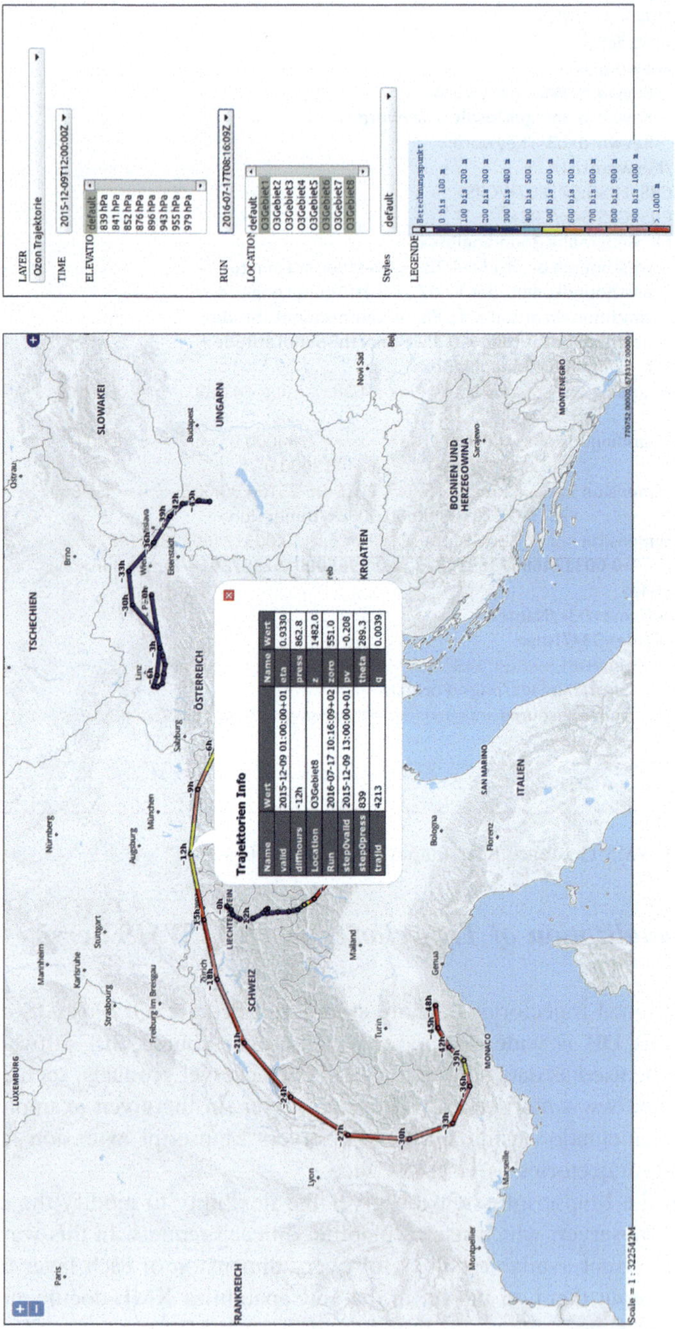

Fig. 13.11 Illustration: Visualization of Trajectories via WMS in Web-application

trajectory structure

DATE: 20151209 TIME: 120000 STOP INDEX: 1 # OF POINTS: 17

SECS	LONGIT	LATIT	ETA	PRESS	Z	Z-ORO	PV	THETA	Q
-21600	11.3618	47.3161	0.9864	852.4	1579.6	108.3	0.474	288.1	0.34E-02
-32400	11.3708	47.2551	0.9828	846.3	1638.3	136.0	0.444	288.3	0.35E-02
...	...	...	...	...	...	...	...	...	...
-43200	11.2792	47.2379	0.9770	830.9	1790.2	181.7	0.183	290.0	0.33E-02

Fig. 13.12 structure of Trajectories content

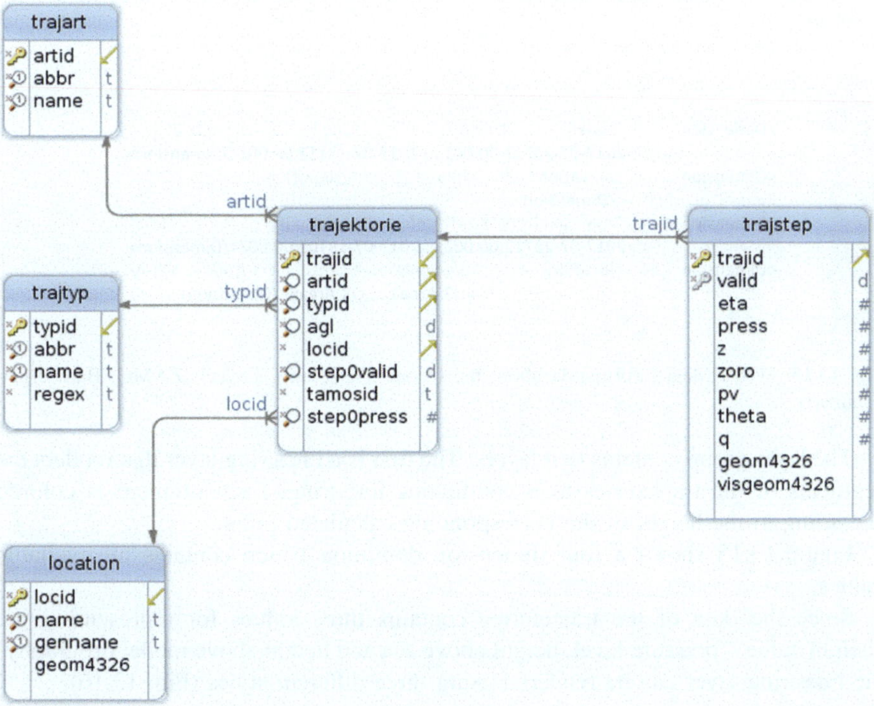

Fig. 13.13 Trajectories Data-Model EER Diagram

To achieve the desired visualization, a line string and a point layer were combined. The line string layer receives a continuous path with segments that are colored according to the height of the air masses. The point layer indicates the exact point from the calculation and places a label with a given offset from step 0 in hourly steps. In addition the two layers are wrapped in a named layer-group. Both can be requested by using the name of the group layer.

Figure 13.14 shows a fragment from the WMS GetCapabilities document which contains the information about the Ozone Trajectories layers.

Ozon Trajectorie Layer

```
<Layer>
  <Name>ozon_traj</Name>
  <Title>Ozon Trajectorie</Title>
  <Abstract>ozon_traj</Abstract>
...
<Layer queryable="1" opaque="0" cascaded="0">
    <Name>ozon_traj_ls</Name>
    <Title>Ozon Trajektorie Linien</Title>
```

Fig. 13.14 WMS Meta information about the Ozone Trajectories Layers (ZAMG, Bamberger, Schubert)

Dimensions

```
<Dimension name="time" units="ISO8601" default="2017-07-23T12:00:00Z"
nearestValue="0">2017-07-23T12:00:00Z,...,2017-07-18T12:00:00Z</Dimension>
<Dimension name="elevation" units="hPa" default="ground" multipleValues="1"
nearestValue="0"></Dimension>
<Dimension name="run" units="ISO8601" default="latest" multipleValues="1"
nearestValue="0">2017-07-22T12:00:00Z,...,2017-07-18T00:00:00Z</Dimension>
<Dimension name="location" units="LocationName" default="O3-Area1,...,O3-Area8"
multipleValues="1" nearestValue="0">O3-Area1,...,O3-Area8</Dimension>
```

Fig. 13.15 WMS Meta information about the Ozone Trajectories Layers (ZAMG, Bamberger, Schubert)

The layer group contains two layers. The first is a linestring layer that renders the segments of the trajectories as a continuous linestring. Each segment is colored according to the height of the corresponding calculated point.

Figure 13.15 shows a four dimension definition which contains all available values.

Since the data of the trajectories contains three values for representing the "height value" (pressure level, height above sea and height above model orography) the linestring layer can be rendered using three different styles (Fig. 13.16).

Fig. 13.16 Example of a linestring layer with styles for Ozone Trajectories. The group contains a layer where the calculated points of the trajectories are rendered (ZAMG, Bamberger, Schubert)

Style description

```
<Style>
  <Name>trajektorie:pressure_level</Name>
  <Title>trajektorie:pressure_level</Title>
  ...
</Style>
<Style>
  <Name>trajektorie:z_height</Name>
  <Title>trajektorie:z_height</Title>
  ..
</Style>
<Style>
  <Name>trajektorie:zoro_height</Name>
  <Title>trajektorie:zoro_height</Title>
  ..
</Style>
```

The aforementioned layer is available for the values defined in the four dimensions. Since this layer only renders points, one single style is needed. The amount of data of the trajectories is quite small. But because the calculation uses NWP data, which is described in the previous use case, there are some characteristics that prevent from a straightforward use of these data in the WMS layer.

13.5 Conclusion

The sharing of weather and climate data via spatial data services require complex IT systems, infrastructures and software tools for considering all requirements and caching meteorological data streams. ZAMG started in 1865 with daily weather maps, which were hand drawn on a printed paper template. These drawings included all observed parameters, meta information as well as a narrative description of weather situation. 1877, the daily telegraphic weather report was issued with a synoptic map and the forecast for the following day. With the introduction of computing infrastructures, more and more automated running sensors record data and allow for precise models. Especially the usage of the World Wide Web, digital data management, interoperability requirements and standard conformity are essential for information and data sharing. Incoming observed and recorded measurements do not naturally have the same structure, the same attributes or the same dimension. In the technical environment of high performant operational tasks, unremitting efforts for software development is a key component of each weather service.

With the given three use cases in this contribution, the current development of web based software services could be presented. We presented a data improvement in terms of reliability and interoperability and geo-visualisation within the permanent stream of meteorological recordings.

The developments of the Dynamic Data Citation Service show the collaboration between theoretical approaches and practical implementation. It minimizes redundant data storage of sub sets and makes data available by ensuring good scientific practice, which lowers barriers for re-using data in an automated way.

The Numerical Weather Prediction and Trajectories use cases describe the main issues for raster information, the multi-dimensional path ways and the best practice as well as limitations to adopt OGC services for the meteorological community.

References

Baumann P (2014) INSPIRE MIG workshop on WCS-based INSPIRE download services. https://iessvn.jrc.ec.europa.eu/projects/download-services-tg/wiki/MIG_workshop_on_WCS-based_INSPIRE_download_services

GeoSolutions—SpatioTemporal Data Handling With GeoServer [GeoSolMD] (2017) http://geoserver.geo-solutions.it/edu/en/multidim/

Open Source Geospatial Foundation, GeoServer Documentation—Image Mosaic [GSDocIM] (2017) http://docs.geoserver.org/latest/en/user/data/raster/imagemosaic/index.html

Rauber et al (2015) Data citation of evolving data. https://www.rd-alliance.org/system/files/documents/RDA-DC-Recommendations_151020.pdf

Rauber et al (2016) Identification of reproducible subsets for data citation, sharing and re-use. http://www.ieee-tcdl.org/Bulletin/v12n1/papers/IEEE-TCDL-DC-2016_paper_1.pdf

Schubert (2017) RDA—Data Citation WG, NetCDF Pilot Implementation of Austrain Climate Scenarios at CCCA Climate Change Centre Austria - Data Centre. https://www.rd-alliance.org/group/data-citation-wg/wiki/wgdc-pilot-ccca

Voidrot-Martinez et al (2014) OGC best practice for using web map services (WMS) with time-dependent or elevation-dependent data, MetOcean domain working group–WMS 1.3.0. Best Practice [MetOcBP]. https://portal.opengeospatial.org/files/?artifact_id=56394

Chapter 14
Datacubes: Towards Space/Time Analysis-Ready Data

Peter Baumann, Dimitar Misev, Vlad Merticariu and Bang Pham Huu

Abstract Datacubes form an emerging paradigm in the quest for providing EO data ready for spatial and temporal analysis; this concept, which generalizes the concept of seamless maps from 2-D to n-D, is based on preprocessing incoming data so as to integrate all data form one sensor into one logical array, say 3-D x/y/t for image timeseries or 4-D x/y/z/t for weather forecasts. This enables spatial analysis (both horizontally and vertically) and multi-temporal analysis simultaneously. Adequate service interfaces enable "shipping code to the data" to avoid excessive data transport. In standardization, datacubes belong to the category of coverages as established by ISO and OGC. In this contribution we present the OGC datacube data and service model, the Coverage Implementation Schema (CIS) and the Web Coverage Service (WCS) with its datacube analytics language, Web Coverage Processing Service (WCPS) and put them in context with further related standards is provided. Finally, we discuss architectural details of datacube services by way of operational tool examples.

Keywords Datacube · Coverage · OGC CIS · OGC WCS · OGC WCPS
ISO 19123

14.1 Motivation

Sensor, image, image timeseries, simulation, and statistics data contribute significantly to the Big Data challenge in the Earth sciences. Serving them at a high service quality and at high speed is one of the key challenges for modern technology. In particular, more and more it becomes clear that a zillion of single scenes is not the appropriate granularity for user-friendly offerings. A first step has been done some time back when seamlessly mosaicked maps enabled a smooth zoom and pan experience for users. Effectively, this has opened Earth Observation data a

P. Baumann (✉) · D. Misev · V. Merticariu · B. P. Huu
Jacobs University, rasdaman GmbH, Bremen, Germany
e-mail: p.baumann@jacobs-university.de

J. Döllner et al. (eds.), *Service-Oriented Mapping*, Lecture Notes in Geoinformation and Cartography, https://doi.org/10.1007/978-3-319-72434-8_14

much larger, new community than just geo experts—we all embrace Google Maps and similar services today as part of our common Web experience. The same paradigm shift is now heralded by the datacube concept, but along the time axis: organizing all data from one satellite instrument into a single x/y/t datacube has the potential of simplifying access to multi-temporal data. Similar arguments hold for the vertical spatial axis, spectral bands lined up as an axis, and further dimensions that may occur.

Generally, a datacube represents a multi-dimensional (usually: spatio-temporal) object which, when offered through some service, can be sliced and diced, aggregated, and analyzed. The value of a datacube unfolds when compared with classical archives where masses of data are provided as masses of files in some given granularity of partitioning and using cumbersome identification mechanisms, often with metadata encoded in file and directory names following some ad hoc convention. In a datacube such files are homogenized and integrated logically into larger spatio-temporally extended objects, such as all Level 2 products of some particular satellite sensor. The data homogenization work—which traditionally has been with the data user—is shifted to the data center where it can be done once and for all. With an appropriate datacube access and processing language, it can be avoided to download excessive data; actually, data download often can be avoided at all by letting the server do the filtering, processing, and fusion of datacubes. Altogether, datacubes effectively contribute to the goal of presenting data in an analysis-ready manner.

Experts and purists may argue that the long processing pipeline through which pixels have to go change their values in an inacceptable way; furthermore, in case of overlapping images several pixel values might be available for the same position, and these should be retained while for a datacube a choice will have to be made. It can be argued, though, that this is relevant only for a comparatively small domain of experts; for a wide range of user communities these intricacies are negligible.

This tells us that it is the operations a service offers that ultimately define usability. The paradigm of "ship code to data" is well known, but there is less agreement on what code to ship. Standard program code is not what users want to write, and what administrators want to let through the firewall. Declarative query languages, as demonstrated in business world by SQL, might be a good way forward towards shipping simple, high-level code to the data (written by humans or, more often, generated transparently by clients during interaction)—among others, this opens up opportunities for automatic server-side optimization, parallelization, etc.

In this contribution we adopt a standards-based perspective and focus on data and service models provided by adopted Earth datacube standards—while acknowledging that there is a body of research on the conceptual modeling of datacubes in sciences and engineering (Baumann 1994; Pisarev 2003; Boulil et al. 2015) as well as implementation issues (RDA 2017; Andrejev et al. 2015; Baumann et al. 2013; Cheng and Rusu 2015; Furtado and Baumann 1999), not to speak about sparse datacubes in OLAP (see Blaschka et al. 1998 for a survey).

The remainder of this chapter is organized as follows. In the next section, we discuss the logical model of datacubes and the requirements that arise from a user's quality of service perspective. After that, we present the OGC data and service model for datacubes, the Coverage Implementation Schema (CIS) and the Web Coverage Service (WCS) together with its datacube analytics language, Web Coverage Processing Service (WCPS). After that, we give a brief outlook on further standards in the field and their relationship to the OGC datacube standards. An architectural perspective complements this, based on the OGC reference implementation, rasdaman. The last section summarizes the plot.

14.2 The Datacube Conceptual Model

Recently, the term datacube is receiving increasing attention as it has the potential of greatly simplifying "Big Earth Data" services for users by providing massive spatio-temporal data in an analysis-ready way. However, there is considerable confusion about the data and service model of such datacubes.

A first condensed clarification of the term has been provided with The Database Manifesto which establishes six requirements on the conceptual data and service model (Baumann 2017a). A more architectural view, among others also taking account stewardship and infrastructure aspects, has been published subsequently (Strobl et al. 2017). From a standardization perspective, though, the conceptual model is of prime interest so we focus on this for the remainder of this section.

The Datacube Manifesto (Baumann 2017a) is a crisp 2-page summary of the core properties of a datacube structure and service model. Based on a definition of gridded multi-dimensional data it proposes six requirements; we inspect each of these in turn.

Definition. A datacube is a massive multi-dimensional array, also called "raster data" or "gridded data"; "massive" entails that we talk about sizes significantly beyond the main memory resources of the server hardware—otherwise, processing can be done satisfactorily with existing array tools like MatLab (2017) or R (2017). Data values, all of the same data type, sit at grid points as defined by the d axes of the d-dimensional datacube. Coordinates along these axes allow addressing data values unambiguously. A d-dimensional grid is characterized by the fact that each inner grid point has exactly two neighbors along each direction; border grid points have just one (see Fig. 14.1). Point clouds, e.g., are not grids.

Based on these definitions, the Datacube Manifesto establishes the following requirements on datacube services.

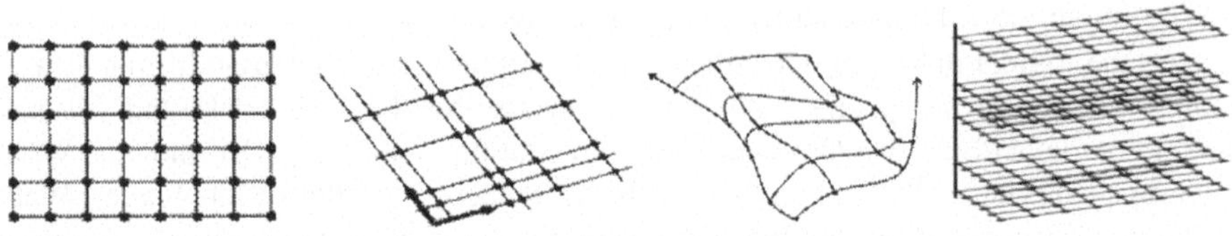

Fig. 14.1 Sample regular and irregular grids (Baumann et al. 2017a)

14.2.1 Spatio-Temporal Gridded Data

Although the name "datacube" suggests a 3-D model, the datacube model is not restricted to that. The spectrum includes 1-D sensor timeseries, 2-D imagery, 3-D x/y/t image timeseries and x/y/z subsurface voxel data, 4-D x/y/z/t climate and ocean datacubes, and even 5-D atmospheric data with two time dimensions, as mandated by climate modelers. Besides these use cases support for lower dimensions is already a must because slices extracted from, say, a 3-D x/y/t image timeseries can be 2-D x/y images or 1-D timeseries. Typically, one datacube is constructed from a large number of files, such as all scenes from one satellite instrument. By ornamenting these axes with domain-specific metadata, spatio-temporal and other semantics can be expressed. The grid can be regular, irregular, or even mixed (such as regular x/y and irregular t), cf. Fig. 14.1. Ideally, the datacube is based on the OGC/ISO Coverage Implementation Schema (CIS) standard.

Requirement 1. Datacubes shall support gridded data of at least one through four spatial, temporal, or other dimensions.

14.2.2 Treat All Axes Alike

In the past, most standards as well as implementations had dedicated methods for spatial extraction, and different ones for temporal extraction. Datacubes, on the contrary, offer the single concept of an axis which, through ornamenting metadata, becomes spatial, temporal, or anything else. After such a unifying step, trimming and slicing (see Fig. 14.2) can follow the same syntax along all axes (see later for the detailed definition of these terms). This is not withstanding different units of measure—Latitude might be measured in degrees like 42°30′, time in days like 2017-06-06, height in feet or in flight levels like FL100.

Requirement 2. Datacubes shall treat all axes alike, irrespective of an axis having a spatial, temporal, or other semantics.

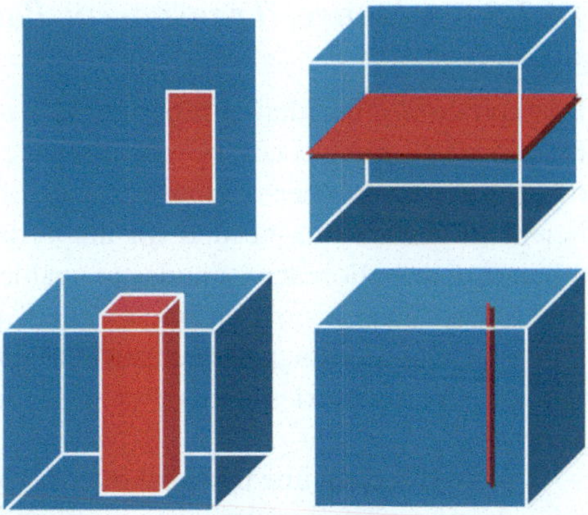

Fig. 14.2 Datacube trimming (left) and slicing (right) (Baumann et al. 2017a; Baumann 2017b)

14.2.3 Efficient Single-Step Multi-Dimensional Extraction

Analysis-ready means: ready for analysis along all dimensions. Inevitably, the first analysis task is spotting the relevant areas in space and time; extracting axis by axis, maybe even transporting intermediate results to the client, leads to excessive copying and transport of data "too big to transport". Therefore single-step multi-axis extraction is a must.

Requirement 3. Datacubes shall allow efficient trimming and slicing from a datacube along any number of axes in a single request.

Ideally, the service interface conforms with OGC Web Coverage Service (WCS) 2 standard—adopted also by EU INSPIRE and foreseen by ISO—as a modular, open, interoperable service interface supporting many protocols (such as GET/KVP and SOAP) and formats (from GeoTIFF over NetCDF to GRIB2).

14.2.4 Fast Along All Axes

Obviously, all datacube trimming and slicing should come at good performance—why is this worth mentioning? Because classic storage is optimized for horizontal access, penalizing temporal and vertical extraction. A stack of GeoTIFF scenes representing a timeseries is not yet a datacube, even when offered through a datacube API—it will be fast in extracting image slices, but convey disastrous performance on timeseries analysis.

Requirement 4. Datacubes shall convey similar extraction performance along any datacube axis.

14.2.5 Adaptive, Transparent Partitioning

It is common sense that partitioning is a must for large datacubes. It does not add functionality, but speeds up performance tremendously if done right because it minimizes the number of disk accesses on server side (ideally, one access is sufficient to get all data needed for answering a particular request), but also for an efficient distribution across nodes to enable parallel processing with minimal traffic.

Notably, there is not one single "good" partitioning; it depends on the individual mix of client access patterns, such as purely horizontal access, purely vertical access, pure timeseries access, or any combination (and variation, over time) of patterns.

A powerful datacube tool will offer partitioning as a tuning option (Fig. 14.3). In this classification, regular aligned tiling is commonly referred to as chunking (Sarawagi and Stonebraker 1994)—obviously, however, more schemes are conceivable. Sometimes it is argued that anything beyond chunking is not worth the effort. To address this, let us look at Open Street Map which also deploys tiling (Fig. 14.4). While tiling is done on vectors the challenge remains the same: in Central New York City users will want to zoom in and have very fine-grain access, reflected by fine-grain tiling, whereas in the Southern Pacific few (or even one) large tiles are adequate. Benchmarks clearly show that the performance impact of tiling can exceed an order of magnitude (Furtado and Baumann 1999).

In any case, partitioning should remain invisible at the access interface—compare it to a database index which resembles some internal tuning option invisible on SQL query level. Having to consider partitions makes analysis queries terribly complicated, in particular on more involved operations like convolution kernels and differential analysis. Offering single data objects analysis-ready is a key concept of datacubes.

These considerations condense into the following requirement:

Requirement 5. Datacubes shall allow adaptive partitioning, invisible to the user when performing access and analysis.

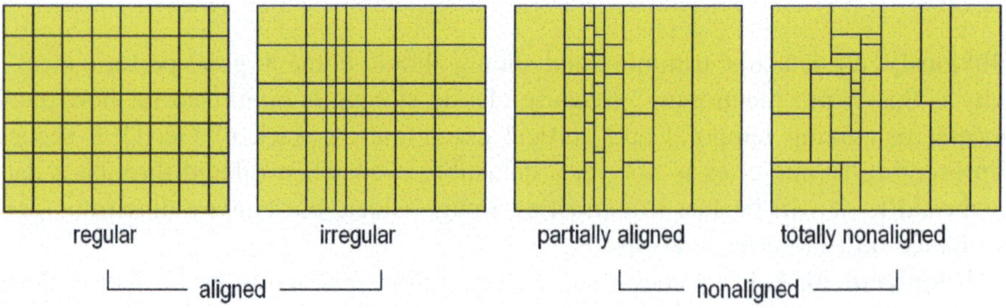

Fig. 14.3 Classification of partitioning schemes (Furtado and Baumann 1999)

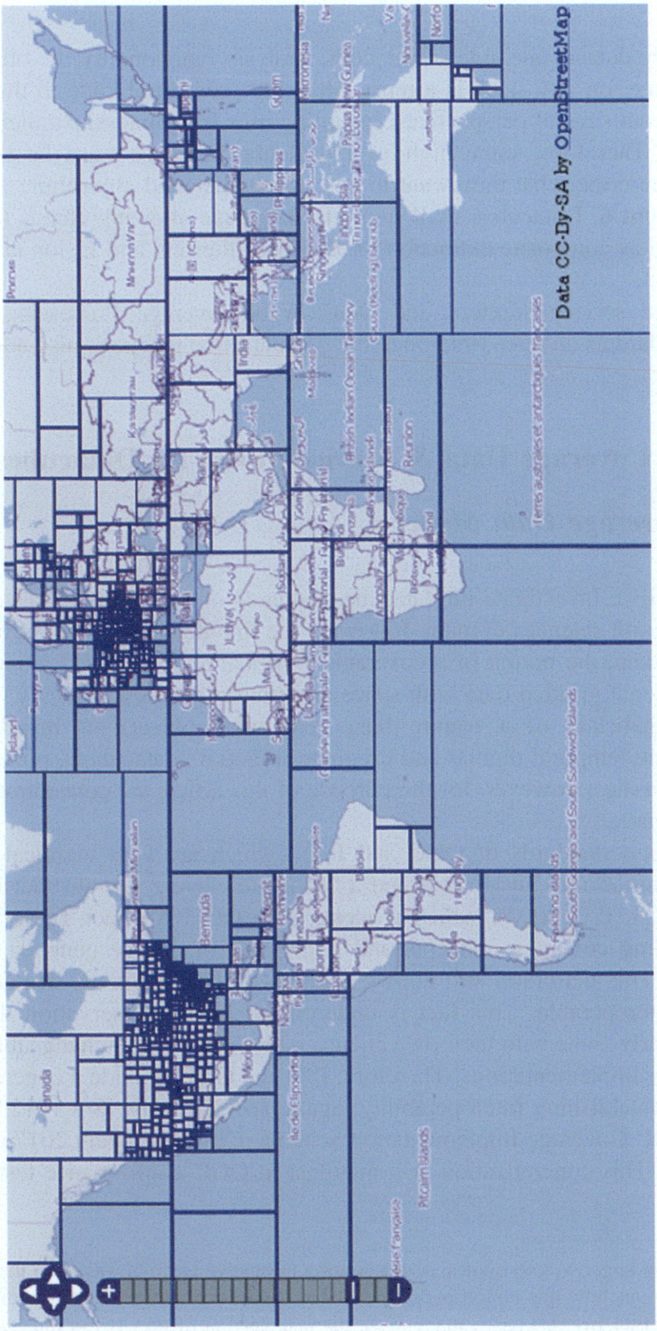

Fig. 14.4 Open Street Map tiling scheme (Open Street Map 2017)

14.2.6 Flexible Access and Analysis Language

Analysis-ready data is one side of the coin, analysis functionality the other; "any query, any time, on any size" is a must. Shipping procedural code to the cube is undesirable from several perspectives, including user friendliness, parallelizability, and security. Therefore, some high-level datacube language must be supported where users describe what they want to get, not the detailed algorithm.

Requirement 6. Datacubes shall support a language allowing clients to submit simple as well as composite extraction, processing, filtering, and fusion tasks in an ad hoc fashion.

Ideally, the service supports the OGC Web Coverage Processing Service (WCPS) standard as an open, interoperable geo datacube language also adopted by EU INSPIRE.

14.3 The Coverage Data & Service Model for Datacubes

14.3.1 Coverage Data Model

In this section we investigate on datacube standards of central importance specifically to the Earth sciences domain, following the line detailed in the introduction.

In this domain, the notion of a coverage is critical as it defines, among others, multi-dimensional gridded data with space and time semantics. Actually, the term coverage—a subclass of a feature (i.e., geographic object)—is broader as it describes spatio-temporal regular and irregular grids (i.e., datacubes), point clouds, and general meshes; however, for the purpose of this article we concentrate on the gridded data part.

The coverage standards of OGC and ISO—which are kept identical—divide coverage definition into an abstract and a "concrete" level.[1] On abstract level we find ISO 19123 (ISO 2004), which is identical to OGC Abstract Topic 6 (OGC 2007), providing coverage concepts and definitions in a very general, comprehensive way. This definition still allows manifold implementations which are not necessarily interoperable. This fact is underpinned by the observation that such services regularly come with their own clients, which are not interchangeable across different server implementations. Therefore, ISO and OGC provide a concretization of coverages establishing interoperability; again, specifications ISO 19123-2 (ISO 2016) and OGC Coverage Implementation Schema (CIS) (Baumann 2017a, b, c, d) are identical.[2] This concretization gets manifest in OGC conformance tests which

[1]We put the word *concrete* into quotes as this is not a normative term; however, in the sequel it will become obvious how this notion is complementary to the (normatively used) term *abstract*.

[2]This naming as ISO 19123-2 paves the way for the new version of ISO 19123 (currently under work) to become 19123-1 so that the numbering reflects the companion nature of both 19123-x specifications.

assess any Web Coverage Service (WCS) and the coverages it delivers down to the level of single pixels; hence, CIS/WCS are interoperable and can be used in any combination of clients and servers.

Since 2000 when the ISO 19123 document was established coverage modeling has undergone some evolution due to the increase in insight, but also in the variety of stakeholder communities with their manifold requirements. At the time of this writing, ISO has adopted CIS 1.0 as 19123-2. In parallel, OGC has been working (and has adopted) CIS 1.1 which will be addressed next by ISO. CIS 1.1 is a backwards compatible extension of CIS 1.0 (i.e., all CIS 1.0 coverages remain valid also under CIS 1.1), but with a modernized grid model plus several representation schemes requested by the communities (Hirschorn and Baumann 2017).

In the remainder of this section we inspect the coverage data structure based on CIS 1.1, reminding again that CIS 1.1 is in accordance with the abstract coverage model of OGC AT6/ISO 19123.

Generally, a coverage is a function mapping location (i.e., coordinates) to data values. This is modeled through the coverage's domain set (at what coordinates can I find values?) and its range set (the values).

The domain set consists of direct positions where values are located; via interpolation, if applicable, inbetween values may be derived. In case of a datacube (remember that coverages can express more than those) the domain set is given by a multi-dimensional grid made up from spatial, temporal, and/or other axes. Coordinates are described in a single Coordinate Reference System (CRS) which may reference a predefined one (such as defined in the EPSG database) or composed; the OGC resolver database hosts all EPSG CRSs, but also spatial and Cartesian ("index") CRSs and offers a composition mechanism to construct CRS URLs. The following example defines a 4D CRS—suitable, e.g., for weather data—via a 3D x/y/z EPSG CRS plus a time axis:

```
http://www.opengis.net/def/crs-compound?
  1=http://www.opengis.net/def/crs/EPSG/0/4979
 & 2=http://www.opengis.net/def/crs/OGC/0/AnsiDate"
```

The range set contains the cell ("pixel", "voxel", etc.) values for each direct position. Depending on the coverage's storage format this may be simple linearization (such as in XML and JSON) or some more sophisticated binary format, possibly including tiling (such as NetCDF).

A coverage's range type captures the semantics of the range set values; its definition is based on SWE (Sensor Web Enablement) Common (Robin 2017) so that sensor data can be transformed into coverages without information loss, thereby enabling seamless service chains from upstream data acquisition (e.g., through OGC SOS) to downstream analysis-ready user services (such as OGC WMS, WCS, and WCPS).

Notably, the range type can go far beyond just a datatype indicator (such as integer versus float); unit of measure, accuracy, nil values, and the semantics (by way of a URL reference), and more information can be provided with a range type, thereby accurately describing the meaning of the values. The following is an example range type definition for panchromatic optical data, encoded in GML:

```
< swe:field name = "green">
 < swe:Quantity
  definition = "http://opengis.net/def/property/OGC/0/Radiance">
   < swe:description > Green Channel </swe:description>
     < swe:nilValues>
     < swe:NilValues>
       < swe:nilValue reason=
       "http://www.opengis.net/def/nil/OGC/0/BelowDetectionRange">
       0
       </swe:nilValue>
       < swe:nilValue reason=
      "http://www.opengis.net/def/nil/OGC/0/AboveDetectionRange">
       255
       </swe:nilValue>
     </swe:NilValues>
  </swe:nilValues>
  < swe:uom code = "W.m-2.sr-1.nm-1"/>
  < swe:constraint>
    < swe:AllowedValues>
      < swe:interval > 0 255 </swe:interval>
      < swe:significantFigures > 3</swe:significantFigures>
    </swe:AllowedValues>
   </swe:constraint>
 </swe:Quantity>
</swe:field>
```

Finally, an optional metadata bucket is part of the coverage which can carry any additional information that may be relevant. Figure 14.5 shows this structure as UML class diagram.

Such coverages can be structured in a variety of representations—including tilings, band-separated coverages, coordinate/value pair lists supporting streaming, as well as nested coverages—and encoded in a variety of formats, such as XML, JSON, RDF, a variety of binary formats (including GeoTIFF, NetCDF, GRIB2, and several more), as well as "containers" (Fig. 14.6). In the latter variant, some suitable container format (such as Multipart/MIME, zip, SAFE, etc.) will contain a canonical header in some information-lossless format (like the abovementioned XML, JSON, and RDF) accompanied by one or more files which typically would

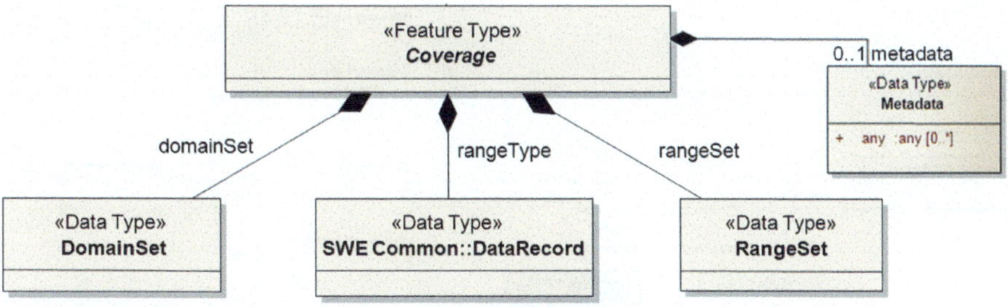

Fig. 14.5 UML class diagram of a coverage (Baumann et al. 2017a)

contain some efficient binary encoding. Hence, tools can request coverages in their favorite format from a server.

In order to keep this wide range of features tractable CIS 1.1 groups them into conformance classes (Fig. 14.7). Any implementation must realize the Core conformance class; aside from that, choice of optional modules is guarded by the dependencies. For example, an implementation might support regular or even irregular grids (green classes), partitioning of coverages (grey class), and any of the formats presented in the yellow encoding class category.

14.3.2 Coverage Service Model

While coverages can be served through a variety of interfaces—including WFS, WMS, and WPS—only Web Coverage Service (WCS) (Baumann 2017b) offers comprehensive functionality. Similar to the modularly organized CIS, the WCS suite also has a mandatory Core around which a series of optionally implementable extensions gathers. WCS Core offers very simple, focused coverage access and extraction functionality. Subsetting of a coverage allows trimming (getting a cutout of the same dimension) and slicing (getting a cutout with reduced dimension) as

Fig. 14.6 Different coverage encodings (Baumann et al. 2017a)

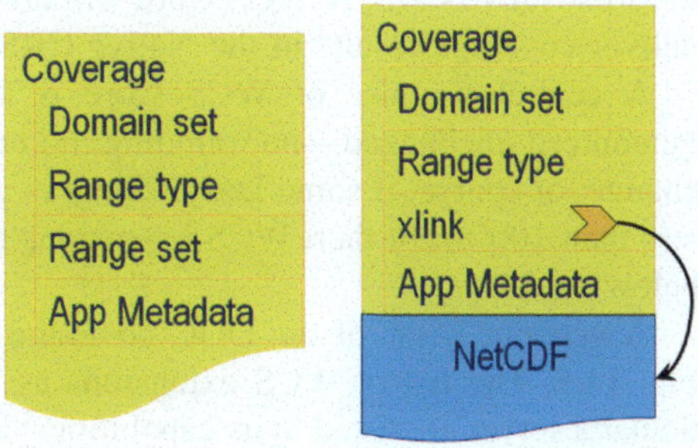

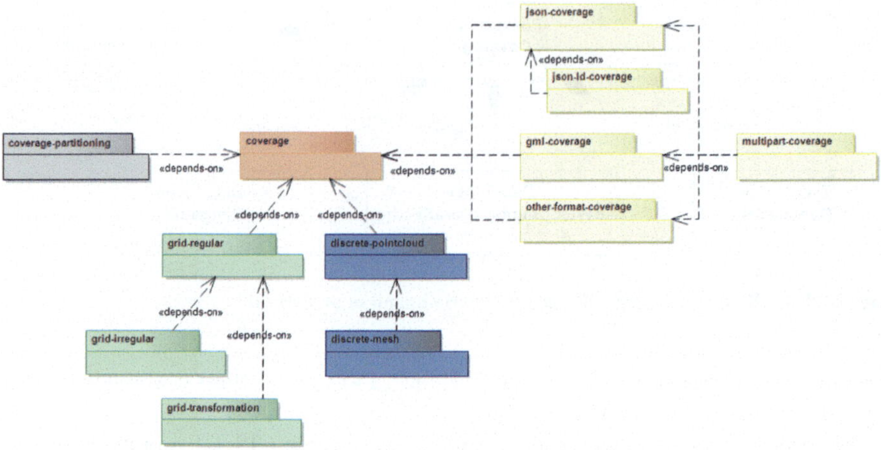

Fig. 14.7 CIS 1.1 conformance classes and their dependencies (Baumann et al. 2017a)

introduced earlier, as well as encoding in some user-selectable format. The following GetCoverage request, using GET/KVP encoding, performs trimming in space and slicing in time, effectively extracting a 2D GeoTIFF image from a 3D timeseries datacube:

```
http://www.acme.com/wcs ? SERVICE=WCS & VERSION=2.0
   & REQUEST=GetCoverage & COVERAGEID=c001
   & SUBSET=Long(100,120) & SUBSET=Lat(50,60)
   & SUBSET=time("2009-11-06T23:20:52")
   & FORMAT="image/tiff"
```

The encoding format in which a coverage is stored in a server is called its Native Format; without any FORMAT parameter in a request the coverage will be delivered in its Native Format—note that this also applies in case of subsetting. The CRS in which the coverage's domain set coordinates are expressed is referred to as its Native CRS. A WCS Core will always interpret subsetting coordinates and deliver coverage results in the Native CRS of the coverage addressed.

A central promise of WCS Core is that coverage values will be delivered guaranteed unchanged—no rounding, no resampling etc. may take place. This may change, of course, if some lossy format is used (such as JPEG with a quality factor less than 100%); further, WCS extensions may change this behavior as we will see below.

A synoptic view of the OGC coverage data and service model is presented in Fig. 14.8. The list of WCS extensions as well as data formats supported by particular a server are listed in its capabilities document which can be retrieved through a GetCapabilities request.

WCS Extensions can be classified into functionality extensions and protocol bindings. The latter allow expressing any request of Core and extensions in one of GET/KVP, POST/XML, SOAP, and (prepared, but still under discussion) REST. The functionality-enhancing extensions provide a variety of useful extras over WCS Core; we will briefly look at each service extension in turn:

- *WCS Range Subsetting* extends the GetCoverage request with an extra, optional parameter to select range components ("bands", "variables"). With RANGESUBSET a selection of bands can be extracted; for example, the following request extracts a false-color image from some hyperspectral scene:

```
http://www.acme.com/wcs ? SERVICE=WCS & VERSION=2.0
   & REQUEST=GetCoverage & COVERAGEID=c001
   & RANGESUBSET=nir,red,green
```

  Generally, one or more bands can be extracted and brought into any sequence desired. For large amounts of bands, intervals can be specified, such as red:blue.
- *WCS Scaling* allows changing resolution of a gridded data set; to this end, an optional SCALING parameter is added to the GetCoverage request. The

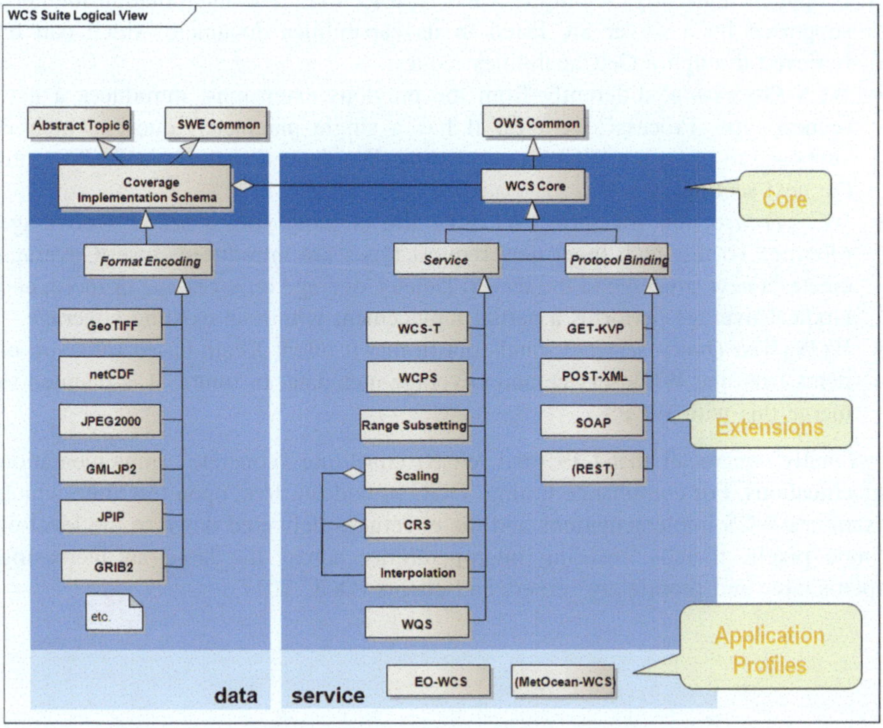

Fig. 14.8 OGC WCS suite: synopsis of data and service model specifications (Baumann 2017d)

interpolation method applied is implementation dependent unless WCS Interpolation is implemented in the server (see below). Obviously, in presence of scaling it cannot be guaranteed any longer that range values remain unchanged.

- *WCS CRS* supports reprojection of a coverage from its Native CRS—i.e., the one in which the coverage is stored in the server—into another one. With the same mechanism as before, an optional parameter OUTPUTCRS lets users indicate the target CRS in which the coverage should be delivered. Following OGC convention, CRSs are expressed as URLs, such as in the following example:

```
http://www.acme.com/ows?SERVICE=WCS & VERSION=2.0
   & REQUEST=GetCoverage & COVERAGEID=c001
   & OUTPUTCRS=http://www.opengis.net/def/crs/EPSG/0/4326
```

  The list of CRSs supported by a server are listed in its capabilities document which can be retrieved through a GetCapabilities request.
- *WCS Interpolation* enables clients to request a particular implementation method—such as nearest-neighbor, linear, or quadratic—whenever interpolation is applied (currently: scaling or CRS reprojection). The interpolation methods supported by a server are listed in its capabilities document which can be retrieved through a GetCapabilities request.
- *WCS Processing*, differently from the previous extensions, introduces a new request type, ProcessCoverages. It has a single parameter, query, which is supposed to contain a WCPS query string. We defer details about WCPS until the next section.
- *WCS Transaction* (or short: WCS-T) allows modifying a server's coverage offering. To this end, three new request types are introduced: InsertCoverage creates a new coverage in the server; DeleteCoverage removes one or more, and UpdateCoverage performs a partial replacement within an existing coverage.
- *WQS* (*Web Query Service*) stands out in that it offers XPath-based selection of items from the WCS service and coverage metadata. In future, it is planned to merge this with WCPS.

Finally, we recall that CIS (and WCS) constitute "concrete" implementation specifications. For compliance testing, OGC provides a free, open test suite which examines WCS implementations and the coverages delivered down to the level of single pixels, thereby ensuring interoperability across the large an increasing open-source and proprietary servers and clients (OGC 2017).

14.3.3 Coverage Analytics

Big Earth Data Analytics demands "shipping the code to the data"—however, the question is: what type of code? For sure not procedural program code (such as C++ or python) which requires strong coding skills from users and developers and also is highly insecure from a service administration perspective; rather, some safe, high-level query language is of advantage, in particular as it also enables strong server-side optimizations, including parallelization, distributed processing, and use of mixed CPU/GPU, to name a few. Further, to provide data ready for spatial and temporal analytics, a semantically adequate integrated space/time coordinate handling is necessary. Already in 2008, OGC has standardized the Web Coverage Processing Service (WCPS) geo datacube analytics language which today is part of the WCS suite (Baumann 2010, 2017c). Today, WCPS is being used routinely in Petascale operational datacube services, e.g., in the intercontinental EarthServer initiative (EarthServer 2017).

In a nutshell, the WCPS language works as explained by the following example; see (Baumann 2010) for a comprehensive presentation. Consider the following WCPS query which expresses the task "From MODIS scenes M1, M2, M3, deliver NDVI as GeoTIFF, but only those where nir exceeds 127 somewhere":

```
for $c in ( M1, M2, M3 )
where some( $c.nir > 127 )
return encode( ($c.red−$c.nir)/($c.red+$c.nir), "image/tiff" )
```

The for clause defines iterations over coverage objects; the where clause allows filtering of coverage combinations to be retained, and in the return clause, processing (including coverage fusion) of the coverages is specified, including an indication of the delivery format of the result. This syntax has been designed tentatively close to XQuery/XPath so as to allow integration with XML or JSON based metadata retrieval; Kakaletris et al. have implemented a coupling this language with XPath, leading to unified data/metadata queries (Liakos et al 2015).

At the heart of WCPS are coverage expressions. We have seen already so-called induced operations where any unary or binary function defined on the cell types can be applied to all coverage cells simultaneously. This includes all the well-known arithmetic, comparison, Boolean, trigonometric, and logarithmic operations, as well as a case distinction. The following example performs a color classification of land temperature (note the null value 99,999 which is mapped to white) (Fig. 14.9).

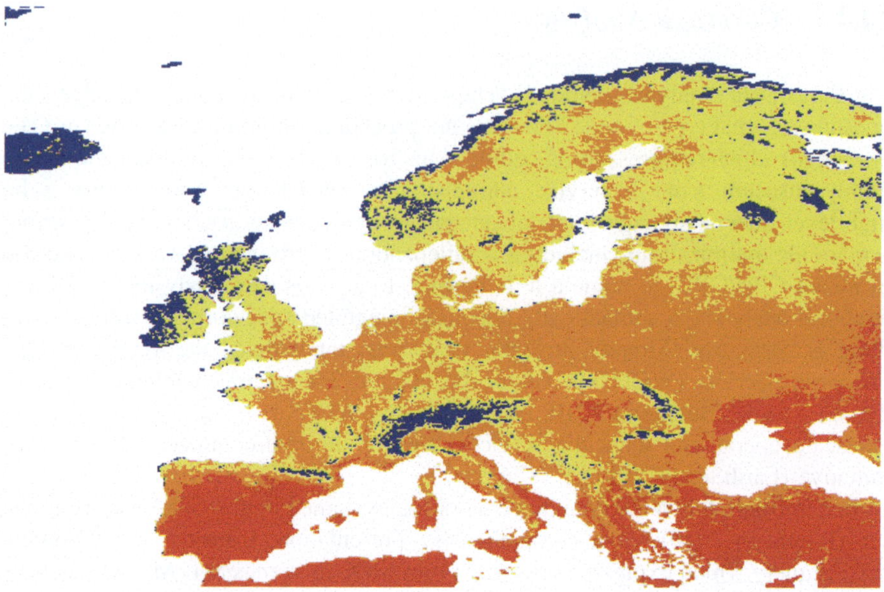

Fig. 14.9 Result of WCPS classification query yielding a color image (*source* rasdaman)

```
for c in ( AvgLandTemp )
 return
  encode( switch
   case c[date("2014-07"), Lat(35:75), Long(-20:40)] = 99999
    return {red: 255; green: 255; blue: 255}
   case 18 > c[date("2014-07"), Lat(35:75), Long(-20:40)]
    return {red: 0; green: 0; blue: 255}
   case 23 > c[date("2014-07"), Lat(35:75), Long(-20:40)]
    return {red: 255; green: 255; blue: 0}
   case 30 > c[date("2014-07"), Lat(35:75), Long(-20:40)]
    return {red: 255; green: 140; blue: 0}
   default return {red: 255; green: 0; blue: 0},
   "image/png"
  )
```

In the next example we combine a satellite image with an elevation model; by copying elevation data (which need to be scaled to the resolution of the scene) into a PNG image, WebGL can render this in 3-D in a browser, thereby enabling terrain draping (Fig. 14.10). The corresponding query looks as follows:

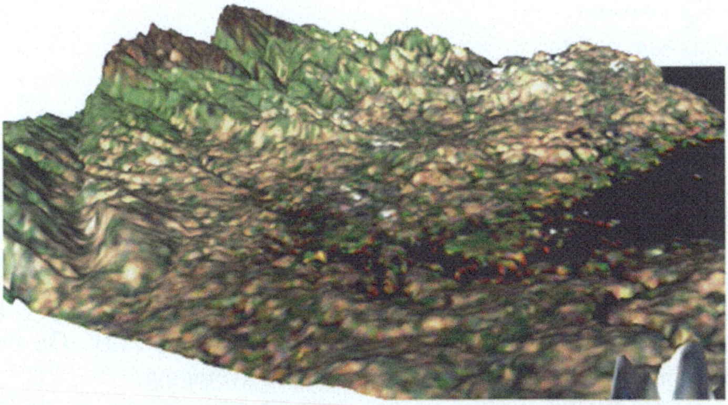

Fig. 14.10 Rendering of a WCPS terrain draping query result through WebGL (*source* EarthServer)

```
for $s in (SatImage), $d in (DEM)
where $s/metadata/@region = "Glasgow"
return
  encode(
    struct {
      red:  (char) $s.b7[x0:x1,x0:x1],
      green: (char) $s.b5[x0:x1,x0:x1],
      blue:  (char) $s.b0[x0:x1,x0:x1],
      alpha: (char) scale( $d, 20 )
    },
    "image/png"
  )
```

These example shows that WCPS can generate visualizations as a special case of processing, specifically: if the result structure is a 2-D 3-band (sometimes 4-band) integer image. In general, we need to emphasize that WCS is a data service delivering data that can be processed further, as opposed to WMS and WMTS which deliver images for human consumption.

All these coverage-combining expressions form part of a larger expression type where a new coverage is built from one or more existing ones; in particular, the combination of two or more coverages constitutes an array join which poses specific implementation challenges (see later). Creating a coverage is accomplished through an expression like the following one which delivers a coverage containing the average value over the pixel history for each location):

```
coverage myCoverage
over   $y imageCrsDomain( $cov, Lon ),
       $x imageCrsDomain( $cov, Lat )
values  avg( $cov[ $y, $x, *:* ] )
```

A coverage with identifier myCoverage is built from a coverage identified by variable $cov which has dimensions Lon, Lat, and date; iteration variables $y, $x, and $t are bound to each dimension, respectively, assuming all possible coordinates as defined in the coverage domain set. The values clause itself can contain any expression, thereby allowing nested queries of any complexity. On the fly we already have introduced aggregations, in this case averaging.

The result of a WCPS request can be of completely different dimensionality as compared to the input coverage(s). For example, the following query retrieves a histogram represented as a 1-D CSV array:

```
for $s in ( LandsatScene )
return
  encode(
    coverage $bucket in ( 0:255 ) values count( $s.red = $bucket ),
    "text/csv"
  )
```

As a further example, the following request performs a matrix multiplication over two 2D coverages A and B with sizings m × n and n × p, resp.:

```
for $a in ( A ), $b in ( B )
return
  encode(
    coverage matMult
    over $i in [0:m], $j in [0:p]
      values condense +
          over    $k in [0:n]
          using   $a [ $i, $k ] * $b [ $k, $j ],
    "text/json"
  )
```

The usual aggregation operations (count, sum, avg, min, max, and the all and exists quantifiers) are mere shorthands of a general aggregator, called condenser, which follows the scheme indicated by the following example:

Fig. 14.11 Kaleidoscope of multi-dimensional WCPS query results (*source* rasdaman)

```
condense min
over    $d in domain($cov)
using   $cov[ $d ]
```

The result of this expression will be a single scalar value representing the minimum of all values in the target coverage; incidentally, this expression is equivalent to min($cov).

A more complex example would perform an edge filtering on a 2D image using a 3 × 3 filter matrix:

```
coverage filteredImage
over    $px x ( x0 : x1 ),
        $py y ( y0 : y1 )
values  condense +
        over    $kx x ( -1 : +1 ),
                $ky y ( -1 : +1 )
        using  $c.b[ $kx + $px, $ky + $py ] * k[ $kx , $ky ]
```

Figure 14.11 collects sample results all coming from WCPS queries, thereby demonstrating the broad range of applications.

14.3.4 Datacube Clients

In the past sections we have seen WCS requests and WCPS queries. None of these should be interpreted as an end user interface, and definitely datacubes do not enforce 3D visualization or any other visualization method.

Actually, OGC datacubes constitute nothing but a logical model with convenient APIs for rapid application development; thanks to the standards, any client

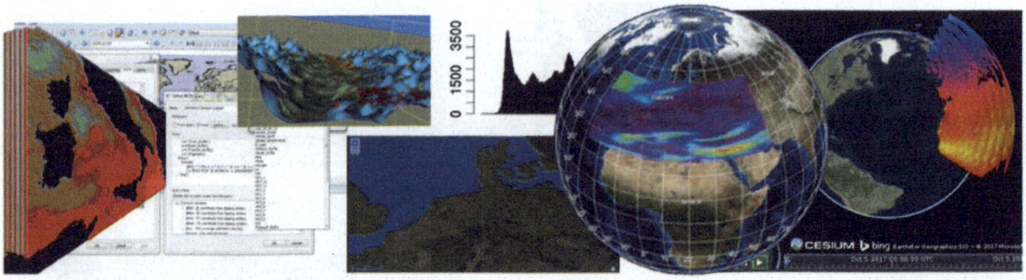

Fig. 14.12 Selected WCS/WCPS clients (from left to right): DLR 3D client; Envitia Chartlink; WebGL; OpenLayers; python; NASA WorldWind; Microsoft Cesium (*source* rasdaman/EarthServer)

supporting WCS or WCPS can interface to any service supporting the same. The contents of a datacube maybe aggregated into 1D timelines, or 2D horizontal slices may be extracted, surface data might get draped over terrain data yielding a 3D representation in the browser, etc. In the end, the range of possible visualizations is as broad as the extractions and aggregations that can be performed. By way of example, Fig. 14.12 lists some of the clients used routinely in EarthServer.

14.4 Related Standards

In this section, we address related data and service standards for datacubes; for a tool survey see, e.g., RDA 2017.

14.4.1 OLAP and Statistics Datacubes

Actually, datacubes are used since long in statistical databases and Online Analytical Processing (OLAP) (Blaschka et al. 1998). Such datacubes share some properties with spatio-temporal raster data (mainly, their multi-dimensional nature) but also differ in some aspects. Scaling of OLAP datacubes ("roll-up") is done only in discrete steps which are modeled in advance through dimension hierarchies (such as days–months–years) whereas scaling in raster data is continuous, possible for any real-numbered scale factor. Further, OLAP data are sparsely populated, typically three to five percent, whereas raster data typically are densely populated: in a weather forecast, for example, every cell contains a value. This has given rise to completely separate technological developments, leading to OLAP databases and ad hoc storage of raster date in file systems, more recently: in Array Databases.

Let us discuss this briefly. In Relational OLAP (ROLAP), every cube cell is represented as a tuple in a so-called fact table. This is efficient on sparse data, and

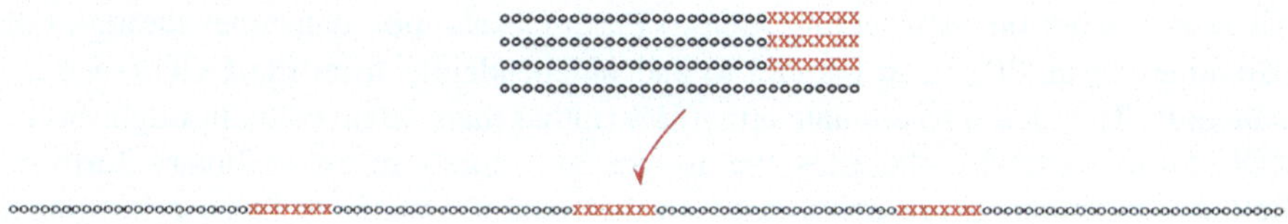

Fig. 14.13 Loss of nD spatial clustering during linearization for storage

actually resembles an effective compression scheme. However, it has been shown (Cheng and Rusu 2015) that on dense data a representation of raster data as tuples in a relational database is prohibitively inefficient for storage and processing by orders of magnitude as compared to dedicated array engines. One reason is that spatial clustering of cube cells is lost in relational storage (Fig. 14.13).

14.4.2 Datacubes in SQL

ISO following research progress in Array Databases has established SQL/MDA ("Multi-Dimensional Arrays") as a candidate standard currently under final adoption vote (ISO 2017). These array extensions are generic and need to be ornamented through spatio-temporal meta information, such as axis definitions and extents, to be used in an Earth science context. The following example, showing a table definition and a query on it, may serve to illustrate the flavor.

```
create table ThematicScenes(
  id: integer not null, acquired: date,
  scene: integer mdarray [ 0:4999,0:4999] )
select id, encode((scene.band1-scene.band2)
            / (scene.band1+scene.band2)), „image/tiff" )
from  ThematicScenes
where acquired between „1990-06-01" and „1990-06-30" and
     avg((scene.band3-scene.band4)/(scene.band3+scene.band4)) > 0
```

Aside from the high-level datacube language, SQL/MDA will be a revolution in that, for the first time in history, data and metadata retrieval can be combined in one and the same query. This effectively is going to abolish the age-old data/metadata chasm, establishing a common integrated data modeling platform for all kind of Earth data.

14.4.3 Datacubes in RDF/SPARQL

W3C RDF Data Cube Ontology (QB) is an RDF-based framework originally devised for sparse statistical data (Cyganiak and Reynolds 2014); recently, W3C has started work on representing (dense) Earth Observation datacubes through QB (Brizhinev et al. 2017), in parallel to and independently from the OGC coverage standards. The idea is to emulate array tiles (rather than single cells) through sets of RDF triples whereby the tiles themselves are stored in some binary format.[3] Retrieval examples given concentrate on metadata it remains to be seen how such services can be designed to perform, e.g., image and timeseries processing involving coordinate addressing in a simple and efficient manner. In this context, the array/RDF integration work with SciSPARQL needs to be mentioned (Andrejev and Risch 2012) which injects arrays into the RDF data model; in its implementation, a mediator approach is employed where array subqueries are redirected to a dedicated engine for an efficient evaluation.

14.4.4 Further Data Standards

NetCDF, HDF, and GRIB are well known examples of data format standards for exchanging multi-dimensional data. While they all offer an array model at their core, metadata management differs significantly. Mapping of OGC CIS to these and further formats are described through CIS extension specifications; for a guaranteed lossless use they can be combined in CIS containers where a canonical header ensures that all metadata are retained and available in some canonical format like XML, JSON, or RDF. Conversely put, CIS critically relies on such encodings.

OGC TimeseriesML is a data standard for organizing data along the time axis, effectively yielding a 1D sequence. By storing 2D images as cell values a 3D datacube can be emulated on principle. When used as a data organization for a datacube service extraction of image slices will be efficient; however, inefficiencies will be encountered when doing timeseries extraction and analysis as every image slice needs to be loaded for extracting the history of a particular pixel location, as illustrated in Fig. 14.14. No corresponding service model is provided; theoretically, though, TimeseriesML could act as an output format generated by a WCS server.

CoverageJSON is early-stage work on a JSON encoding for multi-dimensional data (CoverageJSON 2017). It represents an interesting approach from a scientific perspective. However, while it adopts some design ideas from the OGC coverage standard CIS it is not compatible through its basically different structure and information content. While this is not a problem from a research perspective it can lead to significant confusion in practice; for example, a WCS as per OGC, ISO, and INSPIRE standard cannot operate on W3C CoverageJSON due to its incompatibilities. Note also that OGC CIS already offers a JSON encoding, in addition to XML, RDF, and binary formats.

[3]This is similar to PostGIS Raster and Teradata ASTER where users need to reassemble tiles explicitly in their queries.

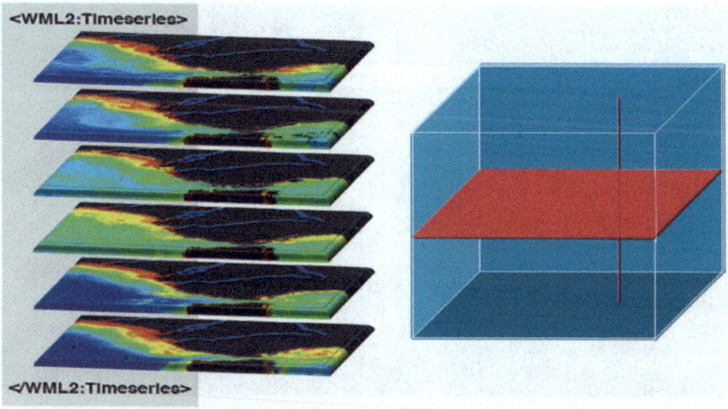

Fig. 14.14 WaterML/TimeseriesML slice stack and CIS coverage cube

14.4.5 *Service Standards*

While coverages long have been neglected in the orchestration of geo services currently a significant increase in interest can be observed.

OGC Sensor Web Enablement (SWE) is a suite of standards focusing on any kind of sensor data. SWE and coverages are harmonized in particular through the shared definition of the cell semantics: the CIS schema reuses the SWE Common DataRecord for its range type definition. This ensures that sensor data collected through, say, Sensor Observation Service (SOS) (Broering et al. 2017) can be exposed without loss of information by a WCS. In general, SWE and its SOS are suitable for upstream data collection and harmonization whereas the WxS services provide functionality for down-stream user services (Fig. 14.15).

Web Processing Service (WPS) defines a generic interface mechanism for offering any kind of service via an OGC Web API (Mueller and Pross 2017). Due to this genericity, WPS per se does not define any particular semantics—the invocation syntax is given as a machine-readable process definition while the semantics is provided human-readable in the title and abstract of the process definition (Fig. 14.16). WCPS, conversely, establishes semantic interoperability based on the concise, formal definition of the query language—a query like

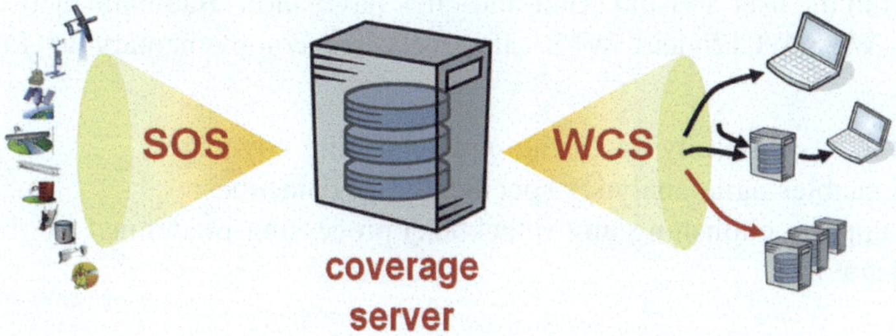

Fig. 14.15 OGC service pipelines: upstream SOS and downstream WCS

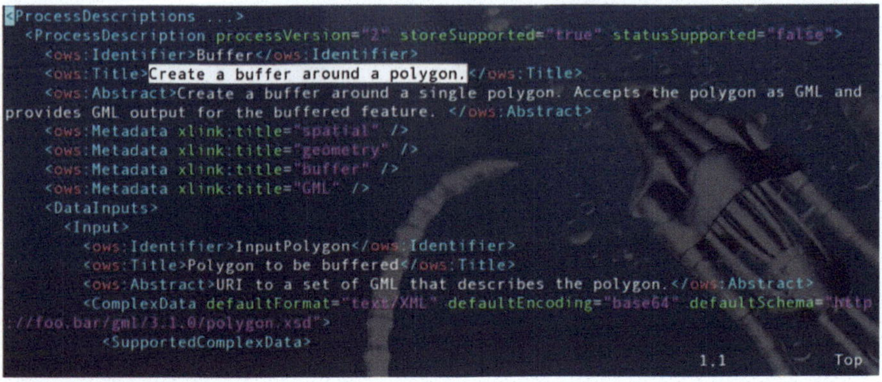

Fig. 14.16 Sample WPS process definition excerpt

Fig. 14.17 Complementarity of WCS, WCPS, and WPS

```
for $c in ( M1, M2, M3 )
return encode( abs( $c.red - $c.nir ), "application/x-hdf" )
```

will always be understood identically on client and server side. WPS is particularly useful in cases where the semantics is too complex for formalization (and for WCPS query formulation), such as in offering simulations over the Web.

However, it is well possible to define a WPS process as a façade for WCS and WCPS. For example, a WCPS process would receive a character string as input (the query) and return a set of coverages or, in case of an aggregation query, a set of scalar values. This has been demonstrated successfully in the VAROS project where the Envitia ChartLink client automatically generated WPS-embedded WCPS requests from the user's visual point-and-click navigation (Baumann et al. 2011). In summary, WCS/WCPS and WPS effectively are complementary as Fig. 14.17. illustrates:

- WCS focuses on easily accessing and subsetting coverages;
- WCPS enables agile analytics specifically on datacubes;
- WPS supports publishing any (blackbox) processing predefined by the service administrator.

14.5 Architectural Support for Datacubes

By providing analysis-ready data to users these get released from the classical workload items of selecting, homogenizing, and integrating data—this functionality now has to be provided by the service provider as an enhancement of its quality of service. In other words, this particular—usually time consuming—work needs to be accomplished by those offering data access, naturally enhancing it with analysis capabilities. One might ask, then, to what extent technology can support service providers in their new situation.

Indeed, methods and tools for datacube handling have been and are being developed in several domains, as discussed in the introduction. For our discussion of tool support and suitable architectures we focus on the OGC Reference Implementation of WCS and its extension, rasdaman ("raster data manager"), which has been developed over two decades into a cross-domain datacube engine (Baumann 1994; Baumann et al. 2013; Baumann and Merticariu 2015; Dumitru et al. 2016; Furtado and Baumann 1999) which today is in operational use by smart farming startups (Karmas et al. 2015; Tzotsos and Karmas 2017) and on Petascale research data (Baumann et al. 2017b; Marco Figuera et al. 2017). A general survey of Array Databases and related technology on datacubes is in progress by RDA (2017).

The rasdaman engine resembles a complete software stack, implemented from scratch to support management and retrieval on massive multi-dimensional arrays in a domain agnostic way. Its raster query language, rasql, extends SQL with declarative n-D array operators; actually, rasql has been selected by ISO as the blueprint for the forthcoming SQL Multi-Dimensional Arrays (MDA) standard (ISO 2017).

The overall system architecture, shown in Fig. 14.18, centers around the rasdaman worker processes which operate on tiled arrays. Storage can be in some relational database system (such as PostgreSQL), directly in some file system (which is about 2x faster than using a DBMS), or by accessing pre-existing archives without copying data into rasdaman.

Data ingestion is done based on WCS-T, but extending it with an Extract/Transform/Load (ETL) tool which homogenizes data, addresses defaults and missing data and metadata, as well as the tiling strategy to be followed. In addition, a variety of tuning parameters is available to the database administrator, including adaptive data partitioning ("tiling", cf. Fig. 14.19), lossless and lossy compression, indexing, cache sizings, etc. To this end, the insert statement has been extended with a storage layout sub-language detailing the tiling strategy to be chosen, etc. For example, the following excerpt would choose strategy "area of interest", enumerating the areas of interest and providing default sizings outside these, plus more:

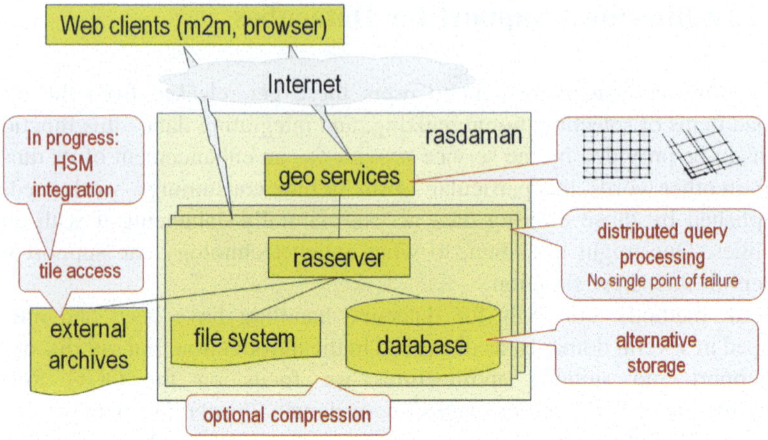

Fig. 14.18 High-level architecture of the rasdaman datacube engine (Baumann et al. 2013)

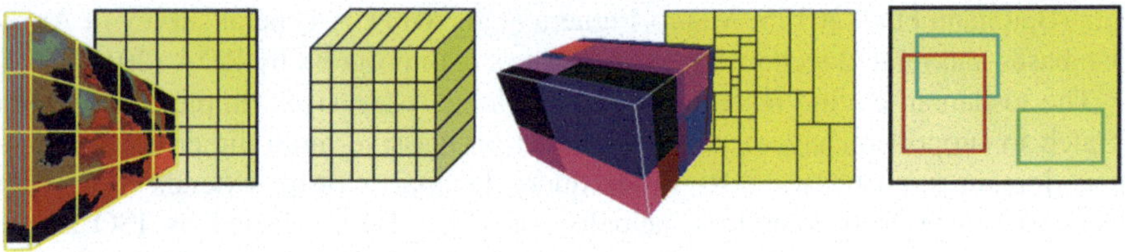

Fig. 14.19 rasdaman datacube partitioning strategy examples (*source* rasdaman)

```
insert into MyCollection
 values ...
 tiling
   area of interest [0:20,0:40], [45:80,80:85]
   tile size 1000000
   index d_index storage array compression zlib
```

During connect, each client gets assigned a dedicated process for a complete session, thereby effectively isolating different sessions against each other. Internally, each worker process is multi-threaded to make use of multiple CPU and GPU cores. Incoming queries undergo a variety of optimizations, including query rewriting (including common subexpression elimination), cost-based optimization, join strategy selection, semantic caching, and mixed multi-core CPU/GPU processing. In a rasdaman federation, worker processes can fork subqueries to other nodes for load sharing and data transport minimization (Fig. 14.20). Queries have been successfully distributed across more than 1000 Amazon cloud nodes (Dumitru et al. 2016).

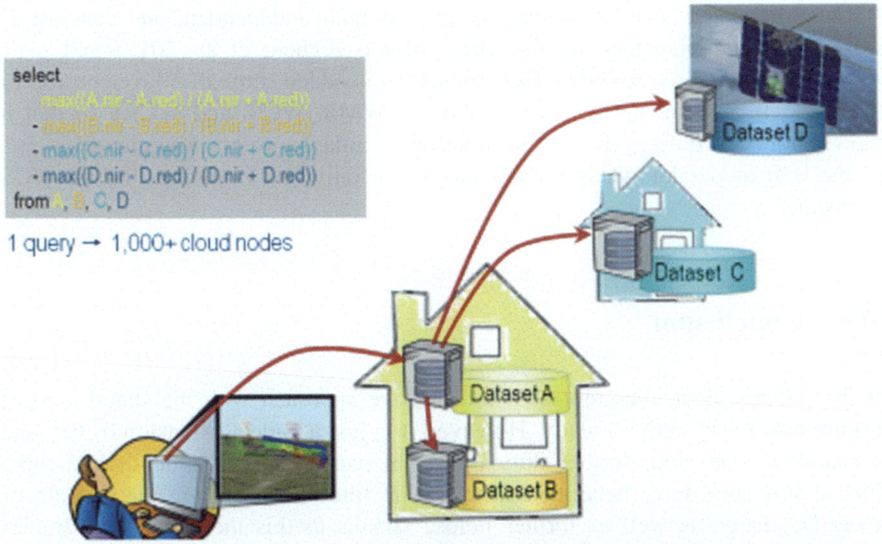

Fig. 14.20 rasdaman transparent distributed query processing (*source* rasdaman)

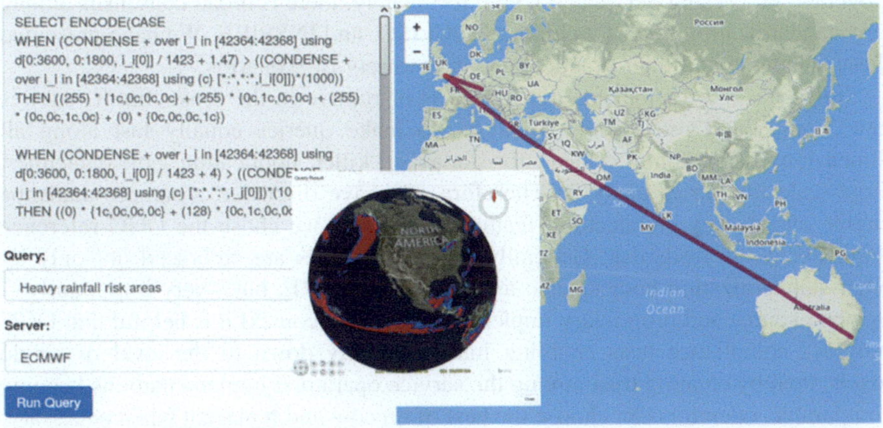

Fig. 14.21 Sample rasdaman distributed query processing in a federation: query sent from Germany to ECMWF/UK which forks off a subquery to NCI Australia, returning a precipitation/ Landsat8 fusion product; result displayed on a virtual globe in center (*source* EarthServer)

By using the same principle across data centers, federations can be established. Figure 14.21 shows a visualization of actual federated query processing between the European Centre for Medium-Range Weather Forecast (ECMWF) in the UK and National Computational Infrastructure (NCI) in Australia—both running rasdaman—for determining heavy rainfall risk areas from precipitation data at ECMWF and the Landsat8 datacube at NCI.

The rasdaman system described so far is domain independent and can just as well serve, say, planetary science data (Marco Figuera et al. 2017) and gene expression data (Pisarev 2003). Geo semantics is added through a layer on top of this stack; it implements the OGC standards WMS, WCS, and WCPS so that it knows about coordinate reference systems and regular as well as irregular grids, etc. To the best of our knowledge, rasdaman is the only tool implementing all WCS extensions.

14.6 Conclusion

At first glance, datacubes represent a disruptive approach with the stated goal of making data more analysis ready. However, it is just a logical extension of the idea of seamless maps into further dimensions, in particular: height/depth and time. Current standards have benefitted from a long line of research in the domain of Array Databases as well as further fields. Thanks to this theoretical and implementation work a set of Earth datacube standards is in place today, encompassing the OGC Coverage Implementation Schema (CIS) accompanied by the OGC Web Service (WCS) service model together with its datacube analytics language, Web Coverage Processing Service (WCPS); fortunately, there is broad consensus among key standardization bodies including OGC, ISO, and INSPIRE. We have presented these standards and their practical use in a variety of scenarios.

Still, datacube implementation indeed is disruptive in the sense that it requires new storage and processing paradigms to make queries equally fast along all dimensions—something critical for today's "killer application" of datacubes, large-scale timeseries analysis. Therefore, we have discussed a sample datacube engine and its key architectural features; naturally, we chose the OGC reference implementation, rasdaman. Generally, OGC datacubes are embraced not only by other standardization bodies, such as ISO and INSPIRE, but—very importantly—by open-source and proprietary implementers. In this context it is helpful that OGC provides compliance tests enabling interoperability down to the level of single pixels, thereby creating trust among the service operators: implementations become comparable, operators can choose the best of breed—and replace it when necessary. Altogether, availability of datacube standards and tools is heralding a new era of service quality and, ultimately, better data insights.

Already today, datacube services like the Petascale EarthServer initiative have proven to be significantly more user-friendly, faster, and scalable than typical classical services—massive gridded spatio-temporal data become analysis-ready, they can be exploited better and faster. As the EarthServer example shows, the OGC datacube standards CIS, WCS, and WCPS allow scalable service implementation and constitute flexible, handy client/server programming interfaces.

Standardization work on CIS, WCS, and WCPS is ongoing. The abstract coverage model of ISO 19123 is under renovation to yield ISO 19123-1, tightly aligned with ISO 19123-2 aka CIS. After adoption of CIS 1.0 by ISO version 1.1 is going to follow. In WCS, new functionality extensions are being asked for by stakeholders, such as multi-dimensional raster/polygon clipping. For WCPS, the data/metadata integration achieved with SQL/MDA on relational level is foreseen also for XML-based retrieval; see the work in (Liakos et al. 2015). Further requirements can be expected along with the increasing uptake pace, likely resulting in new WCS extensions; raster/polygon clipping is one candidate already under work.

Acknowledgements Research is partially funded by European Commission EINFRA project EarthServer-2 and Federal Ministry of Food and Agriculture project BigPicture.

References

Andrejev A, Risch T (2012) Scientific SPARQL: semantic web queries over scientific data. In: 3rd International ICDE workshop on data engineering meets the semantic web (DESWEB 2012). Arlington, USA, 1–5 Apr 2012, pp 5–10

Andrejev A, Baumann P, Misev D, Risch T (2015) Spatio-temporal gridded data processing on the semantic web. In: 2015 IEEE international conference on Data Science and Data Intensive Systems (DSDIS 2015), December 11–13. Sydney, Australia, pp 38–45

Baumann P (1994) On the management of multidimensional discrete data. VLDB J 4(3):401–444 (Special Issue on Spatial database systems)

Baumann P (2010) The OGC web coverage processing service (WCPS) standard. Geoinformatica 14(4):447–479

Baumann P (2017a) The datacube manifesto. http://earthserver.eu/tech/datacube-manifesto, seen 28 May 2017

Baumann P (2017b) Web coverage service (WCS) interface standard—core, version 2.0. OGC document 09-110r4,\www.opengeospatial.org/standards/wcs, visited 28 May 2017

Baumann P (2017c) Web coverage processing service (WCPS), version 1.0. OGC document 08-068r2. http://www.opengeospatial.org/standards/wcs, seen 28 May 2017

Baumann P (2017d) OGC coverages domain working group public wiki. http://myogc.org/coveragesDWG, seen 28 May 2017

Baumann P, Merticariu V (2015) On the efficient evaluation of array joins. In: Proceedings of the workshop big data in the geo sciences (co-located with IEEE big data). Santa Clara, US, 29 Oct 2015

Baumann P, Feyzabadi S, Jucovschi C (2010) Putting pixels in place: a storage layout language for scientific data. In: Proceedings of IEEE ICDM workshop on spatial and spatiotemporal data mining (SSTDM'10). Sydney, Australia, 14 Dec 2010

Baumann P, Brackin R, Owonibi M, Cutsworth T (2011) Towards standards-based, interoperable geo image processing services. In: Proceedings of Geoinformatik, vol 41. ifgiPrints, Münster, Germany, 15–17 June 2011

Baumann P, Dumitru A, Merticariu V (2013) The array database that is not a database: file based array query answering in Rasdaman. In: Proceedings of the 13th international symposium on advances in spatial and temporal databases—(SSTD'2013). Munich, Germany, 21–23 Aug 2013

Baumann P, Hirschorn E, Maso J (2017) Coverage implementation schema, version 1.1. OGC document 09-146r6. www.opengeospatial.org/standards/wcs, visited 30 June 2017

Baumann P, Rossi AP, Bell B, Clements O, Evans B, Hoenig H, Hogan P, Kakaletris G, Koltsida P, Mantovani S, Marco Figuera R, Merticariu V, Misev D, Pham Huu B, Siemen S, Wagemann J (2017b) Fostering cross-disciplinary earth science through datacube analytics. In:

Mathieu PP, Aubrecht C (eds) Earth observation open science and innovation—changing the world one pixel at a time. International Space Science Institute (ISSI), (in print)

Blaschka M, Sapia C, Höfling G, Dinter B (1998) Finding your way through multidimensional data models. in: DEXA workshop data warehouse design and OLAP technology (DWDOT'98). Vienna, Austria, 24–28 Aug 1998, pp 198–203

Boulil K, Bimonte S, Pinet F (2015) Conceptual model for spatial data cubes: a UML profile and its automatic implementation. Comput Stand Interfaces 38(2015):113–132

Brizhinev D, Toyer S, Taylor K (2017) Publishing and using earth observation data with the RDF data cube and the discrete global grid system. In: W3C working group note, 28 Sept 2017. https://www.w3.org/TR/eo-qb/, visited 28 May 2017

Broering A, Stasch C, Echterhoff J (2017) OGC sensor observation service interface standard. OGC document 12-006. https://portal.opengeospatial.org/files/?artifact_id=47599, visited 28 May 2017

Cheng Y, Rusu F (2015) Formal representation of the SS-DB benchmark and experimental evaluation in EXTASCID. Distrib Parallel Databases 33(3):277–317

CoverageJSON (2017). https://covjson.org/, visited 28 May 2017

Cyganiak R, Reynolds D (2014) The RDF data cube vocabulary. W3C, 2014. https://www.w3.org/TR/vocab-data-cube/

Dumitru A, Merticariu V, Baumann P (2016) Array database scalability: intercontinental queries on petabyte datasets (system demonstration). In: Proceedings of the 28th international conference on Scientific and Statistical Database Management (SSDBM). Budapest, Hungary, 18–20 July 2016

EarthServer (2017) EarthServer—big datacubes at your fingertips. www.earthserver.eu, visited 30 June 2017

Furtado P, Baumann P (1999) Storage of multidimensional arrays based on arbitrary tiling. ICDE'99. Sydney, Australia, 23–26 Mar 1999

Hirschorn E, Baumann P (2017) OGC testbed 11 referenceable grid harmonization engineering report. OGC document 15-065r1. http://myogc.org/coveragesDWG/CoveragesBigPicture/15-065r1_Testbed_11_Referenceable_Grid_Harmonization_Engineering_Report_FINAL.pdf, visited 28 May 2017

ISO (2004) Geographic information—schema for coverage geometry and functions. 19123

ISO (2016) Geographic information—schema for coverage geometry and functions—part 2: coverage implementation schema. ISO New Work Item Proposal 19123-2

ISO (2017) SQL 9075 Part 15: Multi-dimensional arrays (MDA). ISO Committee Draft 9075-15

Karmas A, Tzotsos A, Karantzalos K (2015) Scalable geospatial web services through efficient, online and near real-time processing of earth observation data. In: Proceedings of the IEEE international conference on big data computing service and applications

Liakos P, Koltsida P, Baumann P, Ioannidis Y, Delis A (2015) A distributed infrastructure for earth-science big data retrieval. Int J Coop Inf Syst 24(2)

Marco Figuera R, Pham Huu B, Minin M, Flahaut J, Halder A, Rossi AP (2017) Analyzing CRISM hyperspectral imagery using planetserver. Geophysical research abstracts, #14632

MatLab (2017). https://www.mathworks.com/products/matlab.html, visited 28 May 2017

Mueller M, Pross B (2017) WPS 2.0 interface standard corrigendum 1. OGC document 14-065. http://docs.opengeospatial.org/is/14-065/14-065.html, visited 28 May 2017

OGC (2007) Topic 6—schema for coverage geometry and functions, version 7.0. OGC document 07-011, 2007

OGC (2017) WCS 2.0 compliance test suite. http://cite.opengeospatial.org/teamengine/about/wcs/2.0.1/site/index.html, visited 28 May 2017

Open Street Map (2017). http://www.openstreetmap.org, visited 28 May 2017

Pisarev A (2003) Mooshka: a system for the management of multidimensional gene expression data in situ. Inf Syst 28(4):269–285

R (2017) https://www.r-project.org, visited 28 May 2017

RDA (2017) Array database assessment working group report. https://www.rd-alliance.org/groups/array-database-working-group.html, visited 28 May 2017
Robin A (2017) OGC® SWE common data model encoding standard. OGC document 08-094r1. http://www.opengeospatial.org/standards/swecommon, visited 28 May 2017
Sarawagi S, Stonebraker M (1994) Efficient organization of large multidimensional arrays. In: Proceedings of the 10th international conference on data engineering (ICDE), Houston, Texas
Strobl P, Baumann P, Lewis A, Szantoi Z, Killough B, Purss M, Craglia M, Nativi S, Held A, Dhu T (2017) The six faces of the data cube. In: Proceedings of the big data from space, Toulouse, France, 28–30 Nov 2017 (accepted)
Tzotsos A, Karmas A (2017) A datacube approach to agro-geoinformatics. In: Proceedings of the 6th international conference on agro-geoinformatics (Agro-Geoinformatics2017), Fairfax, USA, 7–10 Aug 2017

Part III
Requirements of the New Map Production Paradigm

Chapter 15
How Standards Help the Geospatial Industry Keep Pace with Advancing Technology

Lance McKee and Joan Masó

Abstract This chapter describes how, beginning in the 1990s, the information technology (IT) industry moved into a new paradigm based on widespread networking of computers. This new paradigm forced map production and geoinformation management system providers and users to adapt accordingly. Widespread networking of computers brings the opportunity for widespread discovery, sharing and integration of data. This cannot happen, however, without widespread agreement on standard protocols, data models, encodings and best practices. The The inherent complexity of geospatial information (Longley et al. 2015) imposes particular interoperability problems that can only be resolved through a geo-focused standards development process. The Open Geospatial Consortium (OGC) emerged to lead this progress, coordinating with many other organizations. It appears that the geospatial standards development process will be ongoing as long as the underlying information technology platform continues to evolve and new spatial technology user domains emerge and evolve.

Keywords Geospatial standardisation · Application needs · Technical restrictions
Open geospatial consortium · OGC · ISO/TC 211 · W3C

15.1 Introduction

In the early 1990s the world was leaving the Computer Age and entering the Communication Age. In the Communication Age, diverse computers operate as clients and servers, interoperating, sending and receiving both instructions and data. Computers themselves have continued to evolve, of course, but the networking of

L. McKee (✉)
Worcester, MA 01603, USA
e-mail: lmckee76@gmail.com

J. Masó
CREAF (Ecological and Forestry Applications Research Centre), Cerdanyola del Vallès, Spain

J. Döllner et al. (eds.), *Service-Oriented Mapping*, Lecture Notes in Geoinformation and Cartography, https://doi.org/10.1007/978-3-319-72434-8_15

computers on a vast scale means that now "the network is the computer." This transition soon began to have extraordinary significance in virtually every domain, from daily life to business, education, news, science, culture, politics and national security. Geoinformation is important in all these domains.

Map production and geoinformation management systems had evolved in a time when sharing and integrating data involved batch data storage and batch transfer of data between dissimilar computer systems. Working with geospatial data required special training in data formats. Map production was mostly a matter of painstakingly turning digital data into paper maps. All of this has changed dramatically.

The IT industry trend toward service-oriented architectures (SOA) and distributed computing began in the 1980s (Townsend 2008). In the late 1980s, there was a growing realization that bridging an enterprise's "islands of automation" could be done most efficiently by decomposing systems into discrete functional components that could be shared by multiple systems using common sets of application programming interfaces (APIs). This followed logically from the use of subroutines in programming and from computer architecture design principles. Various providers of networks and middleware products competed to provide the necessary network protocols and bridging software or "glue" not provided by operating systems.

In Service-Oriented Architectures (SOA), diverse distributed systems act as clients and servers, sending and responding to service instructions, enabling the diverse connected computers to communicate. Communication means "transmitting or exchanging through a common system of symbols, signs or behavior." Standardization means "agreeing on a common system."

Enterprises were developing their own private APIs and encodings or else working with companies who provided proprietary APIs and encodings. An enterprise-wide "common system" could be shared across an enterprise, but it was becoming increasingly obvious that what was needed was client/server communication between diverse, independently developed systems inside and outside an enterprise. Such communication would have tremendous value, for example, for companies that were merging, for communication between corporate partners, and for efficiency in government.

This would require open standard APIs and encodings that would enable SOA on top of TCP/IP, the open communication protocol that was quickly becoming the standard for "inter-networking". TCP-IP enabled a network of networks that had become known as the "Internet". The Internet was necessary but not sufficient for widespread communication of service requests and responses.

First the Internet Engineering Task Force (IETF) and later the World Wide Web Consortium (W3C), and then other standards development organizations as well, including both new organizations like OASIS and older "de jure" standards organizations like ISO and ITU, began reviewing the technical requirements for SOA standards. Companies of all sizes were involved in these standards development organizations. Meanwhile, they were also devising strategies for operating and profiting in the new open environment.

Enterprises were discovering that re-engineering their information systems and information flows provided opportunities, and indeed imperatives, to re-engineer their organizational structures. With the emergence of the Web, companies realized in addition that the availability of high-bandwidth TCP/IP networks with free World Wide Web capabilities offered new market opportunities and imperatives. By making data workflows more efficient, network based computing was "disintermediating" not only information workers and whole corporate departments, but also legacy providers. Almost all value chains, including those involving map production and geoinformation management systems, were being disrupted in some way by the rise of networks and network-savvy competitors. New proprietary and open source capabilities layered on top of Internet and Web standards further expanded what was possible and further accelerated the pace of change.

15.2 Mobile Devices Gave Geoinformation a Highly Visible New Dimension

The new market realities boosted market demand for improved geoinformation products. In parallel with the arrival of the Communication Age, semiconductor advances rapidly expanded

- Resolution of cameras, first in Earth imaging cameras, then video and still cameras, and then cameras integrated into telephones
- Precision of optical ranging, LiDAR and development of surveyors' digital "total stations"
- Image processing speed
- Bandwidth for sharing high-volume geoinformation
- Capabilities of global navigation satellite systems (GNSS): the US Global Positioning System (GPS), Russia's GLONASS, Europe's Galileo, and China's BeiDou-2

Advances in solid-state sensors brought improved sensitivity, accuracy, range and collection speeds while dramatically reducing sensors' price, size and energy requirements. Satellite launch capabilities were improving and public and private sector investment in Earth imaging satellites was increasing.

Throughout the 1990s semiconductor advances and advances in microwave technology were also driving the rapid rollout of cell phones, and GPS was becoming a public utility. The first cell phone with GPS was introduced in 1999 (Sullivan 2013). As cellphone networks expanded, the phones' storage, computing and display capabilities were also increasing. In 2007, the iPhone marked the introduction of smart phones, that is, wireless phones with sensors, a developer-friendly platform for applications ("apps"), Internet access, app-supporting links to a "cloud" full of data and processing resources, and location awareness. The iPad, released in 2010, had similar features, including GPS as well as floatability by calculation of proximity to 3 or more cell tower antennas. Gyroscopes and accelerometers were added to phones and to notepad computers to provide additional location and motion data.

The most frequently used sensors in wireless devices are the GPS transponder and the gyroscope. (Kumar 2017). Spatial information adds considerable value to mobile devices. Sensors in wireless devices provide vastly increased opportunities to gather spatial data, often at virtually no cost. As the network grows, the spatial data it provides becomes more abundant and more valuable, resulting in increased investment in products and services involving map production and geoinformation management.

Map production in the new environment came to mean on-demand automated cartography, with maps viewed on digital displays. Google Maps, launched in 2005, and other map and navigation desktop applications and mobile device apps quickly became hugely popular. Makers of handheld devices usually make the phones' GPS coordinates accessible to app developers. Because location is important in so many human activities, location-based apps make up a very large category of apps. In 2014, research firm Ipsos MediaCT (sponsored by Google) discovered that among the substantial subset of those who searched while "on the go," a majority of those searches (56%) were local searches (Sterling and Sterling 2017). The study also showed that almost 90% of mobile users search for local information.

Professional GIS users, image analysts, geoinformation managers and even government cartographic agencies were in some cases being disintermediated, but overall they were benefiting from the technologies that were putting maps on cell phones and desktops everywhere. Besides now being the geospatial "back room" for a rapidly expanding set of information providers, they were enjoying much more efficient data discovery and access, and much more data was becoming available to them. Often without realizing it, as they enjoyed these benefits they were benefiting from geospatial system and data providers' implementations of the new standards that were being developed in the OGC.

15.3 OGC's History and Role

The importance of "location" is immediately obvious to mobile device users, but they see only the tip of the iceberg. The geospatial industry has been transformed by mobile location apps, but today's mobile location capabilities owe much to the pre-mobility geospatial standards platform that was already in place. Looking to the future, overcoming the major limitations of today's mobile apps will depend on the ongoing geospatial standards effort.

Prior to OGC and prior to the Communication Age, virtually all geographic information systems (GIS), Earth imaging systems, navigation systems, automated mapping and facilities management (AM/FM) systems and other spatial technologies were available only as monolithic systems. Basic functions such as data storage, data analysis, and data presentation were bound together in these monolithic systems using proprietary interfaces and encodings. The vector based systems and raster based systems were in most cases not interoperable with each other or with other kinds of spatial systems (such as those for navigation, transportation, civil engineering, simulation & modeling and multidimensional scientific array analysis) or with other vendors' geographic information systems. Sometimes the semantic elements in the data schemas of particular domains were integral with the geometric elements, making whole systems usable only within that domain, and making it difficult to integrate spatial data developed in other domains. Sharing and integrating data was impossible without batch downloads and batch conversion from one system's data format to another.

There were literally hundreds of formats, proprietary and open, and there were hundreds of data transfer filters and conversion utilities to perform the one-to-one format conversions. There were also interchange formats such as SDTS, SAIF, GeoTIFF, the National Image Format Transfer Standard, DXF and Shapefile. An interchange format is an intermediary format that enables aggregation of data from various systems' formats for storage and later conversion to formats used by other systems. This was the approach to geodata interoperability employed by major users such as government agencies. Their only alternative was to adopt one vendor's product set as the agency's standard platform for internal use. External partners were obliged in that case to buy from that vendor's product offerings or rely on interchange formats.

By the mid-1980s, the market for GIS software was growing rapidly, but the expensive software's limited extensibility and flexibility and inability to easily share geospatial data between systems was a serious problem for users. MOSS (Map Overlay and Statistical System), an open source vector GIS, and GRASS (Geographic Resources Analysis Support System), an open source raster GIS, were developed by US federal agencies to reduce GIS software costs and give the agencies more control over development cycles. GRASS had become an international open source development project and it had been successfully integrated with other public domain spatial, statistical and imaging analysis packages. In 1992, under the leadership of David Schell, who later became the founding president and CEO of OGC, the GRASS user community formed a non-profit organization—the Open GRASS Foundation (OGF)—whose mission was to stimulate private sector support for GRASS and create a consensus-based membership process for management of GRASS community affairs. This proved difficult to sustain financially. Also, GRASS and MOSS, though free, modular, and maintained in a process driven by user input, did not provide a full interoperability solution. Each had an open data format, but that was not sufficient to enable interoperability with commercial spatial software products.

From 1992 through early 1994 members of the OGF board of directors organized meetings around the US and Canada that led to the formation of OGC in 1994. Platforms for service-oriented architectures had advanced to the point where it seemed feasible to attempt to achieve OGC's vision: a world in which everyone benefits from geospatial information and services made available across any network, application, or platform.

From 1994 to 2017, OGC's membership grew from 20 to more than 500 government, academic, and private sector organizations. (OGC, membership) Private sector members now include traditional GIS vendors along with technology integrators, data providers, and companies at the cutting edge of location services. Though founded in the US, the OGC now has more members in Europe than in all of North and South America. Many of these are major government agencies. The percentage of members from other world regions continues to grow (OGC, World regions).

At the round of meetings that led to the gathering of a board of directors and the formation of OGC, and at the early OGC Board meetings and Management Committee meetings, a main topic of discussion was how to structure the organization and develop a sustainable business model that would pay for staff, manage and sustain the standards work, fairly represent the interests of all the stakeholders, and maximize the likelihood of vendors' adoption and buyers' deployment of the standards.

The early discussions also included, of course, technical topics such as how to model geospatial features and phenomena in a way that accommodated use cases, vendors' data models and users' data schemas. Multiple pre-Web SOA platforms were beginning to be used in the larger IT world, and members had to learn and agree how to write specifications that could be used with different SOA platforms.

The newly formed OGC Technical Committee reached consensus that both an Abstract Specification and implementation specifications would need to be provided. The Abstract Specification describes a conceptual model that provides for the development of Implementation Specifications. Implementation Specifications are unambiguous technology platform-specific programming specifications for implementation of standard software Application Programming Interfaces (APIs) (see Sect. 15.6) Discussions were encapsulated in *The OpenGIS Guide—Introduction to Interoperable Geoprocessing*. Hundreds of copies were sold and made available to members, and the book was soon put online, free for anyone to download. "The Guide" helped educate people about the technological foundations of the OpenGIS (now OGC) Specifications. "Simple Features" implementation specifications were developed for CORBA and Microsoft's COM platform. (The OpenGIS Guide is no longer available. It has been superseded by the OGC Reference Model (OGC, Reference Model), described in Sect. 15.4).

The World Wide Web Consortium (W3C) was founded in the same year OGC was founded, and OGC members were closely following the nascent Web's progress. The free, open source and vendor-neutral Web appeared likely to overtake other SOA platforms like CORBA and COM. At the same time, members had started dealing in OGC with the diverse technical and institutional standards challenges inherent in Earth Observation, Location Based Services, transportation and civil engineering. None of these types of spatial information systems had an open standard way of sharing and integrating data with the other types or with vector based GIS. All of these spatial information systems are different from GIS, and each has different but overlapping communities of users and providers. At the same time, OGC's senior management and directors were working hard to bring in more international members and to actively engage with other regional and international standards organizations, because it was clear that OGC needed to be an international organization. Developing standards was a slow process. Members were obliged to spend time debating the content of the OGC Policies and Procedures document that serves as a "constitution" for the OGC Technical and Planning Committees. Membership was growing each year, but it was difficult for staff to manage all the activity and for members to commit resources in all these areas of activity.

It was helpful that in 1996 the US Dept. of Defense (DoD) National Imagery and Mapping Agency (NIMA), now the National Geointelligence Agency (NGA), had begun working with OGC to provide input into planning NIMA's next generation distributed geoprocessing approaches. A major user of all types of geospatial software, they were transitioning from very large custom installations to a rapid prototyping, spiral engineering approach that would enable them to take maximum advantage of technology development in the private sector. They needed a cross-DoD standards platform and a public/private process for evolving that standards platform. OGC was positioned to provide that. OGC also provided a way to rapidly insert DoD requirements into private sector product development cycles, which benefitted the DoD, the other government agencies and the private sector providers.

These synergies led to the creation of OGC's now ongoing series of major testbeds and related activities (OGC, Interoperability initiatives). OGC's XML-based Web Map Server (WMS) standard, a product of the first testbed in 1999, quickly became the most widely implemented OGC standard. The OGC Interoperability Program (now called the Innovation Program) was instituted to organize and run testbeds, pilot projects, interoperability experiments, concept development, and OGC Engineering projects. The Innovation Program has made standards development faster, less expensive for members, less dependent on one or two sponsoring agencies, more coordinated and more likely to result in standards that will be implemented by vendors and deployed by users. US agencies such as NASA, NOAA, USGS, Census and others, now including the Department of Homeland Security (DHS), benefit from participation in the strategic and technical discussions that surround each initiative as well as benefitting from the results of the initiatives. The same can be said for European agencies such as ESA, JRC, UK Met Office and national organizations from other world regions who have become increasingly involved as sponsors.

Testbeds bring multiple sponsoring organizations together to create scenarios in which existing and prospective OGC standards play a role. The scenarios are broken down into technically detailed use cases based on real world situations that involve sharing and integrating spatial data. Teams of technology providers collaborate in testing, improving and developing standards that meet the sponsors' requirements. Each testbed ends with a public demonstration in which the scenario is played out and videotaped. Testbeds are particularly useful for major national, regional and international geospatial data integration and data sharing programs like US DoD cross-program interoperability, NIEM, INSPIRE, GEOSS etc.

Also in the late 1990s, the OGC Compliance Program was created to provide members with a way to verify their products' compliance with OGC standards. The Compliance Program also provides a way for product purchasers to have greater confidence in the interoperability of products.

15.4 Collaboration with Other Standards Organizations

Because spatial information is important in so many different areas of activity, many other standards organizations need to address simple location and sometimes other more difficult spatial topics such as volume, density, corridors and change. To ensure consistency across the Internet and Web ecosystem, the OGC has alliance partnerships (OGC, Alliance Partnerships) with many other standards development organizations and industry associations, as shown in Fig. 15.1.

These organizations collaborate with the OGC on a wide range of topics such as indoor/outdoor location integration, the built environment, sensor fusion, urban modeling, location based marketing, aviation information management, meteorology, the Internet of Things, Points of Interest and the Semantic Web.

Very importantly, early in its history OGC began to work with ISO TC 211. This collaboration became formalized as an ISO Joint Initiative. Some of the OGC's most widely implemented standards are also ISO standards. These are listed in Table 15.1. For example, the OGC Web Map Server (WMS) 1.3 is identical with the ISO 19128-2005—Geographic information—Web map server interface.

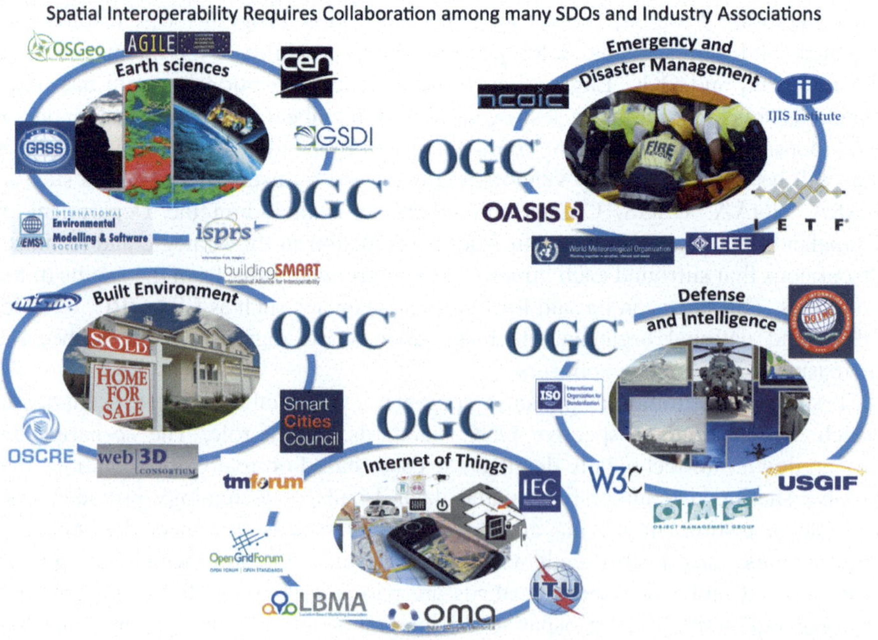

Fig. 15.1 OGC works with many different standards organizations and industry associations

Table 15.1 OGC standards that are also ISO standards

OGC document	ISO document	Current OGC work/notes
OGC Web Map Service 1.3	19128:2005	WMS 2.0 in Progress in OGC
OGC Web Feature Service 2.0	19142:2010	WFS 2.1 in progress in OGC
OGC Filter Encoding 2.0	19143:2010	FE 2.1 in progress in OGC
OGC Simple Features Common Architecture 1.2.1	19125-1:2004	Simple Features 2.0 in progress in OGC
OGC Simple Features Part 2 SQL option 1.2.1	19125-2:2004	Simple Features 2.0 in progress in OGC
OGC Geography Markup Language 3.2	19136:2007	Originated in OGC and accepted by ISO
OGC Geography Markup Language 3.3	19136-2	Originated in OGC and under publication by ISO
OGC AS Topic l: Feature Geometry	19107:2003	Originated in ISO and accepted by OGC. Revision in progress in OGC/ISO
OGC AS Topic 2: Spatial referencing by coordinates	19111:2007	Joint document
OGC AS Topic 2: Spatial referencing by coordinates part 2—parametric	19111-2:2009	Joint document
OGC AS Topic 7: Earth imagery	19101-2:2008	Originated in ISO and accepted by OGC
OGC AS Topic 11: Metadata	19115:2003	Originated in ISO and accepted by OGC. 2014 revision in review by OGC.
OGC AS Topic 12: The OpenGIS service architecture	19119	Originated in OGC and accepted by ISO
OGC AS Topic 18: GeoDRM	19153:2014	Originated in OGC and accepted by ISO
OGC AS Topic 19: Geographic information—linear referencing	19148:2012	Originated in ISO and accepted by OGC
OGC AS Topic 20: Observations and measurements conceptual schema	19156:2011	Originated in OGC and accepted by ISO
OGC rights expression language for geographic info	19149:2011	Originated in OGC and accepted by ISO
Geographic information—well known text representation of CRS	CD 19162	NWIP originated in the OGC but this is a joint project
Geographic information—implementation schema for coverages	NWIP 19123-3	NWIP originated as OGC GMLCOV

Other OGC standards are built to conform as much as possible with pre-existing ISO standards. The OGC Geospatial User Feedback (GUF) set of standards,[1] for example, is based on the ISO 19115-1 and 19157 metadata and data quality models and uses the same encoding model.

Collaboration with W3C is also very important. In 2015 the collaboration between the OGC and the W3C became more formal. The Spatial Data on the Web Working Group was established as a joint OGC/W3C working group. The mission of the Spatial Data on the Web Working Group is to clarify and formalize the relevant standards landscape. The group works to:

- Determine how spatial information can best be integrated with other data on the Web;
- Determine how machines and people can discover that different facts in different datasets relate to the same place, especially when 'place' is expressed in different ways and at different levels of granularity;
- Identify and assess existing methods and tools and then create a set of best practices for their use.

Where desirable, they also work to complete the standardization of informal technologies already in widespread use.

15.5 OGC's Programs

The OGC has a Board of Directors (OGC, Directors) that represents leadership from many disciplines and many regions of the world. OGC Directors provide their professional expertise in advancing OGC objectives. OGC Directors serve as individuals representing the needs of their sectors. They do not represent their organizations of employment. The Global Advisory Council is a committee of the board that functions as a non-executive, "blue-ribbon" panel advising OGC regarding OGC's global outreach and organizational strategies.

The OGC has four core programs (OGC, Core Programs), described below, that deal with different aspects of the standardization process: the Standards Program, the Innovation Program, the Compliance Program, and the Communications and Outreach Program. The Communications and Outreach Program plays a key role in standards adoption, as explained in this section.

The OGC staff manage these programs (OGC, Staff). The OGC also encourages members in countries or world regions to form OGC Forums.

---

[1]The purpose of the GUF is to complement the producer quality documented in the metadata with useful opinions and experiences from users.

15.5.1 Standards Program

In the Standards Program, the Technical Committee and Planning Committee work in a formal consensus process to arrive at approved (or "adopted") OGC® standards. The Technical Committee (TC) is composed of people representing all the OGC member organizations. These member representatives participate in Technical Committee meetings (approximately every 3 months) as well as working group teleconferences. The Planning Committee, which meets on the last day of the Technical Committee meeting week, has ultimate responsibility for approving Technical Committee recommendations for the adoption and release of OGC standards and for Specification Program planning. OGC staff and hosting members attempt to rotate Technical Committee and Planning Committee meetings between venues in different world regions (OGC, Meetings).

Two main types of groups make up the Technical Committee: Domain Working Groups (DWG) ("dwigs") and Standards Working Groups (SWG) ("swigs"). DWGs discuss topics that may be theme based (e.g. the Metocean.DWG), aspect based (e.g. the Data Quality.DWG) or challenge based (e.g. the Big Data DWG) (see Table 15.2).

DWGs often originate as a result of work done in the Innovation program, though they can also originate as a result of discussion among members. A SWG typically spins out of a DWG to focus on detailed technical definition of a standard, a revision of a standard or a profile of a standard.

DWGs and SWGs can also be proposed and chartered by a community external to the OGC that have settled on a community standard or a de facto standard as a way of addressing a problem. That community can bring their standard into the OGC to harmonize it with the OGC baseline and have it become an OGC standard, maintained in OGC's open consensus process. Google did this with KML, their map browser standard API. The climate, meteorology and oceans communities brought in netCDF, a set of software libraries and self-describing, machine-independent data formats that support the creation, access, and sharing of array-oriented scientific data. WaterML and GeoSciML are other examples. Users

Table 15.2 OGC's regional forums

Forum name	Creation date
Asia forum	10/01/2011
Australia and New Zealand (ANZ) Forum	05/07/2012
Europe forum	07/10/2003
France forum (internal)	02/12/2008
Iberian and Latin American forum	22/06/2010
India forum (internal)	16/11/2009
Middle East and North Africa (MENA) Region forum	22/04/2013
Nordic forum	07/12/2011
North American forum	23/03/2012

of contributed standards like these benefit from improved interoperability with the large and diverse global set of geospatial data and services that are accessible through interfaces and encodings that implement OGC standards. They also benefit from the rich geospatial standards experience of OGC members and staff, and they benefit from OGC's disciplined standards maintenance process, Compliance Program and Communications and Outreach Program.

15.5.2 Innovation Program

The Innovation Program (OGC, Innovation Program), described in the OGC history section above, is responsible for scenario-based development and testing of new ideas or new standards. Interoperability Initiatives organized in the Innovation Program build and exercise public-private partnerships that accelerate the development of emerging concepts and drive global interoperability. They motivate the membership to develop specifications, rapidly implement them and then test for interoperability among vendors' implementations. This iterative rapid prototyping process proceeds with the help of the working groups.

Funding for testbeds and pilot projects usually comes from multiple sponsors with similar needs who share the costs. In addition, vendors provide in-kind services. The results of these shared cost, hands-on public/private activities are typically documented in Engineering Reports that are available to the public (OGC, Engineering Reports). Engineering reports may ultimately evolve into standards, standard revisions or best practices.

The important role of the Innovation Program in OGC's standards process makes OGC a unique standards body that directly connects innovation in the industry with the standards. It closes the gap between implementers and the standards process.

15.5.3 Compliance Program

The Compliance Program (OGC, Compliance Program) maintains an online free testing facility (the CITE) that vendors use for testing implementations of standards and for certifying and branding of compliant products. The purpose of the OGC Compliance Program is to increase system interoperability while also reducing technology risks. Vendors gain confidence that they are providing a product compliant with OGC standards, which will be easier for users to integrate and easier to market. Buyers gain confidence that a compliant product will work with another compliant product based on the same OGC standard, regardless of which company developed the product.

15.5.4 Communications and Outreach Program

The OGC Communications and Outreach Program (OGC, Communications) works with OGC members and user communities around the world to encourage "take up" or implementation of OGC standards, as well as to encourage new membership and engagement in OGC programs and initiatives. This program publicizes OGC requests, accomplishments and invitations through publications, workshops, seminars and conferences. It also supports the OGC regional forums.

15.5.5 Standards Adoption

The most important aspect of the standards process is adoption by vendors leading to deployment by users. A standard has little value if it is not implemented in products and used in the real world. Adoption depends on several factors that are not under the control of OGC and mainly depend on the interaction of users and vendors. In some cases, though, a standard becomes a legal mandate ensuring adoption. This has been the case of some OGC standards specified in the European directive for the creation of the European Spatial Data Infrastructure (INSPIRE). OGC's co-branding of standards with ISO TC/211 is important because many procurements require ISO standards.

To support wide adoption, the OGC has a policy that a standard should not be approved by the membership if there is no evidence of implementation. Early adoption is facilitated by interoperability initiatives such as testbeds and plugfests in which vendors develop standard interfaces and test for interoperability between them, resulting in early working implementations.

After a standard has been approved, the Communications and Outreach Program prepares user guides and disseminates information about the new standard in conferences and symposia. The Compliance Program also helps by developing testing platforms and certifying products that have passed the interoperability tests. Despite these efforts, in a few cases, it is clear that a standard has not been implemented by the industry. After verifying the lack of adoption, OGC members may eventually "deprecate" or retire the standard. A recent example is the retired OpenGIS Transducer Markup Language.

15.5.6 OGC Forums

OGC members in most world regions have formed OGC forums to support communication and cooperation among the members in that region. These are listed in Table 15.2.

15.6 OGC Standards

More than fifty approved OGC implementation standards are now freely available to help developers address the challenges of spatial data discovery, sharing and integration. OGC standards, which comprise a platform for geospatial interoperability, have been implemented in hundreds of commercial and open source geoprocessing products and are being deployed in communities and organizations around the world. Today OGC standards are key elements in the geospatial communication interfaces, encodings and best practices for sensor webs, location services, Earth imaging networks, climate models, disaster management programs and national spatial data infrastructures around the world.

In order to provide a structured view of OGC's diverse collection of standards (referred to as the OGC "Technical Baseline"), the OGC provides and maintains the OGC Reference Model (ORM) (OGC, ORM). The OGC Reference Model (ORM) is based on the Reference Model for Open Distributed Processing (RM-ODP, ISO/IEC 10746), a widely used international standard for architecting open, distributed processing systems. The RM-ODP provides an overall conceptual framework for building distributed systems in an incremental manner. The ORM provides an overall conceptual framework for building geoprocessing into distributed systems in an incremental manner. (The above explanation is adapted from Overview of the Abstract Specification Topic 0: Overview of the Abstract Specification (OGC, ORM).

15.6.1 The Challenge of Impermanence

Geospatial interoperability is a moving target. The aspect based, challenge based and theme based topics listed in Table 15.3 are dynamically interweaving, often in unpredictable ways. New topics are added each year, and some drop away. OGC is all about solving complex software architecture and software programming challenges, challenges that are driven by a very broad set of market and institutional dynamics and programming trends that come and go. OGC has evolved a number of strategies for dealing with all this change. The standards development process has become more complex but more ordered, resulting in more successful and easily implemented standards and less wasted time.

15.6.2 Different Kinds of Things Need to Be Standardized

Table 15.3 sorts OGC Domain Working Groups into three different classes based on three different types of standards. Standards Working Groups also fall into these three classes.

Table 15.3 Aspect based, challenge based and theme based domain working groups

Domain working groups	Classification
Coordinate reference system DWG	Aspect based
Coverages DWG	Aspect based
Data preservation DWG	Aspect based
Data quality DWG	Aspect based
Geosemantics DWG	Aspect based
Metadata DWG	Aspect based
Quality of service and experience DWG	Aspect based
Security DWG	Aspect based
Sensor web enablement DWG	Aspect based
Temporal DWG	Aspect based
3DIM DWG	Challenge based
Big data DWG	Challenge based
Catalog DWG	Challenge based
Citizen science DWG	Challenge based
Mobile location services DWG	Challenge based
Oblique imagery DWG	Challenge based
Point cloud DWG	Challenge based
Workflow DWG	Challenge based
Agriculture DWG	Theme based
Architecture DWG	Theme based
Aviation DWG	Theme based
Defense and intelligence DWG	Theme based
Earth systems science DWG	Theme based
Electromagnetic spectrum DWG	Theme based
Emergency & disaster management DWG	Theme based
Energy and utilities DWG	Theme based
Health DWG	Theme based
Hydrology DWG	Theme based
Land administration DWG	Theme based
Land and infrastructure DWG	Theme based
Law enforcement and public safety DWG	Theme based
Marine DWG	Theme based
Meteorology & oceanography DWG	Theme based
Smart cities DWG	Theme based
University DWG	Theme based

Vectors and rasters can be seen as different "aspects" or views of geospatial features and phenomena. In the 1990s, members wanted to expand from GIS to Earth Observation and so they developed an overarching "coverages" concept. The introduction of new aspects of spatial information has continued, resulting in the formation of new aspect-based Domain Working Groups and Standards Working Groups.

The OGC had to address basic mainstream computing platform challenges such as the IT industry's migration from COM and CORBA to the Web. SOAP and XML approaches became mainstream and they continue to be well suited for complex structured data. But programmers in the larger IT world favored JSON, REST and Linked Data for many solutions, which presented OGC with another challenge. Such developments have necessitated the formation of the "challenge based" working groups shown in the table. Others are likely as the IT industry evolves.

Different user domains have seen OGC as a natural forum for addressing their domain-specific interoperability requirements, focusing on semantic interoperability as well as other issues. These user domains form "theme based" Domain Working Groups, as shown in Table 15.3.

15.6.3 Formalization of the Standards Process and Products

In the beginning, OGC standards were monolithic documents typically describing the behavior of a service and the operation it specifies. They described several important aspects that needed to be implemented, mixed with optional details in a descriptive text. Implementers needed to have the ability to find what was mandatory to implement, what was optional and what was an explanation of functionality. It was easy to overlook a mandatory detail. Different standards documents often had different structure.

More formality was introduced by the adoption of UML class diagrams and property tables that list classes and properties and indicate the data type, multiplicity, default value and mandatory or optional character of each parameter. See Fig. 15.2.

This was an improvement that was also complemented by obliging the developers of standards to include an abstract test suite as an annex. Unfortunately, it was not clear how to create a test suite from the standard descriptive text. This resulted in inconsistencies between the main text and the abstract test and in difficulties in creating a concrete test for a standard implementation.

To address this problem, in 2008 the Modular Specification was introduced (OGC, Modular Specification). The Modular Specification increases the level of formality in the way standards should be written. This is another factor that makes OGC a unique standard body. The modular specification introduces the need to enumerate requirements that contain the normative text in short and concrete sentences. Each requirement has a name and a URI that identifies it (transferred and maintained by the OGC naming authority) and has a section specifying its dependencies. Requirements are grouped in requirements classes. A standard can have one or more requirements classes and for each requirement class there is a conformance class with an equivalent list of conformance requirements.

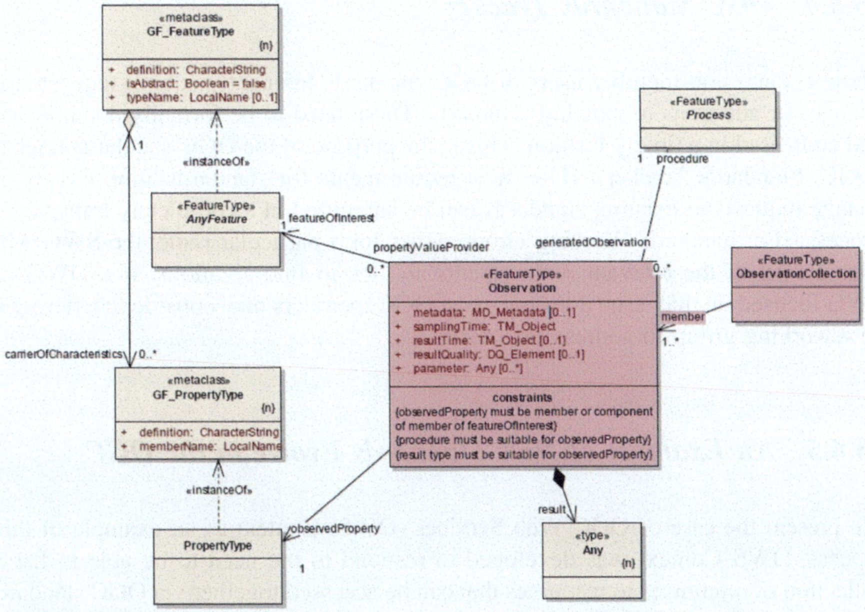

Fig. 15.2 Part of the OGC observations & measurements UML model

Requirement classes (and conformance classes) are indivisible. Implementers of the standard must decide which ones to implement and they must conform to all requirements in the selected classes. Defining a coherent set of requirements is not an easy task and standard writers are now spending more time up front in formalizing the normative text. This increases the quality of the standard. Informative text that precedes or comes after a requirement can help the reader to understand the reason of each normative text. Often provided are practical examples on how to apply the requirement in a particular case.

For each standard there is a requirement class that is more fundamental than others and which should be implemented by all instances. It is called the “core”. All other requirements classes are considered extensions of the core. One of the design rules of the modular specification is that you cannot overrule a core requirement in an extension. In practice, the core should be a very generic class that defines the boundaries of the standard and the kind of resource it is going to deal with.

Often OGC standards documents provide an XML encoding along with a conceptual or abstract standard. In this case, there is a requirement for the XML instances to validate against the corresponding XML Schema. XML Schemas are collected in the OGC schema repository (OGC, Schema Repository) along with application examples. It is foreseen that with the adoption of JSON encodings and the approval of JSON schemas in the IETF, the schema repository will also contain JSON schemas.

15.6.4 OGC Standards Tracker

Members and non-member users of OGC standards frequently submit requests for changes or additions to existing standards. These need to be formally documented and addressed in a timely fashion. This is the purpose of the OGC standards tracker (OGC, Standards Tracker). Here, new requirements for standardization targets or change requests to existing standards can be submitted in a completely transparent process. If a Standards Working Group exists for a particular issue, the SWG will consider all of the relevant requirements and try to find solutions. If a DWG or SWG focused on that issue does not exist, OGC members may consider chartering a new working group to address that need.

15.6.5 An Example of How Standards Evolve in the OGC

We present the case of OGC Web Services (OWS) Context as an example of this process. OWS Context was developed to respond to the need to be able to list a collection of references to resources that can be accessed together via OGC standard services or by other means (e.g. a link to a geospatial file in the web). There was an existing standard that was called WMC (Web Map Context) that had the sole purpose of listing data layers that came from WMS services and that can be shown together in a map browser. With the generalization of other viewing OGC services (such as Web Map Tile Service (WMTS)) and other download services (such as Web Feature Service (WFS) and Web Coverage Service (WCS)) that can be used together in integrated clients, there was a clear need to extent WMC to other OGC services. The process started with experimentation in an OGC testbed that experimented with different possibilities in prototype components and documented the results in Engineering Reports that created an initial set of requirements for OWS Context. Then an OWS Context Standards Working group was created and the first draft of the specification was written while the experimentation on additional testbeds continued. The standard was written under the modular specification and it was divided in two documents: Conceptual and Atom encoding. The Atom encoding was made independent from the conceptual specification to facilitate other encodings that might be required in the future. The conceptual model contains the core class and extensions for each service or format it can include as resources. Recently, a GeoJSON encoding has been approved as a third element of the set.

15.6.6 Addressing Interoperability Among Vendor APIs

The term application programming interface (API) has been in use for decades, but it has taken on a slightly different connotation in today's world of "apps" and the cloud platforms that support those apps. Most apps are written for the platforms provided by major platform providers like Google, Apple and Microsoft. Their goal is to attract many app developers, whose creative apps will attract many more platform users. Other companies, too, from fast food providers to Earth image vendors, provide their own apps, and in some cases they document their APIs for app developers as the major platform providers do. But the app phenomenon is driven mainly by the major platform providers.

Through their APIs, the platform providers give app developers access to resources (sensors, location coordinates, displays, computing resources etc.) available on potentially millions of device owners' devices, and they also provide access to extraordinary processing power and huge stores of data in their platform's cloud. The providers publish the documentation for their APIs in an open way, but their APIs are only useful for accessing their services. Often several different major platform providers' APIs address the same problem using similar solutions based on similar sets of operations, but these are implemented in different ways. Thus apps and scripts using a vendor specific API cannot be applied to another vendor's API which addresses essentially the same problem. Integrating spatial data results obtained through two platform providers' APIs is almost always difficult. This limits the ability of app providers and their users to publish, discover, assess, access, aggregate, analyze and integrate spatial data obtained through two very similar apps that run on two different platform providers' platforms.

The situation is perhaps not as dire as it seems. The OGC® Open Geospatial APIs—White Paper (OGC, APIs White Paper) explains OGC's approach to solving this problem. The API White Paper sets forth a set of "OGC API Essentials", which are basic data structures that are extracted from OGC standards but are not themselves actual standards. They are essentially best practices that a company's geospatial API developers are advised to follow in developing their APIs so that users of APIs will have an easier time using that company's data with data obtained from other sources. Companies providing data want their data to be valued, and it is more valuable if it can be integrated with other data.

15.7 Looking to the Future

Information technology seems destined to advance at an ever faster pace in coming years. We can't know for sure what this will mean for geospatial technology, but we can imagine some possible directions.

Because of the accelerating pace of change, increasing the pace of standards development and early adoption and deployment is a key challenge. The OGC's Innovation Program has proven to be a successful vehicle for meeting these challenges in the past. However, the OGC has continually expanded its range of concerns, as illustrated by the increasing number of "aspect based", "challenge based", and "theme based" Working Groups. This portfolio is likely to continue to grow. More working groups means more competition for time slots at Technical

Committee Meetings, more Innovation Program activities to manage, more members to initiate and counsel, more compliance tests to develop, and more Communication and Outreach responsibilities.

As mentioned above, one key challenge will be helping user communities to deal with the competing giant information technology platform companies whose non-interoperable proprietary spatial apps and cloud resources confound efforts to share and integrate spatial data.

Another challenge is the emergence of the Semantic Web. Geosemantics, already an important topic, is likely to take on new importance and lead to new approaches to data discovery and integration.

The OGC has met basic computing platform challenges, as indicated above in the discussion about migrating from COM and CORBA to the Web, and then the phenomenon of developers frequently choosing JSON, REST and Linked Data instead of SOAP and XML for complex structured data. Both approaches will no doubt continue to be used. Both OGC's technology user members and technology provider members can be expected to continue to bring to OGC's process ideas about new technical innovations and trends. Datacubes, Big Data and crowd sourcing are current examples. In the future, perhaps Blockchain will emerge as the key for tracking and managing geospatial data provenance. Innovators frequently see OGC as an important steppingstone. Discrete global grid systems innovators, for example, have made OGC a key asset in their bid to replace centuries-old coordinate systems with a much more efficient digital approach. Similarly, i-locate has seized upon the OGC IndoorGML standard as a key resource in their mandate to provide a world-leading open platform and tools for indoor location.

The OGC Temporal DWG has been working with ISO to help provide a consistent calendar data model, extending from geological and astronomical periods down to femtoseconds. The Temporal DWG has also generalized the OGC's hydro-based time series model so that it provides a basis for interoperability in other application domains that use time series data. This may inform the time series modeling of high frequency electromagnetic fields, a goal of the Electromagnetic Spectrum DWG. Oscillating electromagnetic fields are a wave phenomenon, and there are other wave phenomena, such as sound, water waves, atmospheric waves and seismic waves. Perhaps a Waves DWG will be chartered to deal with this new aspect of geospatial information.

We are entering an age in which environmental information is becoming increasingly important. Figure 15.3 shows how, in climate science, new sub-disciplines have arisen. The same differentiation or variegation is happening in other domains, many of which need to share and integrate spatial data. Domains with a general focus frequently spawn multiple new sub-domains, each with a narrower, more specialized focus. These represent new information communities: Their data models and semantics differ, but they need to share data and communicate. With this proliferation of domains in relationship, it seems likely that new theme based domains will bring their interoperability requirements into OGC.

Perhaps graph databases, the engine for social networking services, will emerge as environmental science seeks ways to track, understand and manage complex

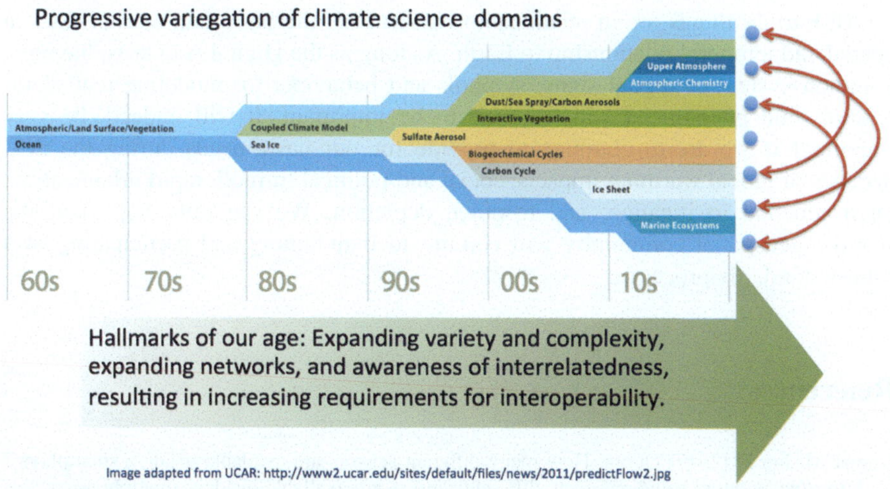

Fig. 15.3 Increasing variety, complexity and interrelatedness are key drivers for standards

ecological networks, which have a strong spatial element. Similarly, environmental health experts seek ways to track and understand the "exposome," the aggregate of environmental health stressors that are co-contributors to diseases and disabilities that are on the rise in our century, diseases such as alzheimer's, autism, obesity and diabetes. Many, though not all, exposures have a spatial element. Like graph databases, array databases provide data management solutions for large, heterogenous datasets that have different underlying data and programming models.

Similarly, array databases (see Chap. 14: Baumann) are promising when it comes to multi-dimensional spatio-temporal raster data, also known as "datacubes". Datacube services aim at doing all of the homogenization work in the server so that, for example, all data from one satellite instrument form one logical object which is ready for spatial and temporal analysis. Graph databases or array databases could help resolve the data/metadata dichotomy underlying many of the difficulties in geoinformation management.

There are other potential trends that will likely be facilitated by the OGC process. OGC and the buildingSMART alliance are working to establish interoperability between geospatial systems and urban 3D models and systems for civil engineering and Building Information Models. Related to this are the OGC IndoorGML standard and i-locate, a project of Europe's ICT-PSP program, which provides the first reference implementation of IndoorGML. I-locate aims to provide an open platform for innovation in integrated indoor and outdoor localisation of people and objects. The effects of electromagnetic fields span all scales, as discussed in the Electromagnetic Spectrum DWG. Perhaps these activities, all bridging geospatial and sub-geospatial scale features and phenomena, are moving OGC in the direction of standardizing the modeling of all spatial/temporal phenomena, from cosmic to mega, macro, meso, micro down to nanoscale features and phenomena.

All Earthly things are in relationship by virtue of the basic fact that all exist in spatial and temporal relationship to Earth. As long as the Digital Age lasts, having a common system of digital signs, symbols and behaviors for modeling real world features and phenomena will continue to be important. It will, indeed, be very important if we are to continue to provide for our basic needs during the next decades of global warming impacts, social and political turmoil, rapid urbanization, environmental toxification, and resource depletion. We can only hope that the world's geospatial community will commit to maintaining and participating in a robust standards process.

References

Kumar Mishra D (2017) Quora. How many different sensors are available inside a smartphone? https://www.quora.com/how-many-different-sensors-are-available-inside-a-smartphone. Accessed 26 June 2017

Longley PA, Goodchild MF, Maguire DG, Rhind DW (2015) Geographic information science and systems, 4th edn. Wiley, New Jersey

McKee L (2017) OGC history. (Note: As of July, 2017, the last update to the OGC History pages was in 2012.) OGC website. http://www.opengeospatial.org/ogc/history adapted from OGC History (long). http://www.opengeospatial.org/ogc/historylong

OGC, Abstract Specification. http://www.opengeospatial.org/docs/as

OGC, Alliance Partnerships. http://www.opengeospatial.org/ogc/alliancepartners

OGC, APIs White Paper. docs.opengeospatial.org/wp/16-019r4/16-019r4.html

OGC, Compliance Program. http://www.opengeospatial.org/compliance

OGC, Communications. http://www.opengeospatial.org/ogc/programs/ocap

OGC, Core Programs. http://www.opengeospatial.org/ogc/programs

OGC, Directors. http://www.opengeospatial.org/ogc/organization/bod

OGC, Engineering Reports. http://www.opengeospatial.org/docs/er

OGC, Innovation Program. http://www.opengeospatial.org/ogc/programs/ip

OGC, Interoperability Initiatives. http://www.opengeospatial.org/projects/initiatives

OGC, Meetings. http://www.opengeospatial.org/event

OGC, Membership. http://www.opengeospatial.org/ogc/members

OGC, Modular Specification. http://www.opengeospatial.org/standards/modularspec

OGC, ORM. http://www.opengeospatial.org/standards/orm

OGC, Reference Model. http://www.opengeospatial.org/standards/orm

OGC, Schema Repository. http://schemas.opengis.net

OGC, Staff. http://www.opengeospatial.org/ogc/organization/staff

OGC, Standards Tracker. http://ogc.standardstracker.org

OGC, World regions. http://www.opengeospatial.org/ogc/regions

Sterling G, Sterling G (2017) 56 Percent of "On The Go" searches have local intent. [online] Search Engine Land. http://searchengineland.com/research-56-percent-go-searches-carry-local-intent-190655. Accessed 26 Jun 2017

Sullivan M (2013) A brief history of GPS. TechHive, PC World. http://www.pcworld.com/article/2000276/a-brief-history-of-gps.html. Accessed 26 June 2017

Townsend E (2008) The 25-year history of service-oriented architecture. An Erik Townsend white paper. http://www.eriktownsend.com/look/doc_download/12-1-the-25-year-history-of-service-oriented-architecture. Accessed 26 June 2017

W3C, Spatial wiki. https://www.w3.org/2015/spatial/wiki/Main_Page

Chapter 16
Standards—Making Geographic Information Discoverable, Accessible and Usable for Modern Cartography

Serena Coetzee, Reese Plews, Jean Brodeur, Jan Hjelmager, Andrew Jones, Knut Jetlund, Roland Grillmayer and Christina Wasström

Abstract Cartography relies on data. Today, data is generated in unprecedented volumes, velocity, and variety. As a result, cartographers need ever more assistance in finding appropriate data for their maps and in harmonizing heterogeneous data before functional maps can be produced. Spatial data infrastructures (SDIs) provide the fundamental facilities, services and systems for finding data. Implementing standards for geographic information and services plays a significant role in facilitating harmonization and interoperability in an SDI. This chapter reviews collaboration between standards development organizations in the field of geographic information, and describes resources available for a model-driven approach in the implementation of geographic information standards. Subsequently, good practice examples from Canada, Denmark, Japan, and Europe illustrate how

S. Coetzee
Centre for Geographic Information Science, Department of Geography, Geoinformatics and Meteorology, University of Pretoria, Pretoria, South Africa

R. Plews
Plews Consulting Co. Ltd, Tokyo, Japan

J. Brodeur
Geosemantic Research, Quebec, Canada

J. Hjelmager
Agency for Data Supply and Efficiency, Copenhagen, Denmark

A. Jones
Jasper Spatial Consulting, Adelaide, Australia

K. Jetlund
Norwegian Public Roads Administration, Oslo, Norway

R. Grillmayer
Environment Agency Austria (EAA), Vienna, Austria

C. Wasström (✉)
Lantmäteriet, Gävle, Sweden
e-mail: Christina.Wasstrom@lm.se

J. Döllner et al. (eds.), *Service-Oriented Mapping*, Lecture Notes in Geoinformation and Cartography, https://doi.org/10.1007/978-3-319-72434-8_16

standards implementation facilitates harmonization and interoperability, and how SDIs make geographic information discoverable, accessible, and usable for modern cartography.

Keywords Standards implementation · Spatial data infrastructure (SDI)
Map production · Cartography · Model-driven approach · Geographic information

16.1 Introduction

To make a map, one needs data. Today, data is generated in unprecedented volumes, velocity, and variety. User-generated content, a plethora of satellite platforms, ubiquitous sensors and smartphones are adding to the ever-increasing volumes of data. Some data is continuously streamed at a rate that requires novel ways to process the data. The diversity of devices generating the data results in vastly heterogeneous data. The challenge for cartographers is to find appropriate data among the vast volumes of data, and to produce functional maps from these heterogeneous data sources using modern cartographic tools. For this, interoperability is required.

Standards are the foundation and building blocks for harmonization and interoperability of geographic information in a spatial data infrastructure (SDI). Through the implementation of standards, an SDI provides the fundamental facilities, services and systems that make geographic information available, accessible, and usable for modern cartographers.

Standardization in the field of geographic information is a complex task that addresses many different aspects of interoperability, including data types for representing geographic and temporal information, data product specifications, metadata, services, and encodings. In today's fast-moving technology-integrated world, a model-driven approach provides implementation efficiency by simplifying the design process through standardized data models and by promoting communication through standardized terminology and descriptions of good practices.

This chapter begins with brief sketches of the background of standards development organizations in the field of geographic information and services. Next, Sect. 16.3 describes the terminology, data models, schemas, and ontologies available from ISO/TC 211 for a model-driven approach in the implementation of geographic information standards. In Sect. 16.4, good practice examples of standards implementation from Canada, Denmark, Japan, and Europe illustrate how standards make geographic information discoverable, accessible, and usable for modern cartography. Section 16.5 discusses alternatives to national SDIs, namely volunteered geographic information (VGI) and open government data, and the influence of these new paradigms on modern cartography. Finally, Sect. 16.6 summarizes and concludes the chapter.

16.2 Standards Development Organizations

ISO/TC 211, Geographic information/Geomatics, is the technical committee of the International Organization for Standardization (ISO) and is responsible for the standardization of geographic information. Established in 1994, its work aims at establishing a structured set of standards for information concerning objects or phenomena that are directly or indirectly associated with a location relative to the Earth. ISO/TC 211 covers semantic, syntactic, and service issues, as well as procedural standards, at various levels of abstraction.

The Open Geospatial Consortium (OGC) is a voluntary consensus standards organization, also established in 1994. The focus of OGC work is to define, document, and test implementation specifications for use with geospatial content and services. OGC specifications leverage the abstract standards defined by ISO/TC 211. Approved OGC specifications, as well as any schemas (xsd, xslt, etc.) supporting them, are published and freely available on the OGC website (e.g. see Chap. 14, Baumann).

The International Hydrographic Organization (IHO) is an intergovernmental consultative and technical organization established in 1921 to support safety of navigation and the protection of the marine environment. Among its main objectives, IHO is tasked to bring about the greatest possible uniformity in nautical charts and documents (i.e. standardization).

OGC and ISO/TC 211 have a long history of collaboration and development of joint standards. For example, OGC specifications have been submitted to ISO/TC 211 for consideration for approval as International Standards. OGC focuses on specifications that can be directly implemented; many of these are based on the conceptual (or abstract) models defined by ISO/TC 211 or jointly between the organizations. The cooperation between IHO and ISO/TC 211 has been driven by the development of standards for digital hydrographical information and products. IHO standards are based on various standards in the ISO 19100 series of standards (ISO/TC 211 2013).

Collaboration with other standards developing organizations ensures that geographic information and services are aligned with state-of-the-art technologies in the era of modern cartography. For example, the Internet Engineering Task Force (IETF) develops and promotes standards for the Internet, which are widely used. Some of these, e.g. IETF RFC 3986, Uniform Resource Identifier (URI): Generic Syntax, are referenced in standards for geographic information. Similarly, standards by the World Wide Web Consortium (W3C) are referenced, such as editions of the W3C Recommendation, Extensible Markup Language (XML). The Geomatics Committee of the International Association of Oil and Gas Producers (OGP), through its Geodesy Subcommittee, maintains and publishes a dataset of parameters for coordinate reference system and coordinate transformation descriptions, collectively referred to as 'EPSG codes'. These codes are used to uniquely identify a coordinate reference system in various IHO, OGC and ISO standards.

16.3 Resources for Model-Driven Implementations of Geographic Information Standards

16.3.1 Terminology

A vocabulary is an important resource for all who seek an understanding of the community's cultures and practices. The terms and definitions identify the concepts that characterize the community's philosophies, technologies and activities. Its structure and supporting information identify relationships between concepts and the context for their intended used.

ISO/TC 211 has compiled a vocabulary for geographic information by aggregating the terminology clauses from each of the ISO/TC 211 International Standards and Technical Specifications. The vocabulary is published as ISO/TC 211 Multi-Lingual Glossary of Terms (MLGT) and is freely available on the Resources page of the ISO/TC 211 website (ISO/TC 211 2017). The MLGT currently includes terminology records in 14 different languages, namely Arabic, Chinese, Danish, Dutch, English, Finnish, French, German, Japanese, Korean, Polish, Russian, Spanish, and Swedish.

The development of the terminology needs to be carefully managed and coordinated. There are terms and concepts in general language dictionaries that correctly designate and describe geographic information concepts. Similarly, there are geographic information concepts that have already been designated and defined in standards or similar documentation. These are adopted whenever possible, avoiding the unnecessary proliferation of terminological entries.

Quite often, however, definitions in general language dictionaries are insufficiently rigorous or concise to describe the concept. In such cases, the concept, term and definition are refined or adapted, as appropriate. To add a term to the vocabulary, the following steps are required: (1) the identification of a concept; (2) the nomination of a designation (usually a term) for that concept; and (3) the construction of a definition that unambiguously describes the concept.

The identification of concepts is arguably the most complex and demanding part of the terminology development process. The complexity stems from the fact that concepts do not exist in isolation but always in relation to each other, giving rise to a concept system—a set of concepts that are distinct but related to each other. Each concept is capable of separate description and may also be capable of further decomposition, as illustrated in Fig. 16.1. The process of decomposition ceases when the concepts become so basic that they do not need to be defined.

Once the concept has been identified, a single term, the '*Preferred Term*', is adopted as the primary designator. Sometimes there may also be a shortened form of the *Preferred Term*, referred to as the *Abbreviated Term*. The role of a definition is to precisely describe the content of an identified concept. It should be as brief as possible, containing only that information that makes the concept unique, focusing on encapsulation rather than exclusion.

The concept "**coordinate reference system**" is defined as

"*coordinate system that is related to an object by a datum*" [ISO 19111:2007. 4.8].

It associates the concepts "**coordinate system**", which is defined as

"*set of mathematical rules for specifying how coordinates are to be assigned to points [ISO 19111:2007, 4.10],*

and "**datum**", which is defined as

"*parameter or set of parameters that define the position of the origin, the scale, and the orientation of a coordinate system" [ISO 19111:2007, 4.14].*

Further decomposition of "**coordinate system**" and "**datum**" into component concepts is possible (for example, into "**coordinate**", "**origin**", "**scale**", "***axis***")

In addition, "**coordinate reference system**" becomes the superordinate concept for more specific concept "**compound coordinate reference system**".

Fig. 16.1 Concept decomposition

In ISO/TC 211, the Terminology Repository is the principal information system for terminology maintenance and harmonisation. It incorporates six registers, shown in Fig. 16.2. The Document Register contains documents that contribute terminological entries to the Terminology Repository. The Terminology Spreadsheet contains the terminological entries from the most recent draft or published document for each ISO/TC 211 work item. The Terminology Archive contains the terminological entries from drafts and published documents that have been superseded. The Symbols and Abbreviations Registers respectively contain symbols and abbreviations from the most recent draft or published document for each ISO/TC 211 work item. The Packages Register identifies the bi-alpha prefixes associated with UML packages in ISO/TC 211 documents. From these registers, the MLGT is generated and published at regular intervals.

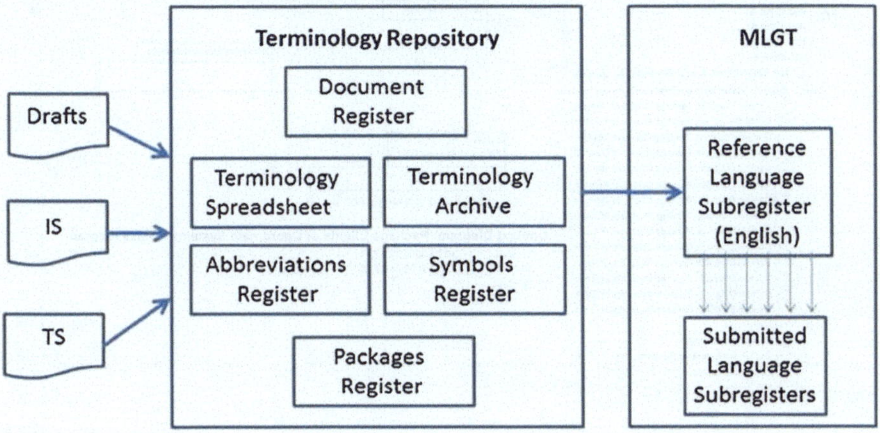

Fig. 16.2 ISO/TC 211 terminology registers (ISO/TC 211 2015)

16.3.2 *Harmonized Model, Schemas, and Ontologies*

Conceptual models in ISO/TC 211 documents are developed in the Unified Modeling Language (UML), based on the ISO/TC 211 profile of UML defined in ISO 19103:2015, *Geographic information—Conceptual schema language*. The UML models included in the different standards are maintained in a single harmonized model repository, commonly referred to as the *Harmonized Model*, under the responsibility of the Harmonized Model Maintenance Group (HMMG).

The GitHub (2017) repository of the HMMG (ISO/TC 211 HMMG 2017) provides guidance for using the *Harmonized Model*. From there, the models can be downloaded as Enterprise Architect (Sparx Systems 2017) projects or as XMI files for packages. In addition, an HTML view of the model is available, as shown in Fig. 16.3.

Implementation schemas for XML and RDF (Resource Description Framework) are generated from the *Harmonized Model*, based on encoding rules defined in ISO/TC 211 standards. To support the Semantic Web community, a suite of ontology standards for geographic information is currently under development. When completed, there will be six parts of ISO 19150, *Geographic information—Ontology*. Based on these standards, OWL (Web Ontology Language) versions of the ISO/TC

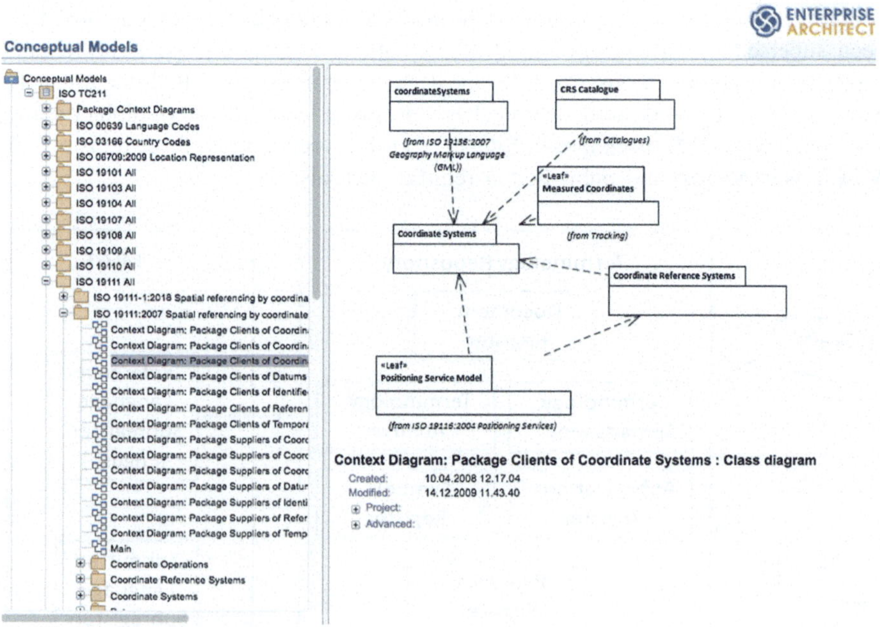

Fig. 16.3 HTML view of the harmonized model in the GitHub repository of the HMMG (ISO/TC 211 HMMG 2017)

211 UML conceptual models have been generated and are published in the GitHub repository of the ISO/TC 211 Group on Ontology Maintenance (GOM).

The models, schemas and ontologies are readily available for anyone who wants to follow a model-driven approach in an implementation of ISO/TC 211 standards. Further implementation guidance is provided on the Standards Wiki (2017).

16.4 Examples of Good Practice

16.4.1 Canada

The Standards user's guide for geographic information published by Natural Resources Canada (2016) was written to support the proper use of geographic information standards. The guide provides background and practical knowledge about standards that support an SDI implementation and, in this specific case, the implementation of the Canadian Geospatial Data Infrastructure (CGDI). The guide addresses three main topics: (1) the establishment of a Geospatial Information Environment, (2) the creation of a Data Product, and (3) the implementation of Web Services.

Geospatial Information Environment. This is the fundamental notion supporting an SDI; it comprises individuals, organizations, and systems that collect, process, disseminate, or act on geographic information, and includes the information itself (Joint Chiefs of Staff 2014). The Geospatial Information Environment, portrayed in Fig. 16.4, includes (1) data structures and schemas described in a standardized application schema; (2) data descriptions and semantics in standardized feature catalogues; (3) standardized metadata; (4) data and metadata capture operations; (5) data (the data elements); (6) data management; (7) discovery; (8) access; and (9) transformation. All these components are well covered by both ISO/TC 211 and OGC standards.

Using a hydrographic network as an example to illustrate a Geospatial Information Environment, first, the required feature types (e.g. river, lake, waterbody, watercourse, stream, creek, bridge, island, obstacle to navigation, geographical name, etc.), their characteristics (e.g. geometry, flow direction, permanency, height, accuracy, etc.), and the associations between feature types (e.g. lake ↔ geographical name, river ↔ bridge, etc.) need to be identified as part of an application schema. Second, these feature types, their characteristics, and the role they play in an association must be defined in a feature catalogue and as appropriate in a concept dictionary to define their semantics for appropriate uses.

Typically, different topics and layers of geographic information are defined and maintained by various stakeholders within a Geospatial Information Environment. Therefore, it is appropriate to set a schema repository that documents all the application schemas and their related feature catalogues, including code lists (permitted attribute values), metadata elements, and encoding types. Additional

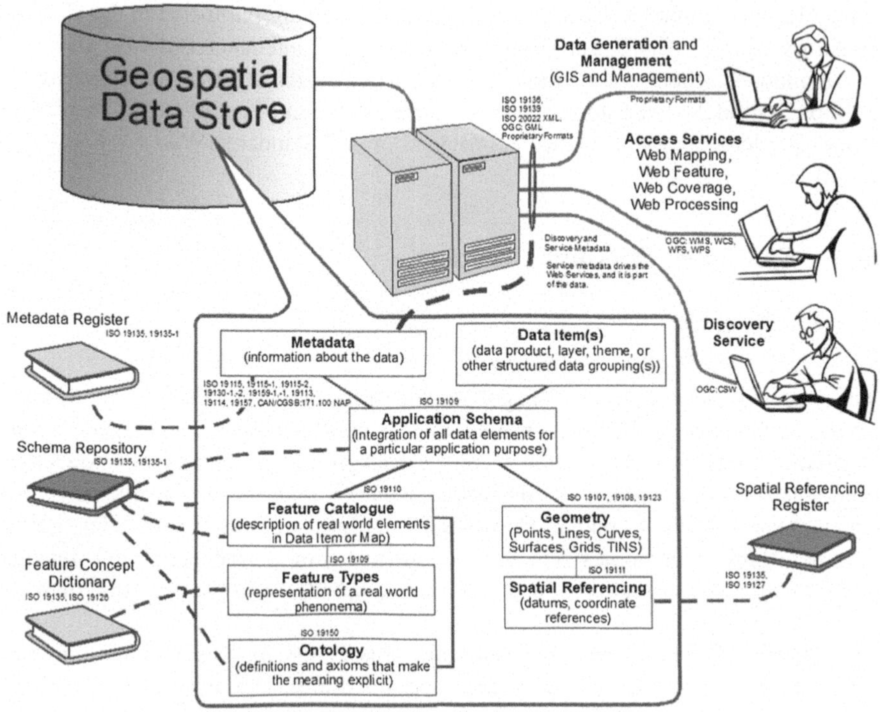

Fig. 16.4 Geospatial information environment (Natural Resources Canada 2016)

information can be included in the schema repository, to support, for instance portrayal application (symbol sets and style sheets), and Semantic Web applications (e.g. ontologies). Data is then acquired and compiled according to the application schema and its feature catalogue. Geospatial data, along with the schema repository, is maintained in a geospatial data store. The geospatial data store supports the management, maintenance, discovery, access, and use of the geospatial data in this Geospatial Information Environment.

Data Products. Geospatial data products are based on the content of the geospatial data store. Ideally, a data product is detailed in a data product specification, comprising application schema, metadata elements, feature catalogues, conformance tests, encoding and portrayal information. The data product specification also provides details on the process for data acquisition (e.g., capture criteria), post processing (e.g., quality control, validation, error correction), and maintenance processes (e.g., revision cycle and methodology).

Printed maps, map series in various scales, GML (Geography Markup Language) files, and ESRI Shape™ files are examples of data products. In Canada, the GeoBase series is an example of an integrated data product. It includes the Canadian Digital Elevation Data (CDED), the National Hydro Network (NHN), the National Road Network (NRN), the National Railway Network (NRWN), Land

Cover—circa 2000-Vector (LCC2000-V), the Canadian Geographical Names, the Aboriginal Lands of Canada, the Municipal Boundaries, the Canadian Geopolitical Boundaries, the GeoBase Orthoimage 2005–2010, the GeoBase Raw Imagery 2005–2010, the Landsat-7 Orthoimage, the Landsat-7 Level 1-G, the RADARSAT-1 Orthorectified Imagery, and the Control Points for Landsat 7 Imagery (Natural Resources Canada 2017). Each of the above is documented in a data product specification compliant with ISO 19131:2007, *Geographic information—Data product specifications*. It is also good practice to include the data product specifications in the schema repository.

Web Services. Nowadays, printed maps have mostly been replaced by digital maps or digital data, which are both accessible online or downloadable via services on the Internet. Relevant web service standards include the Catalogue Service for the Web (CSW) (OGC 2007a) for data discovery by the way of metadata; Web Map Service (WMS) (OGC 2006) for delivering image maps via the Internet; Web Feature Service (WFS) (OGC 2002) for delivering (vector based) geographic feature data via the Internet; Web Coverage Services (WCS) (OGC 2012) for delivering images and other kinds of coverage data via the Internet; and Web Processing Service (WPS) (OGC 2007b) for processing geospatial data, e.g. converting data from one encoding format to another, or transforming it from one projection to another.

As part of the Canadian Geospatial Data Infrastructure, Canada has released Open Maps on its Open Government Portal (Government of Canada 2017b). Various maps can be browsed and data can be downloaded. More recently, a geospatial data extraction tool (Government of Canada 2017a) with similar functionalities has been launched. A screenshot is included in Fig. 16.5.

16.4.2 Denmark

In the Danish eGovernment Strategy for the period 2011–2015 (Digitaliseringsstyrelsen 2011), the vision was to create a simple, more efficient and coherent public sector (government agencies and municipalities). The aim was to achieve this through digitalization and reuse of data across domains. The data covered by the strategy, called *basic data*, originate from five different governmental organizations: the Danish Geodata Agency; the Agency for Data Supply and Efficiency; the Danish Business Authority; the Civil Registration System; and the Danish Taxation Authority.

Over the years, each of the organizations had developed their own environment and conditions for data modelling, information security, distribution of data, and conditions for the access and use of data. To provide their data for reuse across domains, the environments and conditions had to be harmonized. One way of achieving this was through the use of standards, either Danish national standards and/or Danish profiles of international standards.

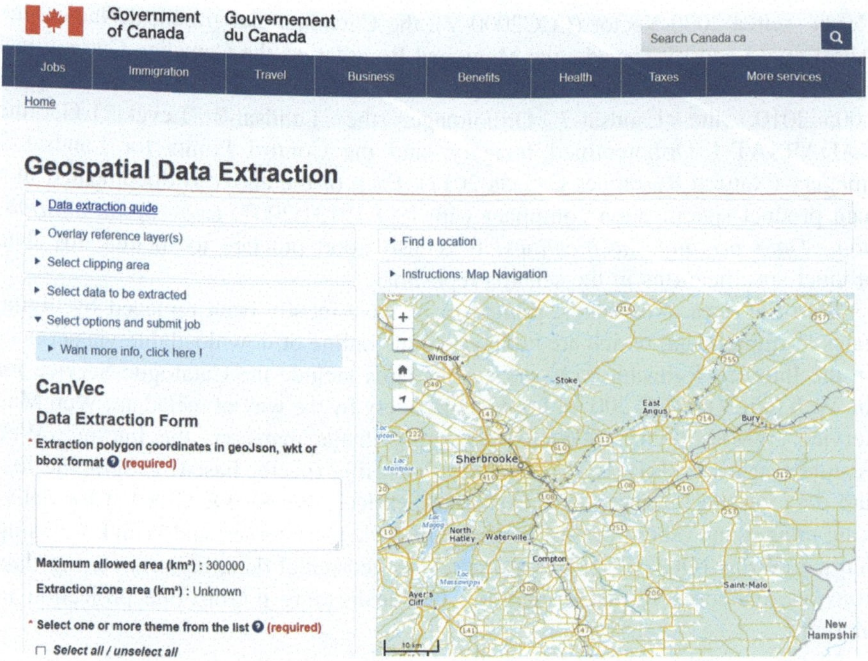

Fig. 16.5 Geospatial data extraction (Government of Canada 2017a)

Data Modelling. The first step in the implementation of the strategy was the development of a single coherent data model of all the basic data. The purpose was to provide an overview and terminology of the basic data and its attribute information, and to show how the basic data fit together across the domains (Datafordeler 2017). Given that the data from the different organizations were historically modelled in different ways and documented in different modelling languages, it was decided to use a common modelling language. UML, or a profile thereof, was selected because it is internationally recognized and used in e.g. ISO/TC211, OGC and INSPIRE (INfrastructure for SPatial InfoRmation in Europe).

In addition to using the same modelling language for all the basic data, various rules for modelling were also developed (OIO Arkitekturguiden 2017a). These rules were based on international standards from ISO, OGC, INSPIRE and Danish national standards (OIO Arkitekturguiden 2017b). The rationale for these modelling rules was to ensure a common approach to the modelling task by the various stakeholders. The rules ensure that the modelling of data objects is based on a common set of guidelines, and that the entire model is based on common normative properties. The result enabled the basic data to be interoperable and re-used across various domains.

Information Security. Since 2014, it is mandatory for all governmental organizations to use a national standard for information security. In 2016, this national standard was replaced by ISO/IEC 27001, *Information security management*.

Among the many reasons for this change, was to ensure uniform and high-level security across the Danish eGovernment and to ensure that Danish citizens trusted the Danish eGovernment solution. In addition, the use of an international standard would pave the way for international interoperability.

Distribution of Data. In order to distribute the basic data, a single point of access website was created, called the *Data Distributor*. This access point contains all the basic data and receives geospatial and other data from the organizations mentioned above in XML and GML formats. These formats imply that the organization responsible for handling the data works with a single standardized format and does not have to handle all kinds of proprietary formats used by owners of the basic data.

For the distribution of the Danish basic data, a service-oriented solution was chosen, i.e. the user can either obtain the basic data via a file download, a web service solution, or an event driven solution. For the file download, the common FTP (File Transfer Protocol) is used.

For the web services, solutions were chosen that relied heavily on standards originating from both OGC and ISO/TC 211. The suite of download services consists of the following: REST (REpresentational State Transfer), WFS (Web Feature Service), WMS (Web Map Service), WMTS (Web Map Tile Service), and WCS (Web Coverage Service). These web services provide online access to the basic data, thereby allowing users to always have access to the latest version of the basic data. The event driven solution is based on an event driven architecture.

Conditions for Access and Use. These conditions are closely related to the distribution of basic data. For some data, users only need to register to gain access to the data, whereas for other types of data, access is restricted due to the sensitivity of the data. Adding to these requirements, Denmark must also conform to the EU Regulation 2016/679 on the protection of natural persons with regard to the processing of personal data and on the free movement of such data, which can be regarded, in some ways, as a standard.

16.4.3 Japan

Japan has been active in ISO/TC 211 as a participating member since 1994, led by the Geospatial Information Authority of Japan (GSI) in the Ministry of Land, Infrastructure, Transport and Tourism (MLIT). Domestically, the organization of TC 211 activities in Japan is carried out under the auspices of the Japan TC 211 National Body. This committee is administered by the Association of Precise Survey and Applied Technology (APA) whose broad membership includes geospatial experts from the public, private and academic sectors.

Recognizing the benefit of using standards for the delivery and exchange of geospatial data, the ISO/TC 211 suite of standards was adopted and translated for domestic use as Japanese Industrial Standards (JIS). The work was guided first by a public-private research project for geospatial standardization and later by the

Advanced level applicants are required to utilize geographic information standards to an even more advanced technical level to create new standards. In addition, the three certification levels have an expiration date of five years, giving applicants an incentive to maintain their certification.

Year on year, the number of certified professionals has increased. In 2014, GSI recognized this certification as one of the prerequisites for bidding on their surveying, topographic map compilation and other cartographic and geospatial related contracts. Currently, professionals who have already obtained beginner level certification are encouraged to progress to a higher level of certification. To facilitate this, opportunities for learning how to utilize geographic information standards and new training techniques to further improve *skill building* are being developed.

Another certification program in Japan was established by the GIS Association (GISA) in cooperation with seven academic associations and non-profit organizations. The GIS Certification Association (GISCA) program (GISA 2017), launched in 2014, organizes educational courses introducing geographic information technology based on geographic information (GI) standards (Ota and Plews 2015; Ota 2017) for industry professionals and students in higher education.

16.4.4 INSPIRE in Europe

INSPIRE is an EU-Directive established by the European Parliament and EU Council in 2007 that addresses 34 spatial data themes needed for environmental applications (INSPIRE 2007, see Chap. 3). The main objective is to facilitate access to cross border harmonized spatial datasets and services at a defined quality level with clearly formulated and concise usage rights.

To support objectives of the Directive, Implementation Rules (IR) were developed for metadata, data specifications, network services, data and service sharing, spatial dataset services and monitoring, and reporting issues. The Implementing Rules have been adopted as Commission Decisions or Regulations and are binding in their entirety. They specify the requirements at an abstract and generic level, while the Technical Guidelines (TG) describe how legal obligations could be implemented based on international standards for geographic information. The relationship between the IR and the TG documents is shown in Fig. 16.7.

Syntactic and sematic interoperability. Data specifications for each theme were developed with the aim to guarantee the best possible level of syntactic and semantic interoperability. For each of the themes, experts developed a UML data model, based on ISO 19103:2015, *Geographic information—Conceptual schema* and ISO 19109:2015, *Geographic information—Rules for application schema*. After a consolidation phase, the UML models were automatically translated into GML application schemas. In various cases, more than one application schema was generated for the same theme, accommodating the different aspects of that theme. For example, five GML schemas are published for the land use data specification, accommodating existing land use, planned land use, gridded land use, sampled land

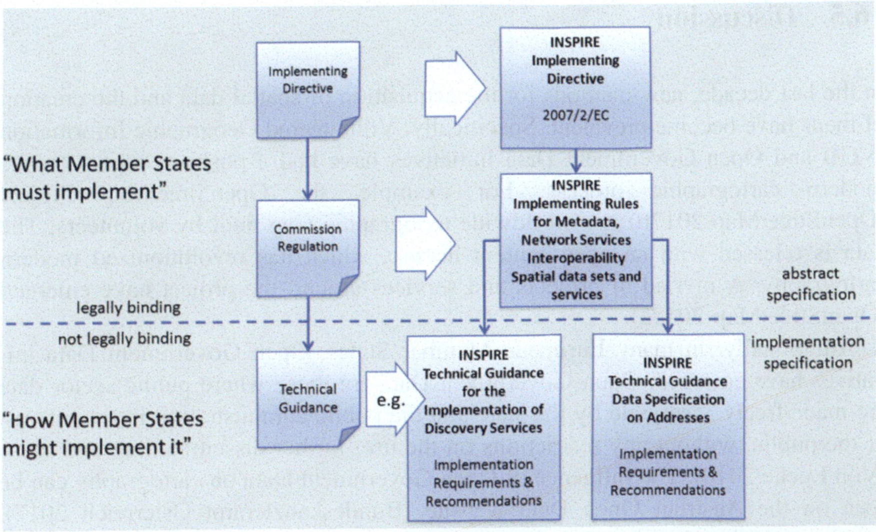

Fig. 16.7 Relationship between INSPIRE implementing rules and technical guidance documents (INSPIRE Thematic Working Group *Addresses* 2009)

use, and land use nomenclatures. The different application schemas incorporate the semantics of the data models, a first step towards syntactic harmonisation of datasets for a specific theme.

To ensure the highest possible semantic interoperability for a specific theme, existing and domain agreed terminology is required. One critical fact within the whole INSPIRE modelling process was that domain specific terminology did not exist for all themes. If a domain agreed terminology was available, semantic interoperability was established by defining a mapping between domain terms and national or local terms used in the source datasets. For all other themes, there is a placeholder for the term in the data model so that source terminology can be considered during harmonization. For such themes, the degree of semantic interoperability is lower and cross-border use of the dataset requires pre-processing.

Data quality validation. Quality issues are addressed by the implementation of a validation procedure. For each theme, conformance classes were defined in the form of Abstract Test Suites (ATS). These were translated into Executable Test Suites (ETS) and implemented in the validation platforms, such as the INSPIRE Validator (INSPIRE 2017). The validation procedure is based on international standards in the ISO 19100 series. When datasets are encoded according to ISO 19136:2007, *Geographic information—Geography Markup Language* (*GML*), a fully automatic validation of many quality aspects is possible, including not only geometry and data consistency, but also semantic issues like the validation of code list values.

16.5 Discussion

In the last decade, new methods for the acquisition of spatial data and the creation of maps have become prevalent. Specifically, Volunteered Geographic Information (VGI) and Open Government Data initiatives have had a positive impact on the modern cartographic process. For example, the OpenStreetMap Project (OpenStreetMap 2017b) is a worldwide topographic map built by volunteers. The data is released with an open-content license, which has revolutionized modern cartography. A myriad of projects and services around the project have emerged (OpenStreetMap 2017a).

Additionally, in many European Member States, Open Government Data initiatives have emerged. Open Government Data are those where public sector data are made freely accessible by the state and the public administration in the interest of the public, without any restrictions on the use, further dissemination, and reuse (Von Lucke 2010). The influence of Open Government Data on cartography can be seen on the Austrian Open Data website (Bundeskanzleramt Österreich 2017), which lists 408 map-related projects, ranging from simple web map visualisations to more complex mobile apps.

The most problematic aspects when working with volunteered geographic information is the reliability of data, which is directly connected to quality issues, such as completeness, semantic heterogeneity and positional accuracy (Dorn et al. 2015; Shudan and Jianghua 2014). These shortcomings restrict the use of VGI to applications.

In contrast, a certain quality level is guaranteed in Open Government Data, because they are produced by governmental organisations with legal obligations. However, quality levels are designed to meet the quality needs of legal obligations (only); no additional processing or harmonization is done. This leads to a low level of semantic interoperability between different datasets of the same theme. The implication is that cartographers are often required to harmonize the data before maps can be produced.

Also, complete national coverage of Open Government Data is not typically provided. Open Government Data for selected themes are commonly made available by federal states, and in many cases themes are available for large cities only. As a result, national- or pan-European mapping applications are not always possible.

16.6 Conclusion

In this chapter we illustrated that standards are the foundation and building blocks for harmonization and interoperability in an SDI, resulting in geographic information being discoverable, accessible, and usable by cartographers.

Through the cooperation between ISO/TC 211, OGC, and IHO, knowledge is shared and re-used, and standards are mostly harmonized with each other. As a result, cartographic tools and technologies can be developed to work with both marine and terrestrial data, and the tools and technologies are usable across many different domains and communities.

Standardized terminology further facilitates work across different domains and communities. ISO/TC 211's standardized terminology achieves this not only in English, but also in the 14 languages into which the terminology has been translated. In a globalized world, a standardized terminology facilitates cartographic communication beyond linguistic boundaries.

The terminology, models, schemas and ontologies are readily available for anyone who wants to follow the proven model-driven approach in the implementation of ISO/TC 211 standards. The Harmonized Model simplifies the design process for standards implementation and for the development of profiles of standards. The examples from the different countries in this chapter show that profile development is often required.

The example from Canada illustrates that ISO and OGC standards are the foundation on which a Geospatial Information Environment is established. Each component of the Environment is supported by specific standards; together the standards facilitate interoperability of geographic information at the syntactic level, as well as at the semantic level.

In Denmark, basic data from various domains was harmonized and integrated. In this way, a single source of data for cartographers was established. The implementation of the Danish eGovernment Strategy would not have succeeded without ISO and OGC standards and profiles.

Japan's continued participation in ISO/TC 211 and the domestic use of standards-based product specifications demonstrates how standards support geoinformation management. It also illustrates how standards are used to make data accessible to developers, cartographers, end users and the public. The certification programs are proof that standards are relevant and important for modern cartographers and geoinformation managers.

The INSPIRE Technical Guidelines are based on international standards published by ISO and OGC. These standards form the backbone of the INSPIRE framework. Implementation of the Technical Guidelines by the member states will maximize cross-border interoperability of the INSPIRE spatial datasets and data services. In this way, diverse data sources are harmonized and available for cartographers to produce maps.

Standards and SDIs have impacted modern cartography in many ways. Amongst others, they have led to quality spatial data of national coverage, available for map production, thus enriching modern cartographic output.

Acknowledgements Serena Coetzee would like to thank the South African Bureau of Standards (SABS) for continued support to participate in ISO/TC 211, Geographic information/Geomatics.

Reese Plews would like to thank JIPDEC for supporting his continued participation in ISO/TC 211, and to GSI, the Japanese Ministry of Economy, Trade and Industry (METI), and APA for their continued support of geospatial standards development in Japan. In addition, Mr. Plews would also like to thank Chikako Kurokawa, Chie Deguchi, and Morishige Ota for providing feedback on earlier drafts of the chapter.

References

Association of Precise Survey and Applied Technology (APA) (2017) Certification of professionals in standards for geographic information (S-GI-Cert) framework (in Japanese). https://www.sokugikyo.or.jp/geographic_information/index.html. Accessed 18 Aug 2017

Bundeskanzleramt Österreich (2017) Applications (in German). https://www.data.gv.at/anwendungen/. Accessed 30 Aug 2017

Datafordeler (2017) Om grunddata, Fælles datamodel, http://datafordeler.dk/om-grunddata/faelles-datamodel/. Accessed 1 Aug 2017

Digitaliseringsstyrelsen (2011) The digital path to future welfare, e-Government strategy 2011–2015. www.digst.dk/~/media/Files/Strategier/Engelsk_strategi_tilgaengelig.pdf. Accessed 24 June 2017

Dorn H, Törnros T, Zipf A (2015) Quality evaluation of VGI using authoritative data—a comparison with land use data in Southern Germany. ISPRS Int J Geo-Inf 4:1657–1670

ISO/TC 211, Geographic information/Geomatics (2015) Report from WG 1—Framework and reference model, and TMG to the ISO/TC 211 plenary, Shenzhen, 2014-11-27/28. ISO/TC 211 document N3945

ISO/TC 211, Geographic information/Geomatics (2017) Website. https://committee.iso.org/home/tc211. Accessed 30 Aug 2017

Geospatial Information Authority of Japan (GSI) (2014a) Fundamental geospatial data (FGD) (in Japanese). Geospatial information authority of Japan. www.gsi.go.jp/kiban/towa.html. Accessed 18 Aug 2017

Geospatial Information Authority of Japan (GSI) (2014b) JPGIS2014 specification document (in Japanese). Geospatial information authority of Japan. www.gsi.go.jp/common/000091216.pdf. Accessed 18 Aug 2017

Geospatial Information Authority of Japan (GSI) (2015) Japan profile for geographic information standards (JPGIS) (in Japanese), Geospatial Information Authority of Japan. www.gsi.go.jp/GIS/jpgis-jpgidx.html. Accessed 18 Aug 2017

Geospatial Information Authority of Japan (GSI) (2017a) Coverage and seamless description of FGD. Geospatial information authority of Japan. www.gsi.go.jp/kokusaikoryu/kokusaikoryu-e30057.html. Accessed 18 Aug 2017

Geospatial Information Authority of Japan (GSI) (2017b) Description of GSI maps available tiles and data layers (in Japanese). Geospatial information authority of Japan. http://maps.gsi.go.jp/development/ichiran.html. Accessed 18 Aug 2017

Geospatial Information Authority of Japan (GSI) (2017c) Experimental vector tiles for GSI Maps platform. Geospatial information authority of Japan. https://github.com/gsi-cyberjapan/vector-tile-experiment/blob/gh-pages/README_en.md. Accessed 18 Aug 2017

Geospatial Information Authority of Japan (GSI) (2017d) FGD dissemination and release schedule (in Japanese). Geospatial information authority of Japan. www.gsi.go.jp/kiban/index.html. Accessed 18 Aug 2017

Geospatial Information Authority of Japan (GSI) (2017e) GSI maps web based map browser (in Japanese). Geospatial information authority of Japan. https://maps.gsi.go.jp. Accessed 18 Aug 2017

Geospatial Information Authority of Japan (GSI) (2017f) Standardization activities. Geospatial information authority of Japan. www.gsi.go.jp/ENGLISH/page_e30210.html. Accessed 18 Aug 2017

Geospatial Information Authority of Japan (GSI) (2017g) Standardization activities (in Japanese). Geospatial information authority of Japan. www.gsi.go.jp/GIS/stdindex.html. Accessed 18 Aug 2017

GIS Association of Japan (GISA) (2017) GIS Certification Association (GISCA) program (in Japanese). www.gisa-japan.org/gisca. Accessed 18 Aug 2017

GitHub (2017) GitHub website. www.github.com. Accessed 30 Aug 2017

Government of Canada (2017a) Geospatial extraction tool. http://maps.canada.ca/czs/index-en.html. Accessed 30 Aug 2017

Government of Canada (2017b) Open government portal. http://open.canada.ca/data/en/dataset?collection=fgp. Accessed 30 Aug 2017

INSPIRE (2007) INSPIRE directive. http://inspire.ec.europa.eu/inspire-directive/2. Accessed 2 Aug 2017

INSPIRE (2017) INSPIRE validator. http://inspire-sandbox.jrc.ec.europa.eu/validator/. Accessed 20 Aug 2017

INSPIRE Thematic Working Group Addresses (2009) D2.8.I.5 Data specification on addresses—guidelines. http://inspire.ec.europa.eu/documents/Data_Specifications/INSPIRE_DataSpecification_AD_v3.0.pdf. Accessed 30 Aug 2017

ISO 19103:2015 Geographic information—conceptual schema language

ISO 19109:2015 Geographic information—rules for application schema

ISO (19131:2007) Geographic information—data product specifications

ISO (19136:2007) Geographic information—Geography Markup Language (GML)

ISO/TC 211 HMMG (2017) GitHub repository of the ISO/TC 211 harmonized model maintenance group (HMMG). https://github.com/ISO-TC211/HMMG. Accessed 30 Aug 2017

ISO/TC 211, Geographic information/Geomatics, in cooperation with the Open Geospatial Consortium (OGC) and the International Hydrographic Organization (IHO) (2013) The UN-GGIM inventory of issues and geographic information standardization. Background document for the third session of the United Nations Committee of Experts on Global Geospatial Information Management (UN-GGIM) in Cambridge, United Kingdom, July 2013. http://ggim.un.org/ggim_committee.html. Accessed 29 June 2017

Japan (2007) Basic Act on the Advancement of Utilizing Geospatial Information (Act No. 63 of May 30, 2007). Ministry of Justice, Government of Japan. Japanese Law Translation Database System. http://www.japaneselawtranslation.go.jp/law/detail/?id=2311&vm=04&re=02. Accessed 18 Aug 2017

Joint Chiefs of Staff (2014) Information Operations. Joint Publication 3–13, Washington, DC. www.dtic.mil/doctrine/new_pubs/jp3_13.pdf. Accessed 27 July 2017

Natural Resources Canada (2016) Standards user's guide for geographic information. Canadian geospatial data infrastructure, information product 46e, 120 pages. https://doi.org/10.4095/297543. Accessed 27 July 2017

Natural Resources Canada (2017) Download directory and documentation. www.nrcan.gc.ca/earth-sciences/geography/topographic-information/free-data-geogratis/download-directory-documentation/17215. Accessed 27 July 2017

OGC (2002) Web feature service 1.0.0. OGC document 02-058

OGC (2006) OpenGIS web map service implementation specification 1.3.0. OGC document 06-042

OGC (2007a) OpenGIS catalogue service implementation specification 2.0.2 - Corrigendum 2. OGC document 07-006r1

OGC (2007b) Web processing service 1.0.0. OGC document 05-007r7

OGC (2012) OGC WCS® 2.0 Interface Standard—Core, Version 2.0.1. OGC document 09-110r4

OIO Arkitekturguiden (2017a) Grunddata Modelregler. http://arkitekturguiden.digitaliser.dk/grunddata-modelregler. Accessed 1 Aug 2017

OIO Arkitekturguiden (2017b) Eksisterende Standarder. http://arkitekturguiden.digitaliser.dk/eksisterende-standarder. Accessed 1 Aug 2017

OpenStreetMap (2017a) List of OSM-based services. OpenStreetMap Wiki. http://wiki.openstreetmap.org/wiki/List_of_OSM-based_services. Accessed 30 Aug 2017

OpenStreetMap (2017b) OpenStreetMap website. www.openstreetmap.org. Accessed 30 Aug 2017

Ota M (2017) Conceptual modeling and its implementation in an education assistance software tool for geographic information technology. Int J Cartography (online: 2017/05/13):1–25. https://doi.org/10.1080/23729333.2017.1308693

Ota M, Plews R (2015) Development of a software tool as an introduction to geospatial information technology based on geospatial standards. Cartography Geogr Inf Sci 42(5):419–434. https://doi.org/10.1080/15230406.2015.1031701

Shudan Z, Jianghua Z (2014) Assessing the completeness and positional accuracy of OpenStreetMap in China. In: Bandrova T, Konecny M, Zlatanova S (eds) Thematic cartography for the society. Springer International Publishing, Berlin. ISBN 978-3-319-08180-9

Sparx Systems (2017) Enterprise architect. www.sparxsystems.com/products/ea/. Accessed 30 Aug 2017

Standards Wiki (2017) http://wiki.icaci.org/index.php?title=Standards. Accessed 29 June 2017

Von Lucke J (2010) Open government. Öffnung von Staat und Verwaltung. Zeppelin University gGmbH, Germany

Chapter 17
From Maps to Apps—Entering Another Technological Dimension Within the Geological Survey of Austria

Christine Hörfarter, Johannes Reischer and Martin Schiegl

Abstract The creation of geological maps and the provision of geological information has a long history at the Geological Survey of Austria (GBA). Geological maps enable the visualization of geoscientific topics of interest to the public and the authorities. As early as the middle of the nineteenth century the Geological Survey of Austria started producing geological maps. From these first individual sketch sheets and the first data sets to digital information provided via web services, information technology now leads us into the world of controlled vocabularies, such as the Thesaurus of the GBA. In the following, the path from individual geological manuscripts via the digitization process to the data administration for apps and semantic web capabilities at the GBA will be described.

Keywords Geological Survey of Austria · Map production · Geological map INSPIRE · SKOS · Linked data · Controlled vocabulary · Semantic web

17.1 Importance and Meaning of a Geological Map

The geological map is of fundamental importance for any geoscientific activity. It allows visualizing various thematic aspects such as lithology, stratigraphy, tectonics, geologic age, mineral resources, mass movements, geomorphology etc. More importantly, questions of daily life which have a geological reference can be answered with it. Although the map is only a two-dimensional portrayal scheme, it is possible for the skilful eye by means of additional information in the map and the legend to obtain a spatial idea of the geological structure in the depth of the area of interest. In addition to the correctness of content the maximum information density, the aesthetic design as well as the perfect readability is an essential requirement for the map (Ebner et al. 2012; Pascher et al. 1996; Schnabel et al. 1999). Printed geological maps are scientific publications: they have one or more authors, an

C. Hörfarter (✉) · J. Reischer · M. Schiegl
Geological Survey of Austria, Vienna, Austria
e-mail: Christine.Hoerfarter@geologie.ac.at

J. Döllner et al. (eds.), *Service-Oriented Mapping*, Lecture Notes in Geoinformation and Cartography, https://doi.org/10.1007/978-3-319-72434-8_17

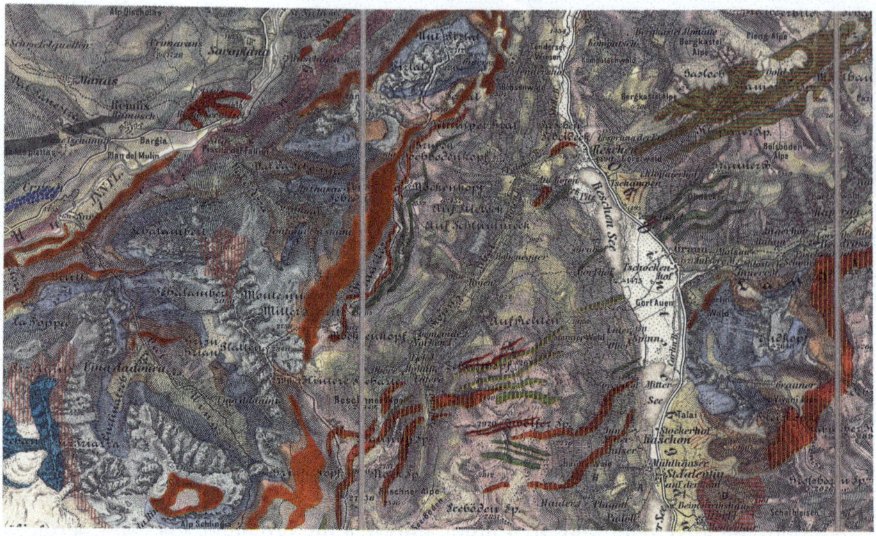

Fig. 17.1 Downscaled example of a hand-colored geological map: Part of the map 5245 Nauders 1:75,000, drawn in the 19th century (Stache 1876)

editorial revision and an exact year of publication (Krenmayr 2010). Not to forget how geological maps fascinate with their colorful appearance and accuracy. Because that the maps were drawn by hand in the early days those maps are often used to be perceived as works of art (Fig. 17.1).

17.2 Map Production at the Geological Survey of Austria

Since the foundation of the Geological Survey of Austria in the year of 1849, the scientific geological survey of the territory is the most important objective of this institution. The knowledge gained is documented in reports and maps. The first sheets were drawn by hand and colored with watercolor. Because of an increasing demand at the end of the 19th century, it was necessary to print the maps (the scale was 1:75,000). In the interwar period and after the Second World War many maps were created, but in different scales, representations and with different coverage of an area—an expression of this time of upheaval. In the 1970s, a standardization of the map scale 1:50,000 was implemented and thus followed—just like the grid—the edition of the maps of the topographic land survey (Schnabel et al. 1999). Because topographic maps pose the indispensable starting point for geological mapping in terrain and are the basis for the final presentation of the results in geological maps (Krenmayr 2014). The production took place in a conventional way, with engraving, strip-mask technique, ink drawing and astralon copies (Laschenko 1999).

In the last decades of the 20th century governmental organizations, scientific institutions and companies all over the world began to shift from analogue to digital map production and so did the Geological Survey of Austria too.

17.2.1 Digital Map Production

In the first years of "computerization" the aim was to build up a database of the contents of geological maps to support the preparatory works for printing in the traditional way and to accelerate the rate of publication of the maps. But the high quality of the printed products should be maintained. To reach this aims the GIS *ARC/INFO* was implemented. The approach was to split the thematic information into several layers separated by the type of the geometry of the graphic element: polygon data for geology, moraines and areal signatures, line data for tectonic structure lines, erosion rims and linear signatures and point referenced data for tectonic symbols, landslides and point signatures e.g. boreholes. Special layers for cartographic purposes were added too. The following workflow was established: fine-drawn and hand-colored sketch maps at the scale of 1:25,000 were checked by the editor and then assigned to particular layers as mentioned. These layers were drawn by cartographers on special films, scanned with 400 dpi, vectorized and saved in separate coverages. After these steps were completed, the coverages were cleaned and edited. A very important step of the editing process was the manual encoding of the polygons, lines and points with ID's (=numbers). In this way the features become intelligent parts of a database. Each object with the same properties gets the same ID. On the whole a legend entry corresponds to a record in the database. The next step was the compilation of the different layers in *ARC/INFO* to a geological map. The legend, the scale bar, title and further symbolization were added for completion. In the step of prepress preparation color separated films were generated with *PostScript* Files. Then the printing plates were prepared and finally the map was printed (Reischer and Stöckl 1998) (Fig. 17.2). In the following years this process was steadily refined and adapted to new technologies: at present the layers are managed in a geodatabase and the number of layers has been reduced to four main feature classes. Also the legend is managed in a table in the geodatabase. *Adobe Illustrator* and PDF are used for prepress preparation (Schiegl 2007).

17.3 Data from Maps

In the course of the further development of digital map production, the Geological Survey tried to manage the information from maps and legends in a database system with the aim not only to archive the data, but also with the medium-term aim to automate the map production—to generate a map from data (geometries, legend information and meta information). However, the time was not ripe for the

Fig. 17.2 Part of the geological map of Austria 1:50,000, sheet "144 Landeck", digitally produced (Krainer et al. 2004)

challenge at this point. Because the digital workflow before the map production—from the surveying in the field to the creation of the manuscript maps—had to be developed and established. But the change to digital methods took place only carefully and step-by-step. Meanwhile the demand for digital geological data, in particular also for vector data, rose rapidly and the provision of digital data come to the fore. Due to the mentioned requirements, following workflow naturalised: first the map is created (in the way as described above), then after completion of a sheet the digital spatial data is obtained. In parallel the development of digital surveying methods with GPS and handheld PCs, the storage of analysis results and the recording of mapping points in geo(databases) and the production of the manuscript maps with GIS is fostered.

The digital data obtained from the maps are constructed in a model and stored in a managed database system and a geodatabase. Two paths are taken: one way is the preparation of the data for sale, the other way is the preparation of the data for internal use and the Internet. The data are centrally stored and managed with a *Microsoft SQL* database with *ArcSDE* as middleware.

17.3.1 Data for Sale

Data for sale use a cartographic model (KM) based on the analogue maps. The information—summarized for printing on a few layers—is now split up to 21 layers. The vector data describe geological objects (surface geology) on the basis of the graphical base elements polygon, line and point, equipped with a coordinate system of course. The contents of the individual datasets are limited to the main map. Overview map of authors, tectonic overview and a graphical visualization of the legend are not incorporated. The digital version may include deviations from the original map because of error correction or updating. The data is neither geometrically nor regarding content harmonized with other sheets. The geological areas and their boundaries as well as tectonic structure lines refer exclusively to one map sheet. The release of the data take place in shape-file format (GBA 2013).

17.3.2 Data for Internal Use and the Internet

The data for internal use and for the provision via Internet are also mapped in a cartographic model too, but with a different focus. For all maps in the scale of 1:50,000 the layers with the same topic are grouped together, e.g. all moraine layers of the maps 1:50,000 are grouped to one moraine layer for the Austrian territory. But these data are also neither geometrically nor regarding content harmonized. The features get a unique number, which consists of IDs for map, feature class and feature. The terms of the text of the legend are analysed for their meaning and divided into different attributes—especially lithology, age and tectonics is extracted. But the symbolization is also stored in the data model: styles, symbols and color values in CMYK and RGB are recorded (Strauß et al. 2004).

17.4 WMS, Image and Map Services

As a governmental research institution, the Geological Survey of Austria has a mediating role between science on the one hand and administrative authorities and the public on the other hand. The Survey is called upon to inform the authorities and the general public about spatial geodata and to facilitate access to it (GBA 2014).

Web services are a great way to provide users with maps and geodata through the Internet. The Geological Survey of Austria has decided to host the services itself and procure the necessary infrastructure. In this context *GeoServer*, *GeoNetwork* and *ArcGIS Server* should be emphasised as the main parts of the system. These servers are applied to share OGC (Open Geospatial Consortium)-compliant WMS (Web Map Services), *ArcGIS* map services and *ArcGIS* image services. Subsequently the services can be added as a layer in a client's GI-System or in a web application.

At the Geological Survey of Austria WMS services are mainly used for the view services for INSPIRE (Infrastructure for Spatial Information in Europe), where OGC standard is supplemented by the required functionalities of INSPIRE. The OGC extension mechanisms will be respected (GBA 2017a). We also host WMS for integration into applications operated by other institutions.

Map services are used to provide vector data in an appealing form. In an *ArcGIS* map document (.mxd) geodata is displayed and processed. Frequently, an attempt is made to produce a map-like appearance of the information, particularly if the data were derived from a printed map.

Most of the printed maps of the Geological Survey of Austria are available as georeferenced images. The data are stored centrally in the file system and are collected in *ArcGIS* mosaic datasets with one dataset covering one theme, for example geologic maps in the scale of 1:50,000. Mosaic datasets also offer the possibility to save attributes to the individual images. The attributes contain the most important metadata of the stored maps as well as links to the library system of the Geological Survey of Austria. To improve display performance of the images down sampled versions are also stored and shown on appropriate display scales (ESRI 2017). With *ArcGIS Server* the mosaic datasets are published as WMS and image services.

17.4.1 Applications

In order to facilitate access to the services and thus to the data, applications were designed. In addition to the general functions for navigation, special queries have been programmed for certain topics, e.g. hydrochemistry or geohazards. In the course of time, numerous applications were created that were basically the same, but deviated from each other in details in programming. With each additional application, the effort of administration and service also increased. The logical step was therefore to reduce the functionality somewhat, but to increase the universality and to create an application currently running under the designation "GBA Map Viewer". With this viewer, services can be easily presented and viewed. At the same time, cloud-based, configurable modules for the creation of applications, as provided by ArcGIS online, are used.

17.5 GBA-Thesaurus and DataViewer Module

The increasing amount of geoscientific spatial data together with the growing demand for interoperability and interdisciplinary usage via the internet requires a sophisticated management of the existing knowledge about basic data information. As geoscientific data cannot be used successfully if the basic structure and interpreted results are not understandable for everyone within a certain community, a comprehensible (transparent) knowledge representation is an important precondition. At the Geological Survey of Austria, the GBA-Thesaurus controlled vocabulary and the DataViewer module based thereon are one approach to face that challenge (Hörfarter and Schiegl 2016).

17.5.1 Initial Situation—The Challenge Called "INSPIRE"

The increase of spatial geodata and the growing demand for interoperability due to interdisciplinary usage needs a sophisticated data management. An important step towards facing that challenge is the INSPIRE[1] directive of the European Union

[1]INSPIRE—Infrastructure for Spatial Information in Europe.

(European Parliament 2007) with the aim to create a standardized spatial geodata infrastructure for the European Union. Especially geology became a crucial theme in the world of geospatial data within INSPIRE. It can be considered as basis for further geoscientific topics, e.g. mineral resources as we can see in the allocation of Geology to Annex II of the directive. The INSPIRE directive was transposed into Austrian law in 2010 (Geodaten Infrastrukturgesetz—GeoDIG; BGBl. I Nr. 14/2010). Therefore, the GBA is legally bound to structure and harmonize the public digital spatial geodata according to a given standard in a semantic and technical way. More precisely, it has to be possible to search, visualize and download the geoscientific data of public authorities throughout Europe in a unique structure and description.

Initially, the enthusiasm was kept within limits because it was recognised early that this implementation requires a lot of effort. Due to the fact, that on the GBA the focus has always been on printed map sheets as a work of authorship, we had no overall data model on Geology. Also, there was no given standard for legend- or feature information. We have, of course, managed the main feature classes in different layers such like planar geology, tectonic structures lines etc., but the main information to be harmonized was just stored as plain text information without the possibility for satisfying queries. Without structuring the geological content of our datasets, processing of the geodata according to a given data model would hardly have been possible.

Therefore, we had to find a solution for a sustainable data structure and started to deal with the subject of knowledge management. It was unavoidable to define a new workflow for our geodata management regarding INSPIRE requirements. First, it was important to have a common understanding of our basic data, the geological information of our maps assimilated in the legend text and legend structure. After being aware of the main themes to cover we took the opportunity to use the INSPIRE core model of "Geology" (European Commission Joint Research Centre 2013: 21) as well as the CGI GeoSciML standard model for Geology (Open Geospatial Consortium 2016: 10) as a basic structure for creating an adapted physical database.

17.5.2 Building up the GBA-Thesaurus

The INSPIRE standard vocabulary, or, to be exact, the INSPIRE code lists, which should be used for encoding the geologic feature objects did not provide a satisfactory solution to assign our dataset objects when we started with the process of harmonization. At the beginning the codes were available only in the form of a PDF with a confusing hierarchical structure and offered too little information output to use them practically. To attribute the geologic feature objects with information focussing on quality and level of detail, the GBA established its own controlled vocabulary—the GBA-Thesaurus—to bridge this gap.

A controlled vocabulary has the purpose to organise information and provide terminology to catalogue and retrieve information. With a thesaurus, in contrast to a lexicon, you create a semantic network of unique concepts, including relationships between synonyms, broader and narrower (parent/child) contexts as well as other related concepts. A thesaurus is much more complicated than a simple list of terms and may contain three types of relationships: equivalence (synonym), hierarchical (whole/part, genus/species, or instance), and associative (Harpring 2010). For creating a controlled vocabulary, it is relevant to consider carefully which relationship type has to be used for modelling. Below, Fig. 17.3a shows an example of modelling a hierarchical relation and Fig. 17.3b an example of modelling an associative relation of a rock.

The GBA-Thesaurus, a bilingual (German/English) controlled vocabulary as knowledge representation of the GBA was realised with the commercial software PoolParty[2]. This thesaurus management software is based on the W3C[3]ontology standard SKOS (Simple Knowledge Organization System) which has been developed to support the use of knowledge organization systems (KOS) such as thesauri, classification schemes, subject heading systems and taxonomies within the framework of the Semantic Web (Isaac and Summers 2009)

Where SKOS is the standard model proposed to structure a controlled vocabulary the Resource Description Framework (RDF) graph based data format behind (W3C, Manola et al. 2004) provides us the possibility to manage information and interchange data in a machine readable way, the basis of the semantic web. In a SKOS based thesaurus each term (SKOS:concept) holds its own persistent unique web address, the so called URI (Unique Resource Identifier). This URI is used for encoding RDF resources. Due to this semantic web capabilities of the thesaurus (SKOS, RDF) it is possible to connect knowledge by linking URIs through a RDF knowledge graph with other resources in the World Wide Web. With realising the linking of open source data based on the described principles then we are talking about Linked Open Data.

With INSPIRE starting to provide their vocabulary code-lists in RDF format—therefore with URIs—we can use the Linked Open Data principle to fulfil the official mandate of INSPIRE compliant data harmonization.

17.5.3 Usability of the GBA-Thesaurus

For the domain experts within the Geological Survey of Austria, the controlled vocabulary of the GBA-Thesaurus provides a source of consistent index terms that spans a comprehensive range of our activities in mapping and research. These

[2]PoolParty—a thesaurus management system distributed by the Semantic Web Company GmbH (SWC).

[3]W3C®—World Wide Web Consortium: https://www.w3.org/.

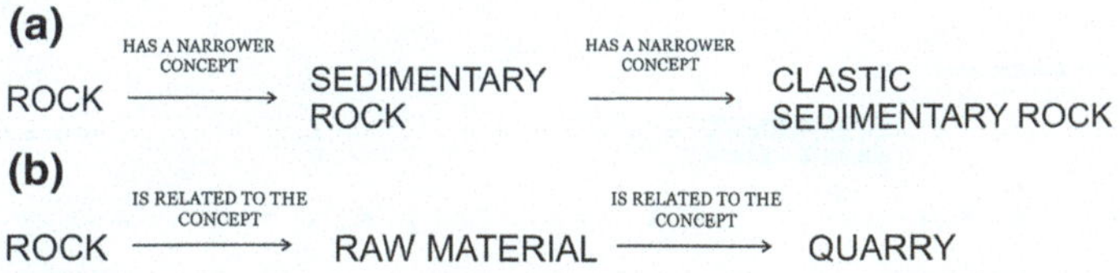

Fig. 17.3 **a.** Example for a hierarchical relationship of concepts. **b.** Example for an associative relationship of concepts

concepts can be used to refine or clarify labels and definitions. Moreover, it is possible to adopt classification models such as proposed from several geoscientific standards (e.g. International Chronostratigraphic Chart).

Based on the published geological datasets and the physical data model for geology, we currently manage the vocabulary of six geoscientific domains within the GBA-Thesaurus:

- *Geologic Units*—lithostratigraphy, lithogenetic units etc.
- *Geologic Structures*—faults, folds and tectonic boundaries.
- *Geologic Time Scale*—international and regional time scales.
- *Lithology*—classification and description of rocks and sediments.
- *Minerals*—rock forming minerals and mineral systematics.
- *Lithotectonic Units*—classification of units defined by tectonic boundaries.

Each topic requires a different approach to the modelling of concepts. When trying to establish the best possible model for structuring the vocabulary one recognises it's a dynamic process and that there is more than one conception of the existing knowledge. In the following, however, the advantages of concept details and GBA-Thesaurus usability will be discussed using a concept frontend example (Fig. 17.4).

GBA-Thesaurus concept—a SKOS concept: In the SKOS primer document describing the SKOS ontology a 'concept' is defined as following:

> The fundamental element of the SKOS vocabulary is the concept. Concepts are the units of thought ideas, meanings, or (categories of) objects and events—which underlie many knowledge organization systems. As such, concepts exist in the mind as abstract entities which are independent of the terms used to label them. (Isaac and Summers 2009).

If we have the idea of something (meaning, concept) and it exists in the world (thing), we try to give it a name (sign, term) (Fig. 17.5). Therefore, a term is something to label a concept (like "Granitic rock" in Fig. 17.4 point 4) and with the description we inform the public what is our basic unit of thought, our interpretation regarding the concept (Fig. 17.4 point 6).

All this information concerning a SKOS concept are stored in form of RDF triples. These triples are composed of a subject, a predicate and an object.

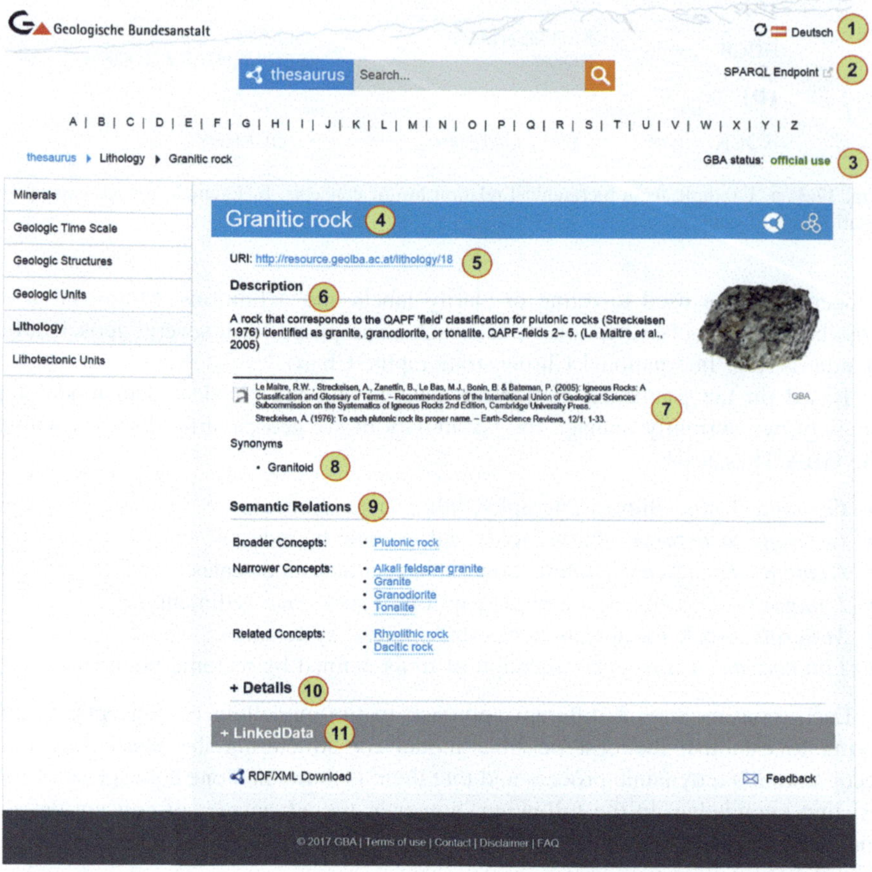

Fig. 17.4 Front-end screen shot of the GBA-Thesaurus showing the search result regarding the concept "Granitic rock": (**1**) language button, (**2**) link to SPARQL endpoint, (**3**) concept status, (**4**) label, (**5**) URI of concept, (**6**) concept description, (**7**) citations, (**8**) synonyms as alternative label, (**9**) semantic related concepts, (**10**) fold-out area for more concept details, (**11**) fold-out area showing Linked Data information

In the following you can see an example of a RDF triple with declaration of subject, predicate and object representing the preferred label information of the concept "Granitic rock":

resource.geolba.ac.at/lithology/18 ➔ skos:prefLabel ➔ "Granitic rock"@en
Subject ➔ *Predicate* ➔ *Object*

The URI is the descriptor, the code of our concept. We use numbers and not the human readable concept label in the URI path, because the URI is persistent and won't get changed even if the label or the interpretation gets modified. This is supposed to happen and is allowed as far as the basic idea and interpretation of the

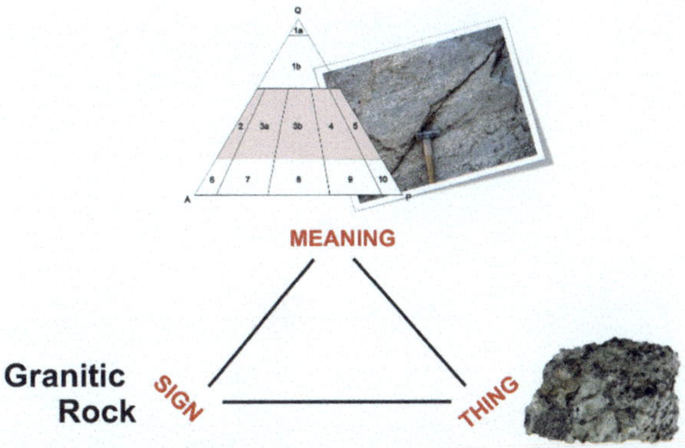

Fig. 17.5 A semiotic triangle relying on Ogden and Richards (1923: 11)

concept does not change. To confirm the interpretation of a concept and to promote acceptance in the geoscientific community each concept description shall be supported with a citation (Fig. 17.4 point 7). In order to facilitate the user's search for literature, the citations are linked to the online library system of the GBA.

Especially, in the scientific area of geology a lot of concepts are historically grown, they get modified in their definition over the time or they change in the range of use. So, the possibility to manage alternative labels like synonyms Fig. 17.4 point 8) and to relate a concept status (Fig. 17.4 point 3) which also describes the processing status, is of great importance.

Semantic relations (Fig. 17.4 point 9): Additionally, the semantic relation of a concept beside a description defines its meaning more concrete within a net of concepts. To assert semantic relationships between the concepts we use the SKOS properties skos:broader, skos:narrower for a hierarchical structure and skos:related for a associative one. Regarding the semantic relations in Fig. 17.4 point 9 the concept "plutonic rock" is a more general concept (skos:broader) and "Granite" is a narrower concept (skos:narrower) of "Granitic rock". The associative link to the concept "*Rhyolitic rock*" asserts a non-hierarchical relationship between these concepts. Below, the brief depiction of triples and the visualization of concept relations (Fig. 17.6):

Granitic Rock	→	skos:narrower	→	Granite	(hierarchical)
Granitic Rock	→	skos:broader	→	Plutonic Rock	(hierarchical)
Granitic Rock	→	skos:related	→	Rhyolitic Rock	(associative)

SPARQL[4]: As mentioned before, one goal of RDF is to provide machine readable data for providing and interchanging of information in the World Wide Web. The other objective is to make these information queryable and available for everyone's purposes. SPARQL is the language to query the graph based database (RDF triplestore) like SQL is the query language for relational databases.

[4]SPARQL—SPARQL Protocol and RDF Query Language (Prud'hommeaux and Seaborne 2008).

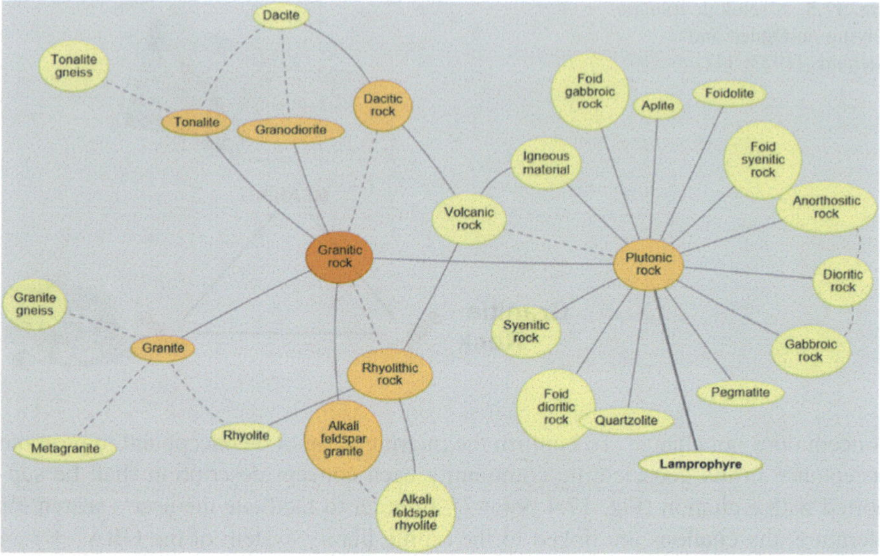

Fig. 17.6 Visualization of semantic relations—a network of concepts (see: http://resource.geolba.ac.at/thesaurus/network.html#lithology/18/en)

With the PoolParty semantic web suite providing a SPARQL endpoint (Fig. 17.4 point 2)—a query interface—it is possible for everybody to download or export individually created lists of information within all concepts of the GBA-Thesaurus like the query example below (Fig. 17.7a, b).

With making our thesaurus vocabulary content available for everyone and the utilization of open formats (RDF, SKOS) the implementation of Linked Open Data (LOD) is feasible. The interlinking of concepts of other controlled vocabularies (via HTTP URIs) and therefore the connection of information increases knowledge enormously. This interlinking of concepts and the SKOS mapping properties within the GBA-Thesaurus is visible in a separate Linked Data information area (Fig. 17.8).

Whenever the URIs of resources are linked by a mapping relationship, the associated information is linked in a comprehensible manner as well. In this case URI links the INSPIRE information to the respective geological features encoded by GBA-Thesaurus URIs of the data set and the requirement of the INSPIRE harmonization is met.

17.5.4 The DataViewer Module—An Extension of the GBA-Thesaurus

Another great advantage is availability of an open access service interface within the PoolParty thesaurus management tool. It allows building external applications

(a)

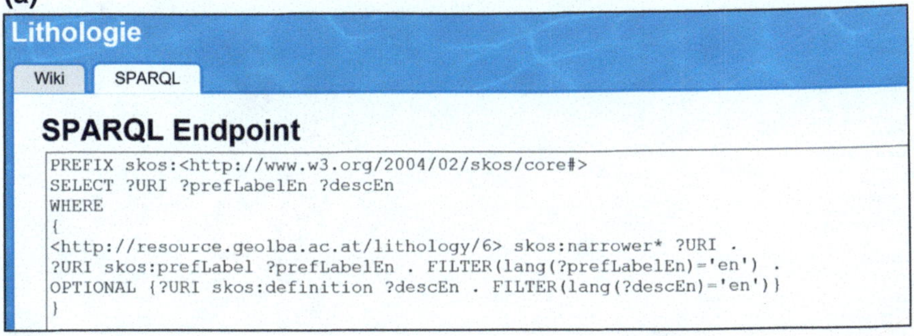

(b)

URI	prefLabelEn	descEn
http://resource.geolba.ac.at/lithology/6	"Volcanic rock"@en	"An igneous rock that can be associated with volcanism and has a relatively fine-grained texture in which most of the individual crystals can not be seen with the unaided eye. The term also includes subvolcanic (hypabyssal) rocks (Le Maitre et al., 2005)."@en
http://resource.geolba.ac.at/lithology/62	"Rhyolithic rock"@en	"A rock that corresponds to the QAPF 'field' classification for volcanic rocks (Streckeisen 1978) identified as rhyolite. QAPF-fields 2-3 (Le Maitre et al., 2005)."@en

Fig. 17.7 **a.** SPARQL query example regarding all rocks with volcanic origin, their English preferred label and English description → GBA-Thesaurus SPARQL bookmarks: http://resource.geolba.ac.at/thesaurus/sparql_bookmarks.html. **b.** SPARQL endpoint query result represented in tabular form—2 examples

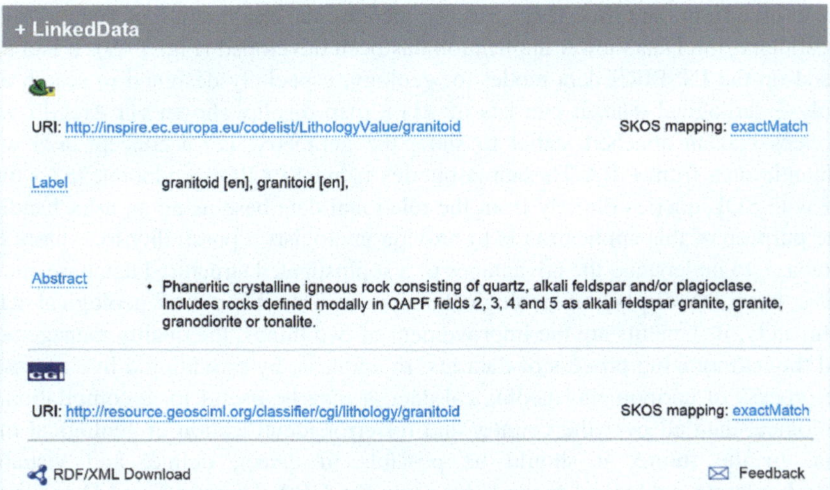

Fig. 17.8 In the linked data area of the GBA-Thesaurus the linking concepts to other RDF resources are indicated as well as the mapping relations (SKOS mapping). As mentioned before, this linked data option is the basis for the GBA to fulfill INSPIRE

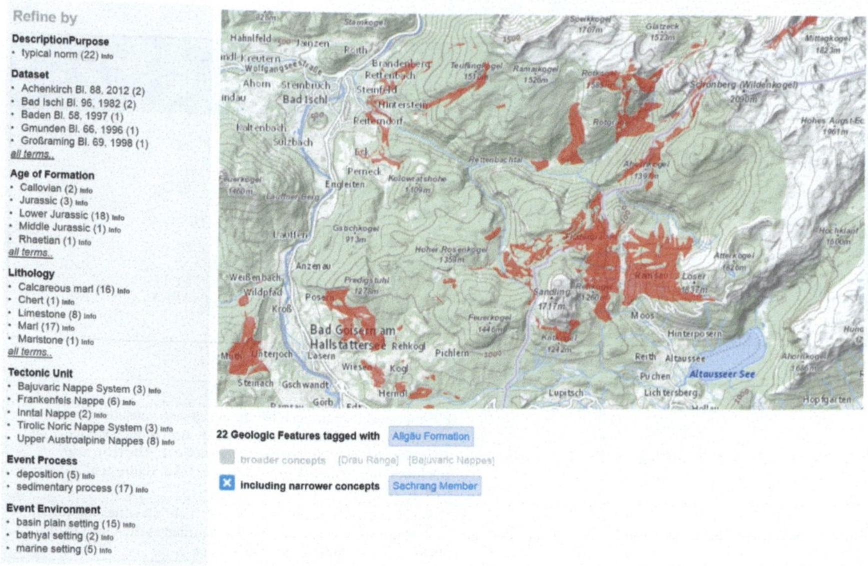

Fig. 17.9 DataViewer Screen Shot which visualizes the distribution of the Allgäu Formation (here limited to the area of Bad Ischl). In addition to the possibility to display the legend text and different base layers, the information already harmonized according to INSPIRE is available to the left of the picture. The DataViewer is accessible via a link in the "Details" fold-out area (Fig. 17.4 point 10)

that evolve from the GBA-Thesaurus. To increase the exploitation of the controlled vocabulary, the DataViewer application has been developed (Fig. 17.9). It is a tool based on the INSPIRE data model for geology, especially designed to search and analyse geological data. It consists of: (1) a map display shown via ArcGIS web services and an attached leaflet to show the geometry, (2) a concept area with semantic data from GBA-Thesaurus queries using SPARQL endpoint, (3) a filter bar with SQL queries directly from the relational data base using an ashx handler. The purpose of this application is to provide geologists a possibility to explore the database, to understand the advantages of a sophisticated structured database and to move from the display of geological maps towards a view of geological data. Ultimately, its benefits are the improvement of two things, the quality management and the harmonizing process of datasets. In addition, by providing a live access to the process of harmonizing geological data, it may be useful for a compilation of geological data all over the country and for error identification in geological map data. In the future, it should be possible to detect, define, and visualize cross-bordering geological features by using the GBA-Thesaurus and DataViewer module to support a common transboundary cooperation regarding geoscientific challenges.

17.6 Future Perspectives in Geoinformation Management and Cartography

17.6.1 Modelling the Information of a Geological Map

The traditional map production as the main field of activity of geological survey organisations (GSO) has now almost completely been replaced by the production of GIS data sets (Ebner et al. 2012). From this point on, GSOs now provide their customers just with data whose queries can be map-like visualized. The process of digitizing a geological map can be compared with a transformation of an unstructured and man-readable text publication into a structured machine-readable table with defined field formats. Some contents can be integrated very well in relational SQL or GIS databases—such as the datatype geometry of a point, polyline, or polygon by storing the coordinates of the vertices. Some other contents like certain vocabulary terms of a map legend, or keywords for tagging metadata, don't fit in SQL very well. Much of these contents have a lot of semantics behind and are only understood in the context of the map and year of publication. Particularly in the case of these text contents of a map legend, a semantic harmonization of vocabulary terms connected with surrounding information is taken into account during data modelling. A paragraph text dedicated to describe a single geologic feature on the map comprises the relational data model for geologic units. Many of the addressed entities within are vocabulary concepts more or less on an atomic level.

In general, a process or a state of semantic harmonization integrates data into one consistent logical view with different approaches to harmonize. It is organised into a knowledge framework like geological data models and enables users to access information in a meaningful use. In practice data models for geological map data are widely covered by existing application schemes like INSPIRE or GeoSciML. But regarding the existing vocabularies and code lists for geosciences, data should be defined more precisely and semantically enriched. Simple terminology services usually contain limited contextual information and cannot easily generalise across disciplines.

So many geological data sets derived from traditional geological maps are simply a combination of data types "Geometry" and "Concept". In that case semantic harmonization issues focus on concepts to put them in the right context. Specifically, Semantic Web technologies (Blumauer 2016; Pellegrini and Blumauer 2006; Pellegrini et al. 2014) (semantic tags or keywords) show a great promise for the semantic interoperability of vocabularies. In addition to the information context we also need to provide source meta-information or a description of the origins of the shared semantics, which is also called knowledge provenance.

17.6.2 Enhance GIS/SQL Data with RDF and Linked Data

In GSOs the application of GIS databases and web services is widely-used and popular to deliver data. Since GIS databases are bound to SQL data storage and modelling there might be interesting approaches to combine SQL technology with RDF capabilities. The comparison of relational SQL databases with RDF triple stores (Wood et al. 2014), i.e. graph databases for the storage and publication of Linked Data (Wood et al. 2014) on the web shows pros and cons likewise on both sides. For example, when storing entities in SQL databases, you get faced with IDs assigned by a "closed" database system while RDF triple stores work with "global" URIs (Unique Resource Identifier, Wood et al. 2014). SQL databases have fixed-defined field formats while RDF triple stores use dynamically type checked attributes. When using RDF instead of SQL a vocabulary concept changes from ID plus text string to an URI plus Linked data concept within the context of a knowledge representation in the web (q.v. open world assumption, description logic). If this concept is migrated to a RDF triple store and published for free and open use, we talk about Linked Open Data (LOD, Wood et al. 2014). Thereby the vocabulary concept gets its own URI web address which should always be resolvable as HTML, and/or RDF/XML. This concept still can be used worldwide by any person to encode other additional data as soon as it has been persistently published on the Web. The creator or publisher does not even need to know who linked to these URIs on the web. Data published this way can be combined and referenced following the principles of Semantic Web and Linked Open Data. On the other hand – storing statically type checked data within clearly defined data models, e.g. geometry polygons or standardized metadata—it is simply a better choice to work with well proven SQL and GIS database technologies. They have been significantly enhanced in terms of performance and stability over the last decades. Here relationships are defined strictly from the start by database tools for referential integrity, which reduces the need for subsequent validation required when working with RDF and Linked Data.

Regarding the database relationships of Linked Data and RDF, they are defined by web ontologies (Wood et al. 2014) analogous to an application schema which can always be extended by custom properties. In doing so, only information of necessary complexity and detail is collected and stored. For SQL databases, any tables are connected with pre-defined cardinality and their fields have to be statically type checked from the beginning, regardless of whether data is present or not. The benefits when working with SQL databases are a homogenous structure and data content.

In summary, original information always should be stored with more advantageous technology (SQL/GIS database vs RDF triple store), and additionally granted access to with the alternate technology. For SQL and GIS, there are standards like OGC (Open Geospatial Consortium 2016) or INSPIRE (European Commission Joint Research Centre 2017) compliant web services (WMS, WCS, WFS,

CSW, and many more) to deliver data. For RDF triple stores (native store or SQL to RDF mapping, van den Brink et al. 2013) usually SPARQL endpoints (Wood et al. 2014) are provided.

17.6.3 Geological Survey—Use Case

Since 2011, the Geological Survey of Austria has been used a thesaurus management system (GBA 2017b; Hörfarter and Schiegl 2016; Ma et al. 2011) for the management of geoscientific terms when processing its map data. This terminological knowledge is published using RDF Linked Open Data (LOD), URI'S and SPARQL endpoints. The published vocabulary concepts are currently linked to resources like DbPedia (Lehmann et al. 2015) (collected knowledge from Wikipedia), INSPIRE code lists (European Commission INSPIRE-Registry 2017), GeoSciML vocabularies (CGI 2017) and BGS rock classification system (BGS 2017). This important step leads to the "fifth star" of Berners-Lee (2015) who pushed the idea "Link your data to other people's data to provide context".

Considering the whole design, including published GIS web map services whose data models are encoded with URI attributes, future web applications offer a wide range of possibilities. Map-like visualizations could be enriched with mash-ups directly from information spread on the web (GBA 2017b). The modern geological map will be situated in the middle of various knowledge representations published on the web.

For example, a geological formation defined in Austria could be easily related to an identical concept of a neighbouring country—just by providing a single triple pointing to the equivalent. By linking to the federated semantic web resources for instance foreign-language synonyms of these concepts could be loaded and displayed like translations at any time. Thus every person, organisation or national authority publishes their associated knowledge domain on the web. Knowledge spaces will grow together only where knowledge is linked because it makes sense. As an example—a knowledge base about glacially affected geologic units from Iceland probably will not be able to link with geological units from Greece formed by tectonics. But vocabulary concepts of Austrian lithostratigraphic units could be easily matched to the cross-border knowledge of our neighbouring countries. This "changes the game" in terms of publishing maps. Maps and information can be shared democratically and used by everyone. Finally, I would also like to point out the many other use cases for publishing vocabularies as Linked Open Data. These mainly focus on moderated (auto complete) and faceted text search where simple keywords are not sufficient to define complex queries. In general constraint satisfaction problems like text search are often modelled and solved using graphs. Recommender systems based on similarity analysis (works like ecommerce product suggestions) or presenting dossiers, fact sheets on your web site, trends by mashing up information are other popular applications (Pellegrini and Blumauer 2006; Pellegrini et al. 2014).

References

Berners-Lee T (2015) 5 stars open data. http://5stardata.info/en/. Created: 2012-01-22 by Michael. Last updated: 2015-08-31 by James

BGBl. I Nr. 14/2010 (2010) Bundesgesetz über eine umweltrelevante Geodateninfrastruktur des Bundes (Geodateninfrastrukturgesetz—GeoDIG). European Legislation Identifier (ELI). https://www.ris.bka.gv.at/eli/bgbl/I/2010/14/20100301. Accessed 29 June 2017

Blumauer A (2016) The semantic puzzle, introducing a graph-based semantic layer in enterprises. https://blog.semantic-web.at/2016/08/15/introducing-a-graph-based-semantic-layer-in-enterprises/

British Geological Survey (2017) BGS rock classification scheme (RCS). http://www.bgs.ac.uk/bgsrcs/

CGI (Commission for the Management and Application of Geoscience Information) (2017) Vocabularies for linked data. http://resource.geosciml.org/. Accessed 05 July 2017

Ebner M, Schiegl M, Stöckl W, Heger H (2012) From printed geological maps to web-based service oriented data products—strategies, foundations and problems. In: Geophysical research abstracts, vol 14

Environmental Systems Research Institute (ESRI), Inc. (2017) ArcGIS desktop documentation. https://desktop.arcgis.com/en/documentation/. Accessed 22 May 2017

European Commission INSPIRE-Registry (2017) http://inspire.ec.europa.eu/registry/. Accessed 29 June 2017

European Commission Joint Research Centre (2013) INSPIRE thematic working group geology D2.8.II.4 data specification on geology—technical guidelines. http://inspire.ec.europa.eu/Themes/128/2892. Accessed 29 June 2017

European Parliament (2007) Directive 2007/2/EC of the European Parliament and of the Council of 14 March 2007 establishing an Infrastructure for Spatial Information in the European Community (INSPIRE). http://inspire.ec.europa.eu/documents/directive-20072ec-european-parliament-and-council-14-march-2007-establishing. Accessed 29 June 2017

Geologische Bundesanstalt (GBA) (2013) Kartographisches Modell 1:50 000. KM50 Geologie. Geologische Bundesanstalt (GBA), Wien. http://www.geologie.ac.at/shop/files/KM50_Geologie.pdf. Accessed 10 May 2017

Geologische Bundesanstalt (GBA) (2014) Programmplan 2014–2017, 104 pages, Geologische Bundesanstalt, Wien

Geologische Bundesanstalt (GBA) (2017a) INSPIRE Darstellungsdienste. https://www.geologie.ac.at/services/inspire/inspire-darstellungsdienste. Accessed 22 May 2017

Geologische Bundesanstalt (GBA) (ed) (2017b) GBA-Thesaurus. http://resource.geolba.ac.at/. Accessed 7 June 2017

Harpring P (2010) Introduction to controlled vocabularies: terminologies for art, architecture, and other cultural works. Patricia Harpring with a foreword by Murtha Baca, 2010, 258 pages. http://www.getty.edu/publications/virtuallibrary/160606018X.html. Accessed 29 June 2017

Hörfarter C, Schiegl M (2016) Thesaurus and dataviewer application tool—knowledge representation to define concepts and visualize geoscientific data. In: Conference Paper 2016 35th international geological congress, at Cape Town

Isaac A, Summers E (2009) SKOS simple knowledge organization system. Primer, World Wide Web Consortium (W3C). https://www.w3.org/TR/skos-primer/. Accessed 29 June 2017

Krainer K, Hauser C, Pavlik W, Pestal G, Nowotny A, Rockenschaub M, Ucik F, Ruthner J (2004) Landeck 1:50.000. In: Geologische Karte der Republik Österreich 1:50.000, Nr. 144, Verlag der Geologischen Bundesanstalt, Wien

Krenmayr HG (2010) Das Geologische Spezialkarten-Programm in Österreich: Umstieg auf das neue topografische Kartenwerk im UTM-System und den Maßstab 1:25.000 – Hintergründe und Strategie. Jahrb Geologischen Bundesanst 150:421–429

Krenmayr HG (2014) Topografische und geologische Landesaufnahme in Österreich: Ein Abhängigkeitsverhältnis. In: Institut für Militärisches Geowesen (ed) 250 Jahre Landesaufnahme. Festschrift, Wien, pp 157–164
Laschenko S (1999) Kartographie und Reproduktion. In: Bachl-Hofmann C, Cernajsek T, Hofmann T, Schedl A (eds) Die Geologische Bundesanstalt in Wien. 150 Jahre Geologie im Dienste Österreichs (1849–1999). Böhlau Verlag, Wien, pp 336–340
Lehmann J, Isele R, Jakob M, Jentzsch A, Kontokostas D, Mendes PN, Hellmann S, Morsey M, van Kleef P, Auer S, Bizer C (2015) DBpedia—a large-scale, multilingual knowledge base extracted from wikipedia. Semant Web J 6(2):167–195. http://wiki.dbpedia.org
Manola F, Miller E, McBride B (2004) RDF primer W3C recommendation of 10 Feb 2004. https://www.w3.org/TR/2004/REC-rdf-primer-20040210/. Accessed 29 June 2017
Ma X, Carranza EJM, Wu C, van der Meer FD, Liu G (2011) A SKOS-based multilingual thesaurus of geological time scale for interoperability of online geological maps. Comput Geosci 37:1602–1615
Ogden CK, Richards IA (1923) The meaning of meaning: a study of the influence of thought and of the science of symbolism. Harcourt Brace Jovanovich, New York
Open Geospatial Consortium (2016) OGC Geoscience Markup Language 4.1 (GeoSciML). http://docs.opengeospatial.org/is/16-008/16-008.html. Accessed 29 June 2017
Pascher GA, Laschenko S, Puhm E, Stöckl W (1996) Die digitale Geologische Karte von Österreich (GÖK) 1:50.000, 1:200.000 und 1:500.000 – Datenbank und computergestützte Kartenproduktion. In: Kretschmer I, Kriz K (eds) Kartographie in Österreich'96. Wiener Schriften zur Geographie und Kartographie, Band 9:37–48
Pellegrini T, Blumauer A (eds) (2006) Semantic web, Wege zur vernetzten Wissensgesellschaft. Springer, Berlin
Pellegrini T, Sack H, Auer S (2014) Linked enterprise data, Management und Bewirtschaftung vernetzter Unternehmensdaten mit Semantic Web Technologien. Springer, Berlin
Prud'hommeaux E, Seaborne A (2008) SPARQL Query Language for RDF. W3C recommendation 15 Jan 2008. https://www.w3.org/TR/rdf-sparql-query/. Accessed 29 June 2017
Reischer J, Stöckl W (1998) Digital map processing at the geological survey of Austria. In: Zepic F, Jagodic A (eds) ICGESA'98 international conference on GIS for earth science applications, Ljubljana, p 243
Schiegl M (2007) Kartenherstellung mit ArcGIS™ an der Geologischen Bundesanstalt Wien 2007. In: Strobl J, Blaschke T, Griesebner G (eds) Angewandte Geoinformatik 2007. Beiträge zum 19. AGIT-Symposium Salzburg, Herbert Wichmann Verlag, Heidelberg, pp 662–669
Schnabel W, Matura A, Pistotnik J (1999) Geologische Landesaufnahme. In: Bachl-Hofmann C, Cernajsek T, Hofmann T, Schedl A (eds) Die Geologische Bundesanstalt in Wien. 150 Jahre Geologie im Dienste Österreichs (1849–1999). Böhlau Verlag, Wien, pp 186–193
Stache G (1876) Nauders 1:75.000, 5245 Zone 18 Col. III., Geological Sketch Map of the First Geological Land Survey of the Austro-Hungarian Monarchy, Wien
Strauß U, Dörflinger EG, Heger H, Reischer J, Stöckl W (2004) Zentrales Geoinformationssystem —ZENGIS. Endbericht für das Projekt ÜLG 47. Geologische Bundesanstalt, Wien
van den Brink L, Janssen P, Quak W (2013) From geo-data to linked data: automated transformation from GML to RDF, TU Delft, linked open data—pilot linked open data Nederland. Deel 2—De Verdieping, Geonovum, pp 249–261
Wood D, Zaidman M, Ruth L, Hausenblas M (2014) Linked data structures data on the web. Manning Publications Co, New York

Chapter 18
Statistical Disclosure Control in Geospatial Data: The 2021 EU Census Example

Fabian Bach

Abstract This chapter outlines challenges and modern approaches in statistical disclosure control of official high-resolution population data on the example of the EU census rounds 2011 and 2021, where a particular focus is on the European 1 km grid outputs derived from these censuses. After a general introduction to the topic and experiences from 2011, the recommended protection methods for geospatial data in the planned 2021 census 1 km grids are discussed in detail.

Keywords Geospatial data · Data protection · Statistical disclosure control Population grids

18.1 Introduction

The decennial EU censuses play a key role in the production of reliable and comparable population statistics for Europe. The EU censuses are governed by law[1] which sets out that data need to be published at the level of administrative regions, down to LAU2.[2] Hence classical census products are sets of large statistical tables which cross-tabulate several population characteristics against appropriate regional breakdowns (up to five dimensions in total); these are usually called "hypercubes". However, it is difficult to compare hypercube data at a given regional level because the definition of administrative regions is subject to different national needs, and to

The views expressed are purely those of the authors and may not in any circumstances be regarded as stating an official position of the European Commission.

[1]Regulation (EC) No 763/2008 of the European Parliament and of the Council of 9 July 2008 on population and housing censuses (OJ L 218, 13.8.2008, p. 14) and its implementing Acts.
[2]"Local Administrative Units level 2", typically representing municipalities.

F. Bach (✉)
European Commission, Eurostat, Luxembourg City, Luxembourg
e-mail: Fabian.BACH@ec.europa.eu

J. Döllner et al. (eds.), *Service-Oriented Mapping*, Lecture Notes in Geoinformation and Cartography, https://doi.org/10.1007/978-3-319-72434-8_18

change over time. Population grids are much more versatile in terms of highly detailed regional and even cross-border analysis: similar to Lego bricks, grid squares can be freely assembled into functional clusters according to user-defined analysis needs (e.g. such as environmental or agricultural aspects), regardless of any administrative patterns.

Along these lines, Backer et al. (2002) proposed statistical grids as a complementary system for the publication of small-area statistics. For European statistics, the GEOSTAT 1 project[3] has recommended statistical grids with 1 km squares. As a result, for the 2021 EU population census Eurostat and the National Statistical Institutes (NSIs) are working together to publish—parallel to the default hypercube programme—13 key census variables at the level of 1 km grid squares. Currently, a planned dedicated legal Act foresees EU-wide information per grid square on the size and basic structure of the population (sex, age group, place of birth, change of place of residence within last year, and whether the person is employed). The census has always been an important source of detailed geographic information. However, this coordinated collection of grid data will be a new departure, allowing new types of statistical and geospatial analyses for areas that can be flexibly defined according to the needs of policy makers and researchers. In the future it will also assure geographical comparability over time.

The present chapter summarizes the ongoing efforts of the European Statistical System (ESS)[4] towards high-quality population data on a common pan-European 1 km statistical grid, as a part of the general EU censuses.[5] There are specific challenges in statistical disclosure control (SDC) of highly detailed geospatial population data, which need to be addressed carefully. Therefore special emphasis is put on legal and methodological aspects of SDC in official population/census statistics: after a general introduction to the topic and various typical methods, the particular scope and challenges of population grid data will be highlighted.

18.2 Statistical Disclosure Control of Population Data

In official statistics within the EU, information considered confidential is protected by law, both at national level as well as EU level. See for instance Articles 20 to 26 of the EU Regulation on European statistics.[6] In particular, statistical results may

---

[3]EFGS/GEOSTAT 1 2012. GEOSTAT 1A—Representing Census data in a European population grid. Final report. http://ec.europa.eu/eurostat/documents/4311134/4350174/ESSnet-project-GEOSTAT1A-final-report_0.pdf.

[4]The ESS is the joint body of Eurostat and the NSIs of all EU countries and Iceland, Liechtenstein, Norway and Switzerland. It is responsible for the development and quality assurance of official European statistics.

[5]For a more detailed discussion on the appropriate/intended framework for the ESS, see Haldorson and Moström (2018), Chap. 9 in this volume.

[6]Regulation (EC) No 223/2009 of the European Parliament and of the Council of 11 March 2009 on European statistics (OJ L 087 31.3.2009, p. 164), amended by Regulation (EU) 2015/759 of the European Parliament and of the Council of 29 April 2015 (OJ L 123, 19.5.2015, p. 90).

not allow the identification of individual statistical units without their explicit permission (cf. Article 20). This is even more crucial in population statistics, where the statistical units are persons.[7] This implies immediately that statistical institutes collecting such data must *physically protect* their databases on individual persons ("microdata"). Consequently, public statistical outputs are usually in aggregated form, meaning that person characteristics are cross-tabulated and only the total counts of persons having the same characteristics are published ("meso-/macrodata" depending on the aggregation level).

However, depending on the design of output tables and details of characteristics asked, even the aggregated table data may still entail a risk of logically disclosing confidential information on individuals or small person groups: for instance, an intruder may use public statistical data and logical inference to first identify an individual person, then learn more (sensitive) details about that person—personal information is disclosed (see Sect. 18.2.1). Therefore, outputs need to be *logically protected* before publication (Sect. 18.2.2). Moreover, the EU countries have different notions and national laws regulating what is considered confidential information in population statistics. For instance, some categories of person attributes may be considered confidential—and hence need to be protected against disclosure —in some countries but not in others (see "attribute disclosure" in Sect. 18.2.1).

18.2.1 Disclosure Risks and Scenarios

The disclosure risk of statistical information considered confidential can be quantified by disclosure risk measures,[8] which help to make decisions about the publication of output data. If the disclosure risk is low, a statistical institute might release the data without any change. However, if the disclosure risk is considered high, the statistical institute is legally obliged to protect this information carefully. In this context, one typically distinguishes in population tables between *identifying* variables (i.e. variables that may lead to the identification of a single person or very small group, such as place of residence, age, sex, …) and *sensitive* variables (i.e. variables containing categories which may disclose sensitive additional information about a person or group, e.g. economic activity—unemployed or criminal record—yes).[9] Accordingly, sensitive cases may include:

---

[7]…or households etc.; for brevity we simply refer to "persons" in the following.

[8]For a general discussion of disclosure risks in frequency tables, see e.g. Hundepool et al. (2010, 2012).

[9]This classification of variables is not rigorous because it depends on the prior knowledge of an intruder, where reasonable assumptions need to be made for a risk assessment. Moreover, some variable categories might be considered identifying *and* sensitive (think about place of birth). Finally, different countries have different perceptions of what is sensitive information—the examples chosen here try to convey an intuitive, common-sense notion of the conceptual difference between *identification* and *disclosure of sensitive information*.

Identification. A small count (e.g. '1' or '2') in a data table means that the respective person may be identified from the identifying variables in that table. Often, it is still impossible to infer sensitive details about this person (no sensitive variables in the table, or no hit in sensitive categories), but generally the risk of attribute disclosure (see below) is higher. Also self-identifications of persons after publication of the data may raise public concerns about sufficient protection, especially in census data where a series of public hypercubes (see Sect. 18.1 and Fig. 18.1) cover the entire population of a country or geographical region. The more cells a hypercube contains (where the cell number is mostly driven by its dimension, i.e. the number of person characteristics that are cross-tabulated), the more small counts will be present.

Attribute disclosure. An attribute of a person, or a small group, can be inferred from the data. This might be considered especially critical if a group included in a subset of the variables defining a table is already rather small. An intruder might first identify this group, and then learn from the other variables in the table that the distribution of certain characteristics among this group is very skewed, i.e. most or all persons in the group have a particular (set of) attribute(s), or conversely none or just a single person in the group has the complementary (set of) attribute(s). Imagine a small group of persons with a specific age/sex combination in a small municipality where the data exhibit that most or all of them are unemployed. Thus, the risk of attribute disclosure is mainly driven by patterns of small or zero counts across the categories of sensitive variables (skewedness). Obviously, risks of attribute disclosure are more direct risks compared to the mere identification risks from small counts. Again, these risks are higher in census data than in data from sample surveys.

In a stricter understanding of attribute disclosure even the information that a non-zero group of persons with a certain attribute combination *exists* (without necessarily being identifiable) is considered risky: This is sometimes referred to as risk of "disclosure of existence" risk.

Disclosure by differencing. This describes a situation where information about one or two very similar (not necessarily small) groups of persons is available from different tables. Subtracting counts from these tables may then disclose information about a very small sub-group of persons: Imagine two tables with identical categories but slightly different age bands, e.g. table 1 starting with the category "0–20" and table 2 with "0–25". Subtracting counts in these first categories then provides information about the potentially small group of persons aged 21–25. Different overlapping but not exactly nested geographical classifications, such as grid squares and LAU2 regions, are another important example (see Fig. 18.2).

18.2.2 Disclosure Control Methods

As briefly explained at the beginning of Sect. 18.2, the production of population statistics typically includes the collection of microdata on individual persons (from

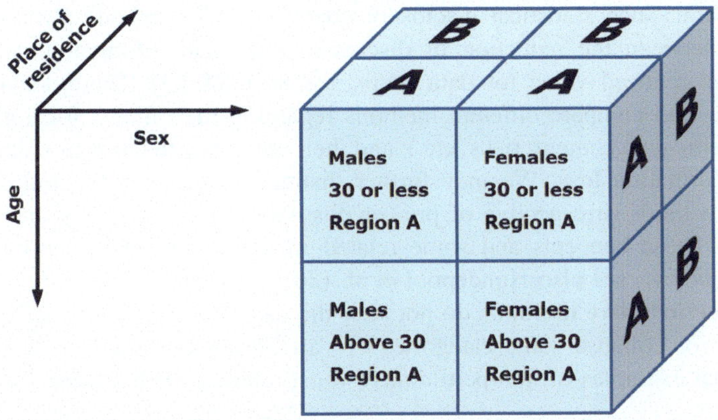

Fig. 18.1 Illustration of a three-dimensional hypercube (census output table), in this case crossing the attributes sex, age and place of residence; up to five dimensions are typical

Fig. 18.2 Illustration of LAU2 municipality boundaries (black lines) overlapping with 1 km grid squares (image produced by Eurostat; credits: map view © OpenStreetMap Contributors and administrative boundaries © EuroGeographics)

population samples or full censuses), and subsequent aggregation into output tables ("counting persons with given characteristics"). Data treatment methods to protect these outputs against logical disclosure of confidential information may be applied either directly to the microdata before aggregation (*pre-tabular* methods), or to the output tables after aggregation (*post-tabular* methods).

In general, such statistical disclosure control (SDC) methods always entail a trade-off between the reduction of disclosure risks and information loss in the output (i.e. reduced value for data users, see Sect. 18.2.3). Therefore, it usually makes sense to compare different methods regarding their fitness for a given purpose (specific data content, risks, etc.), and their efficiency in terms of risk reduction versus information loss. We may further distinguish between the following two types of methods (irrespective of pre- or post-tabular): *perturbative* and *non-perturbative*. These concepts and some related methods applied in census data are presented below; see also Hundepool et al. (2010, 2012).

Non-perturbative methods do not alter the data. Rather, they usually suppress table cells or collapse some categories of variables into a single new one. Two well-known examples of non-perturbative (post-tabular) methods are:

- **Cell suppression** typically consists of two steps: First, a rule is required to determine confidential table cells with unacceptably high disclosure risk. These cells are subject to *primary suppression*. In the second step, some cells are selected for *secondary suppression* to protect the primary cells from disclosure by differencing. Defining a primary suppression rule based on small person counts to protect against identification is simple: e.g., always suppress '1' and '2'. Defining a rule for primary suppression that takes care of attribute disclosure risks (see Sect. 18.2.1) and at the same time does not cause a lot of overprotection due to suppressing also many non-sensitive attributes, is more difficult: one would also have to suppress certain risky patterns of '0' counts. However, a corresponding suppression scheme which is at the same time safe and minimizes information loss is generally very hard to define—for instance, suppressing *all* '0's could be safe but obviously leads to large overprotection, also due to secondary suppression. Indeed, when the goal is to protect *all* low person counts against disclosure (which is required for the census in 9 EU countries, according to a survey that was conducted by the project referred to in Sect. 18.5), the number of secondary suppressions rises drastically with the number of hypercube dimensions. In this situation cell suppression leads to huge information loss. Finally, suppression patterns can differ considerably from country to country, thus jeopardizing severely the comparability at EU level. See Fig. 18.3 for a grid example.
- **Global recoding** means that several categories of a categorical variable are collapsed into a single one, i.e. the "format", or structure, of the protected table may differ from that of the original table. As for the EU census hypercubes, where the formats are legally fixed in advance, global recoding is virtually the same as completely suppressing certain detailed categories of hypercube variables. This entails a huge information loss. In a special grid variant of global recoding called *cell condensation*, grid squares with small counts are joined with neighbouring squares until a certain population threshold is reached (cf. Fig. 18.3); see e.g. Suñé et al. (2017). This method is simple but has the disadvantage that consistency, flexibility and spatial comparability of the grid system is lost.

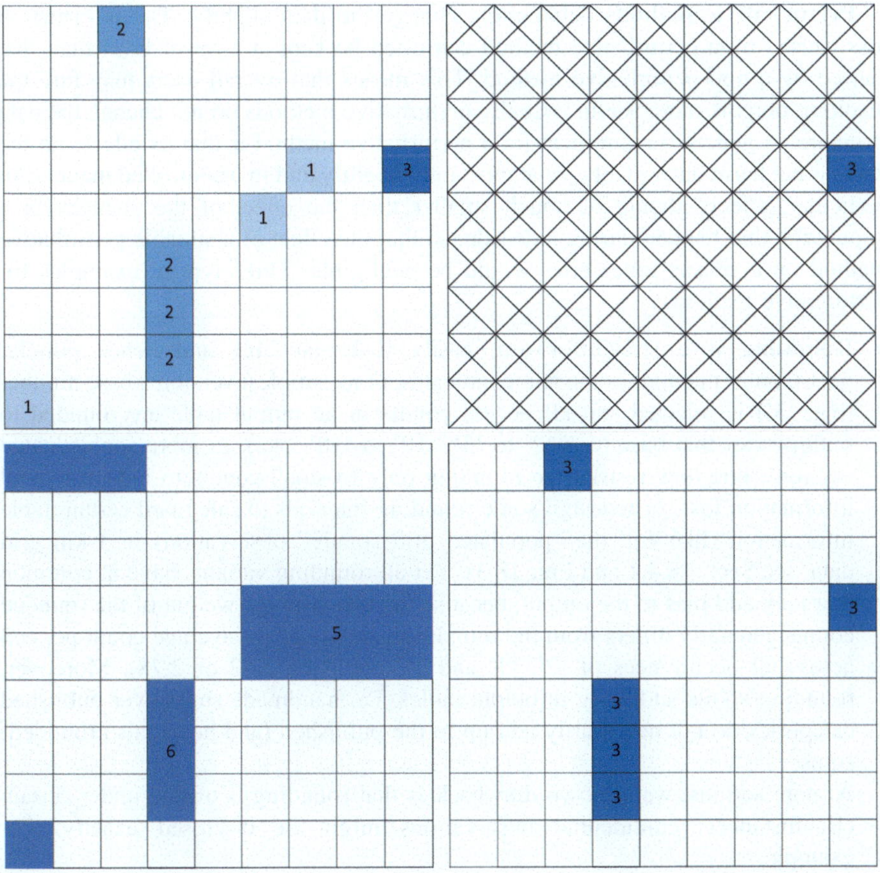

Fig. 18.3 Effects of various SDC methods discussed in the text on a sparsely populated example grid region: the upper left picture shows unperturbed population counts (white squares unpopulated), while in the upper right all counts below ‘3’ are suppressed. The lower left picture illustrates a possible outcome of cell condensation. Finally, the lower right picture shows simple rounding to base ‘3’; note that random noise (cell key method) effects may typically look similar. (The example shows real 2011 GEOSTAT 1 data from Austria; the reported counts may be considered as unperturbed because Austria did not apply SDC measures, cf. Sect. 18.4.1)

In conclusion, while the disclosure risk of data can be lowered by applying a non-perturbative method, this might reduce the information content of the data drastically, leading to unnecessarily high information loss as argued above. Non-perturbative methods can be carried out consistently and tables may remain additive. However, it is very difficult to maintain consistency across large sets of tables based on the same population, like the 2021 EU census hypercube programme consisting of more than 100 output tables, where lack of consistency potentially increases disclosure risks again.

Perturbative methods deliberately change the data slightly. The information loss caused by a perturbative method can often be kept at a lower level than that caused by a non-perturbative method. This means that overall users may find the resulting outputs more useful. Usually perturbative methods do not change the data structure. A general requirement for a perturbative method is that its effects on the data do not harm data quality, or at most only slightly and in a controlled manner. In particular, effects should be much smaller than the effect of the changes in a population that happen during a decade, so that the effect of a suitable perturbative method on a census time series would be negligible. Three typical examples for perturbative methods:

- **Rounding** is a straightforward, easily understandable and hence popular post-tabular method: as the name indicates, in its simplest version a base number (e.g. '3') is selected and all person counts in an output table are rounded to multiples of that base (e.g. '2' to '3', '19' to '18', etc.). In more sophisticated versions, one may restrict the rounding only to small counts to reduce overall information loss, or redesign some rounding intervals to safeguard certain table information (like e.g. the "populated grid square" observations in 1 km grid data, see Sect. 18.4.1 and Fig. 18.3). Yet all rounding variants have in common that they add bias to the output, because the true average weight of the rounded counts generally differs from the rounding value (e.g. the average count per cell across all occurrences of '2', '3' and '4' could be 3.42 or 2.78). Moreover, rounding spoils additivity in output tables, i.e. handmade sums over published categories do not necessarily add up to the published (and hence also rounded) sums.
  A more serious, well-known drawback is that rounding is unsafe under certain circumstances: confidential observations might be disclosed exactly. For example:

	Category A	Category B	Total
Rounded to base '3':	3	3	9
Inferred true counts:	4	4	8

  Gießing (2016) gives a more rigorous formulation of this effect,[10] including a quantitative analysis showing how rounding is generally inferior to random noise (see below) in terms of risk reduction versus information loss.
- **Record swapping** is a pre-tabular method, i.e. applicable directly to the microdata: some pairs of records are selected in the microdata set. The paired persons/households match on some variables in order to maintain the analytical properties and to minimize the bias of the perturbed microdata as much as possible. Record swapping exchanges some of the non-equal variable-values

[10]This risk can be reduced—but not completely eliminated—through *random rounding*, where in our example also '1' and '5' would be rounded to '3' with some probability (and '4' resp. '8' to '6', etc.); see e.g. Hundepool et al. (2010, 2012).

between the paired records. Since this exchange slightly perturbs the microdata, an intruder's inference about a certain statistical unit might not be correct. Record swapping can be random or targeted. In the random case the records to be swapped are selected with equal probability, while in case of targeted record swapping records of high disclosure risk are determined and a matching partner to each of these records is selected for swapping.

- **Random noise** is a technical term denoting the random application of small perturbations to the observed population data, either directly to the microdata, cf. section "noise addition" in Hundepool et al. (2010, 2012), or to the person counts in output tables. As a post-tabular method, it typically consists of a predefined probability distribution for the noise to be added to the tabulated person counts (e.g. ± 1, ± 2, etc.), combined with a mechanism to randomly draw from that noise distribution in a consistent manner. The particular variant considered here is based on the "cell key method" proposed by the Australian Bureau of Statistics (ABS). A generic implementation of this method involves two to three distinct steps or "modules":

  – Cell key module: assigns a fixed random number, or "cell key", to each cell of the output table. The ABS variant ensures by design that the noise added to a specific cell will always be exactly the same, even if the cell appears in different tables—this is meant by *consistency* of the method. In order to guarantee this, one first needs to assign a "record key" to each microdata record. Then each cell,—defining a subset of microdata records with precisely the properties described by this cell,—obtains a cell key constructed from the record keys of all records contributing to the cell. (Note that consistency requires the method to act on microdata—hence *pseudo* post-tabular may be more appropriate.)
  – Noise module: draws the noise for each output table cell from the predefined noise probability distribution, based on the cell's random number (cell key). Essential input parameters for the design of the probability distribution are the statistical noise variance and the maximal noise magnitude (e.g. 1 or 2, etc.), see Gießing (2016).
  – Additivity module (optional!): obviously random noise breaks the additivity of the output tables (cf. rounding above). Additivity can be restored in a post-processing step, but this spoils consistency across several tables. Hence, statistical institutes need to assess carefully which property they consider more important. (In the ESS there is a mixed picture regarding what may be preferred in the 2021 census.)

Table 18.1 summarizes some general properties of all the non-perturbative and perturbative methods described above, while Fig. 18.3 illustrates some of them in an example 1 km grid region. See Hundepool et al. (2010, 2012) for a more comprehensive overview of SDC methods.

18.2.3 Information Loss Assessment

Statistical disclosure control methods affect the output information content. As noted earlier, disclosure risk and information loss are two conflicting concepts that need to be traded off: the lower the disclosure risk, the higher the information loss. The most "efficient" SDC method reduces the disclosure risk to an acceptable level, while keeping the information loss minimal.

Hence, information loss measures are used to quantify and compare the divergence between the data before and after a protection method has been applied. They can help to select between different protection methods, or also between different variants (e.g. parameter settings) of the same protection method, e.g. by ranking protection methods for a fixed acceptable target level of protection. See Shlomo and Young (2006) for a comprehensive overview.

18.3 Census Population Grids: User Requirements Versus Data Protection

It is usually difficult to compare regional census data at the lowest available level (LAU2 or municipalities) because the definition of these regions is subject to different national needs and because the LAU2 units change over time. Data made available on a common European reference grid are much more versatile in terms of highly detailed regional and even cross-border analysis. Moreover, sufficiently small grid squares can be freely clustered into regional aggregates according to individual analysis needs (e.g. such as environmental or agricultural aspects), regardless of any administrative patterns. The provision of reliable census data on a common European grid is therefore an essential complement to the traditional census hypercubes based on administrative areas.

Also key census data users at national and EU level have re-stated the particular importance of grid data in developing and evaluating policy. Serving the demands of such highly experienced long-term users is a core business of the NSIs and the entire European Statistical System (ESS). It is therefore not surprising that many NSIs have already started to develop and publish new products at national level to meet these policy user demands. However, only the ESS can lift such endeavours to the European level to serve supranational and EU-wide user needs: this is a key argument for the joint efforts by Eurostat and its ESS partners to extend, improve and regulate the 2021 census grid outputs compared to the 2011 results—see Sect. 18.4.

One particular policy use of EU-wide census grid data will be in planning and evaluating the distribution of the European Cohesion Policy budget—which roughly amounts to €350 billion over seven years—conducted by the Directorate-General for Regional and Urban Policy of the European Commission (DG REGIO). An important aspect of ensuring the efficient allocation of funds is

Table 18.1 Comparison of the general properties of various disclosure control methods discussed in Sect. 18.2.2

	Non-perturbative methods		Perturbative methods		
	Cell suppression	Global recoding	Rounding	Record swapping	Random noise
Data values changed?	No	No	Yes	Yes	Yes
Information loss?	Often high	Often high	Depends on rounding base	Controllable	Controllable
Table format preserved?	Yes	No	Yes	Yes	Yes
Consistency?	Difficult in high-dimensional and/or linked tables	Possible	Possible	Yes, for tables from the same protected microdata	Yes, if additivity is given up
Additivity?	Mostly yes (suppressed information missing)	Possible	No	Yes, for tables from the same protected microdata	Yes, if consistency is given up

the analysis of populations for small urban and rural areas, and assessing the accessibility of services for citizens. The ability with grid data from the census to define and analyse populations for flexible areas will be of great importance in improving the allocation of funds for development and infrastructure projects—consider for example people living within a certain distance of a hospital. This analysis requires comparable and accurate EU-wide information on the population distribution at high geographical detail. In this context, sparsely populated areas are of particular interest for the EU and at the same time particularly risky in terms of disclosure.

Hence, in view of the generic situation described in Sect. 18.2, grid data are very different from "ordinary" statistical population tables in terms of structure, scope and intended use, thus posing very specific challenges for data protection. Three particular points:

- Grids are not nested with other regional breakdowns based on administrative boundaries (NUTS, LAU), which may increase the disclosure risk of confidential information by *differencing* between partially overlapping small regions (see Sect. 18.2.1).
- Grid data have a *special scope and purpose*: contrary to census tabulations based on administrative regions, they are more than just another "table" of statistical data—their essential and distinguishing feature is the crossing of statistical census information with rigid and standardized geographical information. In that sense, grids may be seen as a hybrid product between traditional

tables and maps. It follows immediately that the geographic dimension of a grid plays an outstanding role: it is not just “another dimension of a statistical table”.
- Directly emerging from the previous point, the *information value* of individual statistical data contained in a grid dataset (and hence notion of information loss) may need to be re-evaluated, taking into account its particular purpose: several key use cases of EU grid data rely on very accurate information about the *spatial distribution of populated areas* (e.g. the DG REGIO use case outlined above). This assigns an outstanding role to the distinction of populated versus unpopulated areas (i.e. grid squares), which establishes a qualitative difference between the total population observation by grid square (or rather, if this number is zero or not) and other person characteristics collected on the grid.

Effectively protecting confidential information, while at the same time respecting the peculiarities of grid data outlined here, entails an added complexity: we will address this in depth in Sect. 18.5, after briefly introducing the evolution of EU census grid data in the following section.

18.4 1 km Grid Data from EU Population Censuses

18.4.1 2011 Census: The GEOSTAT 1 Exercise

From the beginnings in 2010, the ongoing series of GEOSTAT projects[11] is intended to establish common guidelines for the provision of pan-European grid statistics within the ESS, and from a wider perspective to promote the integration of statistical and geospatial information in a common ESS-wide information infrastructure. In 2013, the GEOSTAT 1 project successfully collected and published 1 km grid data on the total population across 30 out of 32 ESS members. While some NSIs already had considerable prior experience with statistical grid products at the national level, the particular success of this voluntary action was the very high participation across the whole ESS, thus piloting the use of a common, INSPIRE-compliant[12] 1 km statistical grid for pan-European official population statistics. Hence, GEOSTAT 1 paved the way for the planned more comprehensive and more harmonized publication of 1 km grid data from the 2021 EU census round as a part of the European statistical programme, to be fixed by a dedicated new EU regulation (see Sect. 18.4.2).

In terms of data protection, the GEOSTAT 1 situation was rather straightforward because only the total population was published on the European 1 km grid, and

[11]GEOSTAT is a joint cooperation of Eurostat with the European Forum for Geography and Statistics (EFGS); see http://www.efgs.info/geostat/.

[12]Directive 2007/2/EC of the European Parliament and of the Council establishing an Infrastructure for Spatial Information in the European Community (INSPIRE) (OJ L 108, 25.4.2007, p. 1) and its implementing Acts.

this information was not considered confidential in the majority (namely 19) of the participating ESS members.[13] Most of the 11 remaining participants who did consider SDC treatment necessary applied simple rounding variants restricted to small population counts only. The ensuing small biases in population aggregates were generally considered negligible compared to other sources of uncertainty typically afflicting census data. However, most NSIs intentionally avoided rounding towards '0' (i.e., rounding '1' to '10' towards '5', or replacing '1' and '2' by '3', or similar): this was to preserve the most accurate distinction between populated and unpopulated grid squares, and hence information on the spatial population *distribution*, in the output data—see Sect. 18.3 for the use cases of statistical grids justifying this requirement.

18.4.2 2021 Census: Planned EU Regulation

The overall success of the voluntary 1 km grid exercise based on 2011 census data as—outlined in the previous section—encouraged the ESS members to build on this experience and aim for some ambitious improvements for the 2021 census round. In particular, the ESS *Task Force on future population censuses*,[14] (henceforth "Task Force", invested considerable efforts over the recent years in the statistical and administrative preparation of an extended and improved 1 km grid output.

As a first step, weighing user needs against the expected availability and quality of geocoded information as well as basic confidentiality constraints, the Task Force agreed on the following list of categories to be collected for each 1 km grid square:

- total population;
- sex (males, females);
- age (under 15, 15–64, 65 and over);
- employed persons;[15]
- place of birth (in the reporting country, in another EU country, outside EU);
- usual residence 12 months before (unchanged, within reporting country, outside of the reporting country).

It is important to note that no cross-tabulation of the different topics is foreseen, mainly due to confidentiality concerns at this very high geographic detail. In effect, 13 numbers would be collected per grid square, representing the total number of persons in each of the categories listed above.

[13]For a comprehensive quality assessment, including a detailed documentation of SDC treatments in the various participating countries, see: http://www.efgs.info/wp-content/uploads/geostat/1b/GEOSTAT1B-Appendix17-GEOSTAT-grid-POP-1K-ALL-2011-QA.pdf.

[14]Bringing together around 20 NSI representatives of census teams.

[15]To be provided only "as far as possible", i.e. the planned Regulation (see below) would oblige EU countries to justify the non-provision of these data (e.g. non-availability).

While the 2011 exercise had been conducted voluntarily as a part of the GEOSTAT 1 project (Sect. 18.4.1), the Task Force concluded that a legal basis for 2021 would have several distinct advantages: in addition to considerable efficiency gains at ESS level, the resulting census grids would meet much higher quality standards than the 2011 ones, especially in terms of accuracy, comparability and completeness of the output data across the whole ESS.[16] At the time of writing this chapter, an advanced draft for a dedicated EU Regulation had been proposed by Eurostat for general discussion at ESS expert level.

After 2021, the infrastructure can then be reused for more frequent and timely population grid products that are foreseen in the post-2021 census strategy. INSPIRE standards also facilitate the integration of possible future changes of the grid resolution, for instance going below 1 km, or even dynamic adaptations of the resolution depending on local population densities (e.g., recursive quartering of grid squares down to a set threshold).

18.5 Recommendations for Data Protection in the 2021 EU Census Including 1 km Grid Data

In 2016 and 2017, Eurostat hosted a dedicated project "Harmonized Protection of Census Data in the ESS"[17] (henceforth "SDC project") to investigate on good confidentiality practices to be recommended for the next EU census round in 2021. One of the key topics was the protection of the planned 1 km grid data described in Sect. 18.4.2, particularly in view of the general considerations outlined in Sect. 18.3: this issue will be addressed in detail in Sect. 18.5.2, directly after a comprehensive summary of the generic project results and recommendations in the following section.

18.5.1 Generic Recommendations for 2021 Census Data

As outlined in Sect. 18.2.2, there is a large variety of methods to protect confidential information in the census outputs, and many of them have already been applied in production by one or more ESS members in the past. In their assessment for 2021, the project team considered it most important that a harmonized SDC approach should:

---

[16]Common quality principles for European statistics—including accuracy, comparability and completeness mentioned in the text—are established in the revised "European Statistics Code of Practice" adopted by the ESS Committee on 28 September 2011; more details on http://ec.europa.eu/eurostat/web/ess/about-us/quality.

[17]https://ec.europa.eu/eurostat/cros/content/harmonised-protection-census-data_en.

- retain the fixed structure of output tables;
- address attribute disclosure risks;
- minimize the information loss, also for very detailed tables;
- be applicable in as many countries as possible, perhaps with slightly different parameters.

Based on these four priorities, the project team concluded that perturbative methods are superior to non-perturbative methods for the scope of the project. In particular, it was concluded that cell suppression is not a good candidate for a harmonized approach, even though it is an established method in some countries: on top of the lack of comparability, applying it to more than 100 census hypercubes is technically not possible in a consistent way with reasonable information loss. On the other hand, a harmonized method should offer enough flexibility for countries to adapt it to their specific needs and expectations regarding an acceptable balance between residual disclosure risk and level of information loss. Thus, it should provide various parameters for simple adaptation, and consist of separate components that can be combined, but may as well be "deactivated" if a country judges that a specific component is not necessary. From that perspective, the project considered it an added advantage to include components both for pre-tabular as well as for post-tabular perturbation.

Resulting from all these considerations, the project team recommends the following two perturbative protection methods and corresponding setups:

- **Record swapping** (pre-tabular): households containing at least one high-risk person (i.e. person with rare characteristics) are identified and swapped with similar, low-risk households (targeted swapping). Technically this is done by exchanging the geographic location of the paired households. In the output tables, this leads to small random variations in the counts of person characteristics—but not in the total population[18]—inside a given geographic area (NUTS, LAU or grid squares).
- **Cell key method** (post-tabular): random noise is added to some population counts in the output tables, where the recommended setup includes a fixed global design variance of 1 and a maximum absolute deviation of 3 from the original counts, while original '0' counts are not perturbed at all. (Other possible setups have different global variance and/or absolute deviation parameters, or they also perturb original '0' counts if this is desired.)

They can be applied either separately or in combination. Both methods explicitly avoid cell suppression, so that national outputs can be combined straightforwardly into European-level data. As further argued in Sect. 18.2.2, such perturbative methods can significantly reduce the overall information loss in the outputs at a comparably low disclosure risk level. Moreover, the parameters of both methods are not fixed, so the countries can adapt them to specific needs.

[18]Only if equal person count is a necessary requirement for pairing similar households, but this is the default in the recommended method variant.

In conclusion, these methods are specifically aimed at

- overcoming the typical drawbacks of traditional methods like suppression or simple rounding, and thus reducing the information loss through the treatment;
- flexibility: various parameters are available that can be adapted to particular national requirements in an EU country;
- straightforward usage, by providing respective SAS (already available) and R (foreseen) implementations out of the box.

If many ESS members apply the same methods—though perhaps in different flavours—this will help prepare European-level data in a more straightforward way. It will also be an advantage when it comes to tool development for actual data production, teaching and documentation of the methods, as well as preparation of user communication. Finally, unlike cell suppression, the proposed perturbative methods will make data available for *all* hypercube cells. This will be a great advantage for all users, significantly improving on the completeness and comparability of the data between countries with respect to the 2011 situation.

18.5.2 Grid Data

Section 18.3 outlined how grid data typically have different scopes and user requirements compared to “traditional” census outputs (hypercubes), and hence also specific quality priorities and potential confidentiality challenges. Moreover, as summarized in Sect. 18.4.1, during the 2011 census grid exercise the overall majority of GEOSTAT 1 participants did not even consider ‘total population’ sensitive information on the 1 km grid. Most of the remaining participants applied simple rounding variants on small counts, but keeping the ‘0’ counts apart to maintain the distinction between unpopulated and populated cells.[19] The 2021 situation is more complicated, though, due to the additional person characteristics collected by grid square (cf. Sect. 18.4.2): this additional information may cause more countries to protect small ‘total population’ counts. Moreover, simple rounding of small ‘total population’ counts would likely lead to a rounding (or suppression) also of many other person characteristics, thus leading to an unnecessarily large information loss.

All this was duly acknowledged and addressed by the SDC project: while the cell key method provides an effective protection against the risk of disclosure by differencing (point 1), some challenges were noted with respect to the other two points highlighted in Sect. 18.3. In particular, a certain tension was identified between the cell key method in its more common parameter setups and the requirement of accurately separating unpopulated from populated grid squares, essentially because the method treats all table cells in the same way. Hence it cannot

[19]Cf. the ‘Confidentiality’ field of the GEOSTAT 1 quality assessment (Footnote 13).

distinguish between 'total population' and other person categories, and there is a considerable share of small counts—typically in the range of '3' or less—being perturbed to '0', thus turning populated grid squares into unpopulated ones.[20]

Based on the 2011 grid data, the project team found that a naïve application of the cell key method in its recommended setup (and, in fact, in most other feasible setups) would jeopardize the accuracy on the total population *distribution* that was mentioned as a benchmark by key grid users, most notably DG REGIO (cf. Sect. 18.3). Several approaches were discussed to alleviate this tension. However, it was judged that going back to the simpler rounding approaches of 2011 would entail a high risk of unnecessary information loss due to secondary rounding and/or excessive suppression also of other person categories on the grid. On the other hand, all of the discussed modifications of the cell key method itself specifically for the EU grid datasets would violate the *consistency* paradigm of the method, and thus potentially increase disclosure risks again.

The 'populated' flag. Ultimately the SDC project proposed the following procedure in order to maintain the distinct advantages of the cell key method, while simultaneously meeting the requirement to accurately distinguish populated from unpopulated grid squares: introduce a flag which marks 'total population' data items as 'populated' if and only if their observed value *before the application of any SDC method* is greater than '0'. The desired information—accurate population distribution—can thus be extracted from the metadata to the 1 km grid dataset (i.e. distribution of the 'populated' flag across the grid squares), see Fig. 18.4 for an illustration. As this flagging is completely independent from the actual 'total population' output data, the recommended cell key method and setup can be applied to the grid dataset in a straightforward way, without touching the accuracy of population distribution information.

The proposed use of the 'populated' flag in combination with the recommended cell key method has distinct advantages from a user perspective:

- **Minimal and controlled information loss**: the complete statistical data (total population plus characteristics) are made available for each grid square with a fixed absolute statistical variance on every data value, throughout the entire 1 km grid data of a country. This means that each and every published number is affected by a design SDC uncertainty of $\pm\sqrt{V}$, where V is the variance parameter of the cell key method (typically around 0.5–2).[21]

[20]Some particular method setups (not recommended by the project for various reasons partly connected to the hypercubes) would also include perturbations of some '0' counts into non-'0', i.e. turning originally unpopulated grid squares into populated ones (cf. Sect. 18.5.1).

[21]Note that record swapping (cf. Sect. 18.2.2) may further increase the SDC uncertainty on counts of person *characteristics*. And of course, all this refers only to the SDC uncertainty: naturally there are additional uncertainties (variance and/or bias) stemming from other sources such as data collection and processing.

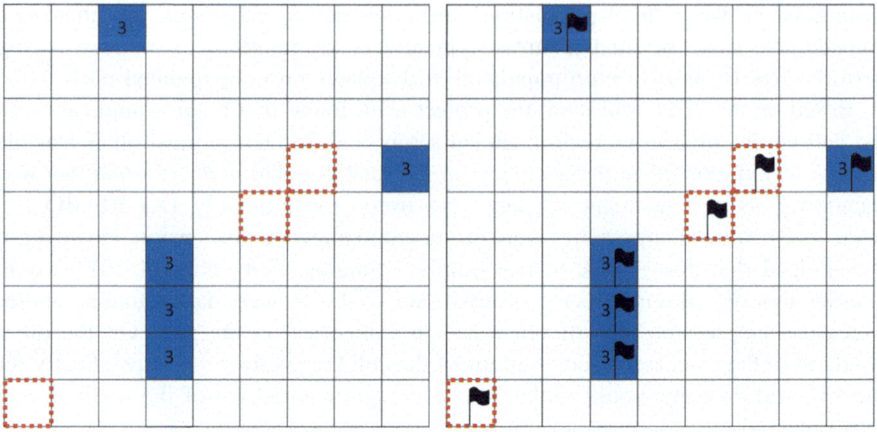

Fig. 18.4 Effect of the 'populated' flag (see text) on the sparsely populated example region of Fig. 18.3: The left picture shows again a possible outcome of the cell key method (without going into detail on the method parameters), but this time populated areas turned into unpopulated ones are marked by a red frame. The right picture shows how the 'populated' flag marks all originally populated areas, while 'total population' would be reported as '0' in the red-framed cells

Most accurate distribution of populated areas:

- can be readily obtained from the distribution of the 'populated' flag across the grid squares. If the value of a 'total population' data item flagged as 'populated' equals '0' (e.g. as appearing in Fig. 18.4), some simple heuristics, taking into account available data on characteristics in this grid square and/or adjacent grid squares, can be applied to arrive at a fair estimate of the total population. For instance, the category sums for each of the four attribute topics (sex, age, place of birth, residence one year before) could be averaged on that grid square to convey a decent first approximation of the total population.

It might be seen as a drawback that a given grid square with 'total population' equal to '0' but flagged as 'populated' (or vice versa in some possible setups, cf. Footnote 20) could be perceived as confusing or unintuitive by "the user". However, it was judged that the advantages above outweigh this by far. Moreover, it may be reasonably assumed that any serious user of the grid data would (and should!) inquire about their scope and usage, for instance in the provided metadata. On the other hand, cursory data users will likely not even be aware of the 'populated' flag in the first place.

18.6 Conclusion

After a brief outline of the evolution of population grids from census data between 2011 and 2021, this chapter describes the results of the dedicated ESS project "Harmonized Protection of Census Data in the ESS". It was conducted between 2016 and 2017 to work out recommendations for good statistical disclosure control practices in the 2021 EU census round. Overall the project team favours perturbative methods over "traditional" non-perturbative methods such as cell suppression, because they are typically much more efficient in terms of information loss versus residual disclosure risk. In particular, the project recommends a combination of pre-tabular *record swapping* and the post-tabular *cell key method* to protect all output tables (including grid data) of the 2021 EU census round: this harmonized approach would greatly improve the comparability and usability of European-level census outputs with respect to the 2011 situation, while providing sufficient flexibility for the EU countries to adapt to national needs (e.g. through method parameters). In order to simplify and thus spread the use of these recommendations as much as possible across the whole ESS, software implementations of both methods will be made publicly available, both in SAS (ready) and in R (planned).

Moreover, in this chapter it was argued that the specific use cases of key users of the European 1 km grid require accurate information on the spatial population *distribution*, and thus an accurate separation between populated and unpopulated areas (i.e. grid squares). The SDC project took special care in addressing this particular scope of grid data vis-à-vis the recommended unified data protection approach: a special challenge was identified in maintaining the generic advantages of the recommended cell key method while simultaneously serving the distinct user need for accurate population distribution. The proposed solution is to encode the required population distribution in a dedicated metadata flag indicating whether a grid square is 'populated' or 'unpopulated', so that the cell key method can be run in a consistent, global setup on the grid dataset just like on any other census output. Ultimately the idea is to provide an efficient, methodologically safe and simple, out-of-the-box treatment to protect census grids along with all other census outputs: in this sense, the 2021 EU census is a pilot exercise to see if this proposed approach is successful, in terms of acceptance across the EU countries as well as eventual user satisfaction.

Finally, it is worth noting that variants of the recommended perturbative protection methods—in particular the cell key method—have a large potential for application in other domains of highly detailed regional or grid statistics: basically any type of sensitive statistical information can be confidentialized using appropriate random noise, while particular implementations profit from generic advantages such as built-in consistency (through the cell key paradigm), or protection on the fly in dynamic publication/access tools (through straightforward automatization).

Acknowledgements The author would like to thank Eric Schulte Nordholt, Sarah Gießing, and all other participants of the ESS project "Harmonized Protection of Census Data in the ESS" for many fruitful discussions, and for their kind permission to use the project deliverables as a key reference for this chapter. The author further owes his gratitude to Ekkehard Petri, who provided valuable expert knowledge and explanations about all geo-referencing aspects, in particular about the GEOSTAT project series.

References

Backer L, Tammilehto-Luode M, Gublin P (2002) Tandem_GIS_I. A feasibility study towards a common geographical base for statistics across the European Union. Eurostat Working papers. http://ec.europa.eu/eurostat/documents/3888793/5816633/KS-AN-02-001-EN.PDF

Gießing S (2016) Computational issues in the design of transition probabilities and disclosure risk estimation for additive noise. In: Proceedings of the 2016 international conference on privacy in statistical databases. https://link.springer.com/book/10.1007/978-3-319-45381-1

Hundepool A, Domingo-Ferrer J, Franconi L, Gießing S, Lenz R, Naylor J, Schulte Nordholt E, Seri G, de Wolf PP (2010) Handbook on statistical disclosure control. Version 1.2, ESSnet centre of excellence for statistical disclosure control

Hundepool A, Domingo-Ferrer J, Franconi L, Gießing S, Schulte Nordholt E, Spicer K, de Wolf PP (2012) Statistical disclosure control. Wiley, Chichester (Wiley Series in Survey Methodology)

Shlomo N, Young C (2006) Statistical disclosure control methods through a risk-utility framework. Privacy in statistical databases. Springer, New York, pp 68–81

Suñé E, Ibáñez D, Rovira C, Farré M (2017) Statistical disclosure control on visualising geocoded population data using a structure in quad trees. In: NTTS 2017 proceedings. https://ec.europa.eu/eurostat/cros/ntts2017programme/data/abstracts/abstract_286.html

Chapter 19
Supply Chains and Decentralized Map Productions

Peter Schmitz

Abstract Map production is a process in which data is collected and production materials are sourced from suppliers; transformed into maps and made available to a customer base by the mapping entity. This production process is a basic supply chain namely there are suppliers that provide the data as well as the production material; the mapping entity is the firm that produces the maps; and the customers of the mapping entity obtains the maps. Maps are available as digital maps that can be viewed on a computer, tablet or on a smart phone or as a paper map. The production process can be totally in-house or it can be decentralized, meaning various entities may collect the data, another entity can collate the data and make it publication ready and other entities can distribute the maps digitally or print the maps for distribution to customers. This chapter looks at the use of supply chain management to manage and improve decentralized map production and to exceed customer expectations using the Supply-Chain Operations Reference (SCOR) model as the supply chain modeling tool.

Keywords Supply chains · Supply chain management · Decentralized production SCOR model

P. Schmitz (✉)
CSIR Built Environment, Pretoria, South Africa
e-mail: schimpmu@unisa.ac.za

P. Schmitz
Department of Geography, Geoinformatics and Meteorology, University of Pretoria, Pretoria, South Africa

P. Schmitz
Fakultät für Vermessung, Informatik und Mathematik, Hochschule für Technik, Stuttgart, Germany

P. Schmitz
Department of Geography, University of South Africa, Florida, South Africa

J. Döllner et al. (eds.), *Service-Oriented Mapping*, Lecture Notes in Geoinformation and Cartography, https://doi.org/10.1007/978-3-319-72434-8_19

19.1 Introduction

Map have been part of humanity over the centuries starting with mapping important stars and maps for hunting purposes such as the Pavlov maps which have been drawn on a rock face between 24,000 and 25,000 BC (Wolodtschenko and Forner 2007; Utrilla et al. 2009). Figure 19.1 illustrates the evolution of maps as indicated briefly in the aforementioned paragraphs. Figure 19.1 is based on the concept showing the evolution of manufacturing paradigms by Mourtzis and Doukas (2014). According to Mourtzis and Doukas (2014) the manufacturing paradigm shifted from farming to small workshops, the industrial revolution bringing the factory concept to mass production and global manufacturing with decentralized production activities. Figure 19.1 illustrates the growth in map production from single items to mass production with decentralized workflows.

The early maps were single items and were either etched onto stone, clay, bronze, silver or papyrus. Later maps that were drawn on paper, parchments and silk may have been copied, thus having the original and a few copies available for use. This changed during the renaissance when printing was invented and maps were commonly available (Pusic 2012). This was seen as bulk production as shown in Fig. 19.1. Mass production in this context is seen as high volume production of maps in various forms as used during the 20th and 21st century ranging from topographic maps, nautical charts, school atlases to folded maps of various places of the world. This continued until the onset of electronic data and spatial and n0n-spatial software that enables decentralized map production.

This chapter focuses on supply chains and decentralized map production which is part of contemporary cartography as shown in Fig. 19.1. The chapter is divided as follows: Decentralized map production is discussed in the next section, followed by a discussion on supply chains and supply chain management which includes the Supply-Chain Operations Reference (SCOR) model. The next section illustrates the use of supply chains and supply chain management in decentralized map production. The chapter ends with the conclusion and recommendations.

19.2 Cartographic Process

This section gives an overview of the cartographic process which will guide the map production process whether the production is completely in-house or decentralized. The cartographic process according to Stevens et al. (2012) starts at the environment that needs to be mapped for a specific purpose, next step is to collect data using various sources such as remote sensing, data from other entities and fieldwork that will best describe the selected environment. The collected data is then prepared for use in the creation of the map. The second process is the translation of the prepared data to be represented visually through generalization, selecting appropriate symbols, etc. Once the cartographer is satisfied that these

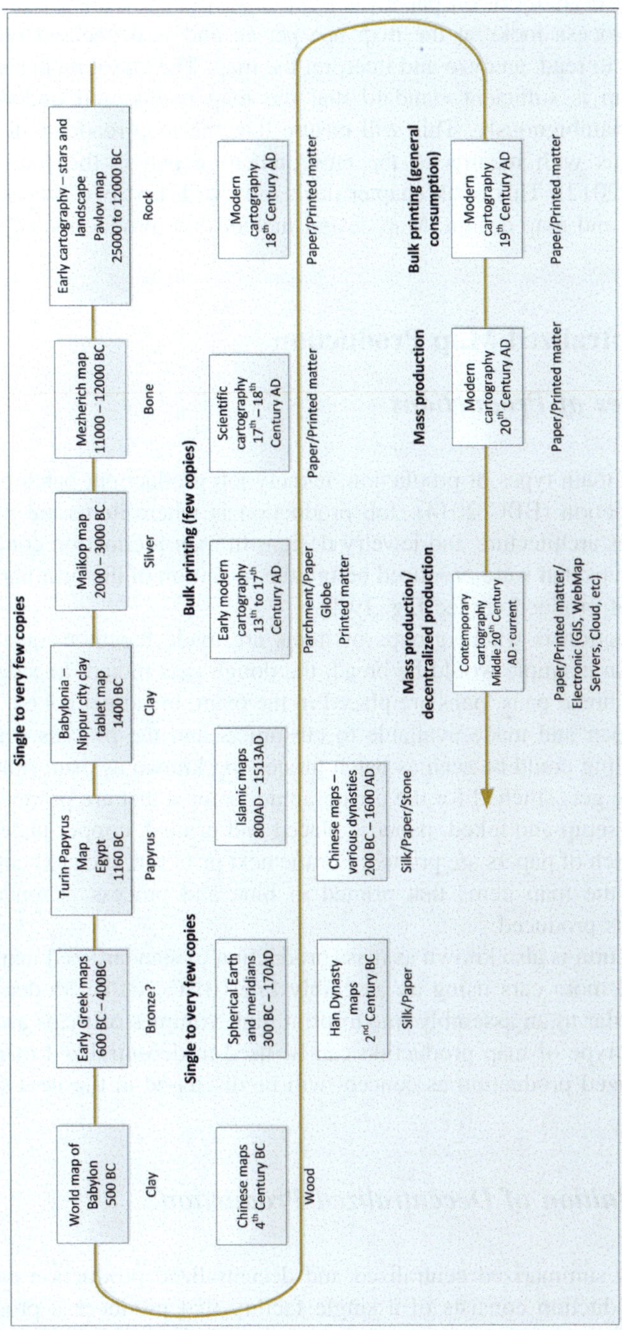

Fig. 19.1 The evolution of maps and map production

represents the selected environment sufficiently for the selected purpose, the map is produced either digitally or on paper.

The third process looks at the map use per se and is dependent on the map reader's ability to read, analyze and interpret the map. The cartographer's role is to ensure that map is sufficient standard that the map reader will understand and interpret the unambiguously. This will ensure that the map reader will make the correct decisions with regards to the environment which is the fourth process (Stevens et al. 2012). This book chapter deals only with first two process namely, data collection and map design. Map design includes the production of the map.

19.3 Decentralized Map Production

19.3.1 Types of Productions

There are three main types of production, namely job production, batch production and flow production (BBC 2014). Job production is where items are made individually such as architecture and jewelry design. In map production context, most of the earlier maps that were produced before the invention of the printing press fell into this type of production (see Fig. 19.1).

Batch production is where groups of items are made together stage by stage (BBC 2014). An example would be bread, the dough gets mixed, kneaded, placed in a number of bread pans, pans are placed in the oven, bread get baked, taken out of the baking pan and made available to customers and the process starts again. Early map printing could be seen as batch processing known as print runs, namely the copper plate gets stitched for the cartographic features that are printed in black, copper plate is setup and inked, paper is placed and printed, copper plate is inked, etc. until the batch of papers are printed. For the next print run, the next copper plate is stitched for the map items that printed in blue and process is repeated until complete map is produced.

Flow production is also known as mass production of standardized items such as soft drinks and moto cars using an assembly line (BBC 2014). Modern printing presses are similar to an assembly line to print high volumes of maps as shown in Fig. 19.2. This type of map production can be used in decentralized map production. Decentralized production as concept will be discussed in the next section.

19.3.2 Definition of Decentralized Production

Garrehy (2015) summarized centralized and decentralized production as follows: Centralized production consists of a single facility that produces a product. The distribution of the manufactured product can be from a single distribution facility or

Fig. 19.2 Example of a modern printer used to print maps (Grough 2006)

through several distributed distribution points. Decentralized production facilitates distributed production facilities and distribution over a large geographic area. This allows for the production and distribution of the products close to the customers.

Benefits for decentralized companies according to Stein (2002), Whitford and Zeitlin (2004) and Garrehy (2015) are flexibility; efficiency in decision making at line level; increased motivation, creativity and research at lower levels of the company which results in faster implementations of improvements in production and supply; individualization of products; rapid response to changes in products; access to regional differences in labor costs; and close to its customer base.

Digital manufacturing furthermore allows for the above and reduces labor costs in "rich" countries and improved collaborative manufacturing services online which is the backbone for decentralized production and dispersed manufacturing (The Economist 2012). The Economist (2012) put the emphasis on the impact of 3D printing will have in decentralized production of specific parts using various types of materials ranging from medical grade epoxies to titanium printing. Linking this concept to mapping agencies, they can have various small printing houses or facilities distributed around a country and print maps on demand using digital prepared maps from a single source.

Decentralized production has its disadvantages as well such as higher setup costs for the various distributed production plants; higher ramp-up costs for new products owing to higher variety of products available to the market as well as the short life cycle of these products; higher costs in quality control and maintaining consistency in production processes, especially when two or production units at various locations produce the same product (Minhas et al. 2012; Garrehy 2015).

19.3.3 The Map Production Process

Gruver and Dutton (2017) the high level map production process starts with data compilation which is the first part of the cartographic process as discussed in Sect. 19.2. This is followed by the map design using in most instances a Geographic Information System, the map or the layers that are created in a GIS is then exported into a software that is designed to enhance the design. Several mapping design software can accommodate spatial data thus being able to create and design a map in such software. Examples of map design software are 1Spatial Management Suite (www.1spatial.com), Esri Production Mapping (www.esri.com), MapMaker 4 (www.mapaker.com), and MAPublisher (www.avenza.com). Map design software range from high-end designing software to entry level map design software. The type and cost of map design software depends on the need of the map producing organization. The process discussed in this chapter is not software specific.

Once the map has been designed it is made available for publishing either digitally; printed on paper or other media using plotters, inkjet or laser printers; or offset printing (Gruver and Dutton 2017). The map design, map enhancement and publishing are part of the second process of the cartographic process discussed in Sect. 19.2. Figure 19.3 shows a typical map production workflow. This workflow will be expanded to illustrate decentralized map production.

The output to inkjet/laser printers, plotters and digital devices are straight forward. The resolution of the final image to digital devices may be reduced to accommodate bandwidth and the speed at which the image will be rendered on the device. However with offset printing there are few more steps that are required before final printed map is used by the map reader (Gruver and Dutton 2017).

Figure 19.4 illustrates these steps. The preparation part has been discussed above and illustrated in Fig. 19.3 and ends with the output. The output for offset printing is the color separated image for a specific map. The map is color separated into black, cyan, magenta and yellow (Gruver and Dutton 2017). The color separation can be done the cartographic entity that created the map themselves or it can be done at the printing entity. This the preparation of the digital positive image part in Fig. 19.4.

Before any offset printing can be done, there several steps that needs to be followed that is known as the pre-press part of the printing process. There are two streams available for pre-press activities as shown in Fig. 19.4.

The image-setter route is where the image setter is used to create a negative image for each color. This is used to create a color proof that is used for reviewing and finally approved for printing. Once it is approved plates are made for each color and the final maps are printed.

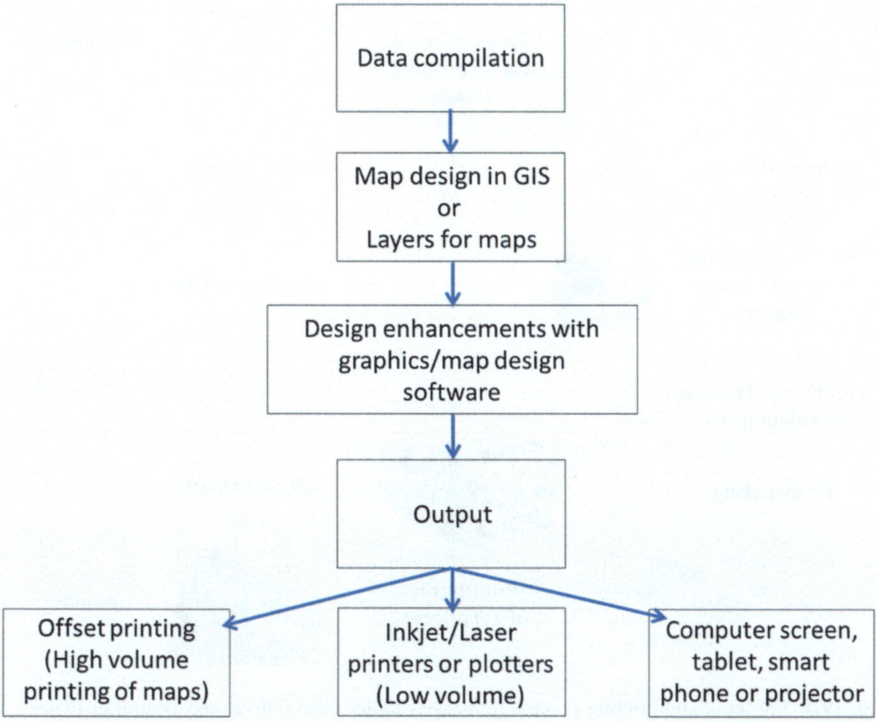

Fig. 19.3 High level map production process (enhanced figure from Gruver and Dutton 2017)

The direct-to-plate route is where the proofed digital image is directly created on a printing plate for each color and used to print the final map. It is important to note that the direct-to-plate route does not allow for final color proofing before the printing run commences (Gruver and Dutton 2017).

The post-printing process involves the preparation of the printed maps for shipping, storing and distribution.

The map production process will be used in a supply chain context to manage and model decentralized map production. The next section gives an overview on supply chains and introduces the SCOR model.

19.4 Supply Chains and Supply Chain Management

19.4.1 Definition

A simple supply chain consists of suppliers, manufacturer and customers (Mentzer et al. 2001). Schmitz (2007) adapted the supply chain definition of Handfields and Nichols (1999) for Geographic Information Systems as follows: "The supply chains

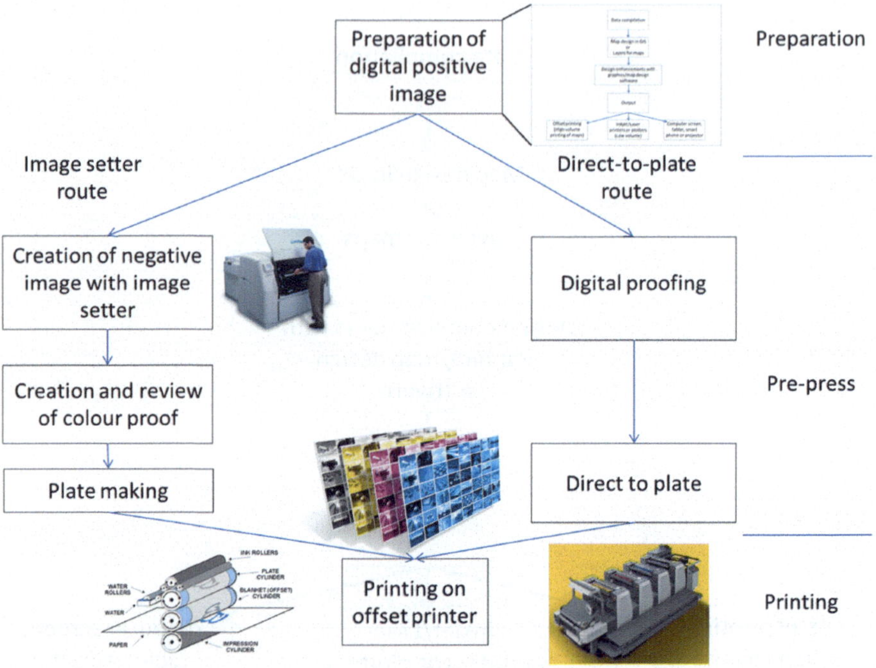

Fig. 19.4 Pre-press and printing process (enhanced figure from Gruver and Dutton 2017)

encompass all activities associated with the flow and transformation of spatial and attribute data from the raw data stage (capturing), through to the end user, as well as the associated information and money flows. Data, information and money flow up and down the supply chain." The same definition applies to cartographic supply chains. Christopher (2016) mentioned that the supply chain became the value chain, meaning value is constantly added to the product as it moves along the supply chain. For the purpose of this book chapter the term "supply chain" will be used to include the value chain. This is also applicable to the cartographic process, namely the value addition occurs along the cartographic supply chain. Figure 19.5 gives an example of a generic high level cartographic supply chain.

The cartographic unit can be its own supplier of data, have its own printing unit, distribution center and retail outlet or it can be completely decentralized or a mixture between the two. This chapter describes the decentralized map production option. Whether the complete supply chain is within the cartographic unit, completely decentralized or mixture thereof, it needs to be managed to function as a whole.

Christopher (2016) defines supply chain management as follows: "The management of upstream and downstream relationships with suppliers and customers in order to deliver superior customer value at less cost to the supply chain as whole".

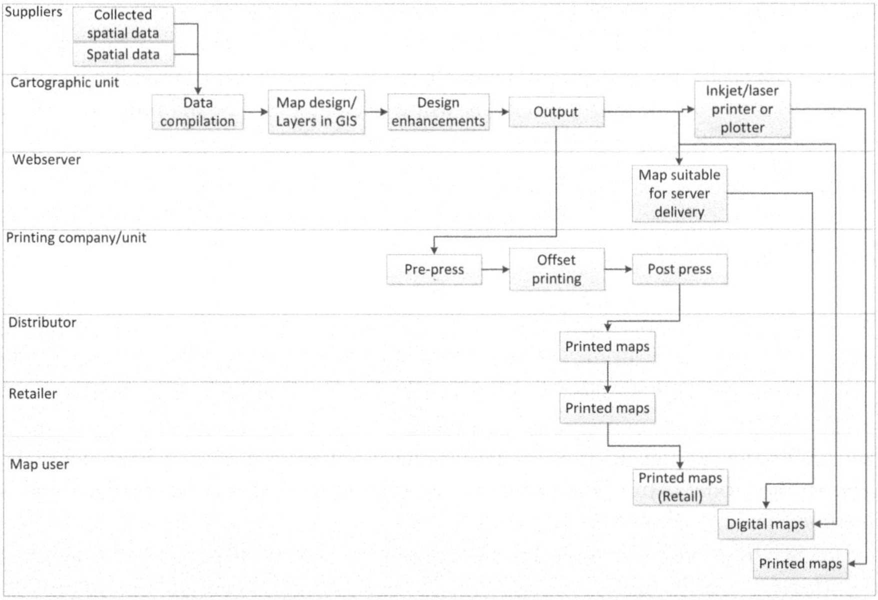

Fig. 19.5 Generic high level cartographic supply chain

In order to do effective supply chain management with the various components in the supply chain as given in the definition and the generic high level cartographic supply chain in Fig. 19.5 it is important understand the supply chain. Modeling a supply chain is one of the methods to understand a supply chain. Bolstorff and Rosenbaum (2003) mention that if you can map (model) the supply chain you can manage and measure the supply chain.

There are several supply chain models available such as the Global Supply Chain Forum's model which ranges from second tier suppliers, to first tier suppliers, to the manufacturing firm, to the customer ending with end user which is the customer's customer. The manufacturer and some of the other members have six components namely purchasing, logistics, marketing, production, research and development and finance which are underpinned by eight supply chain business processes ranging from customer relationship management to returns management (Croxton et al. 2001). Perez (2013) developed the Supply Chain Roadmap encompasses key drivers of supply chain strategies, how these interrelate to provide a competitive framework and to position the business competitively and contains profiles of six different supply chain types namely efficient, fast, continuous-flow, agile, custom-configured and flexible supply chains.

The third example of a supply chain model is the Supply Chain Operations Reference (SCOR) model from APICS (APICS 2014). The SCOR model has six management processes namely plan, source, make, deliver, return and enable and covers the supply chain from the supplier's supplier (second tier supplier) to the supplier (first tier) to the firm, to the customer and the customer's customer (end user). The SCOR model will be used in this chapter since it is a flexible and intuitive model to model a supply chain. The SCOR model can be applied to very simple to very complex supply chains using the same set of definitions (APICS 2014). The SCOR model will be discussed in more detail in the next section.

19.4.2 SCOR Model

The first SCOR model was created in 1996 by the then Supply-Chain Council which was a not for profit organization. The current SCOR model is version 11 and is distributed by APICS. The SCOR version 11 will be used in this chapter.

The SCOR model consists as mentioned in the previous section of six management processes namely Plan, Source, Make, Deliver, Return and Enable and the scope of the model is from the supplier's supplier to the customer's customer and is illustrated in Fig. 19.6 (APICS 2014).

The SCOR model has three levels of analysis namely at process management level (Level 1) as illustrated in Fig. 19.5. These levels are then divided into process categories (Level 2) and each process category is further divided into process elements (Level 3). Level 4 are activity specific and these change from entity to entity and thus outside the scope of the SCOR model (APICS 2014). For the purpose of this chapter to illustrate the use of the SCOR model in decentralized mapping will be up to Level 3 since Level 4 is specific to a specific cartographic unit.

Level 1 gives the scope of the supply chain, in the context of this chapter decentralized map production. The performance targets are set in Level 1 such as costs, reliability, asset management efficiency, responsiveness and agility (APICS 2014). Level 2 process categories are used to define the operations strategy of the cartographic entity and to configure the supply chain. In the SCOR model there are three product streams namely make-to-stock, make-to-order and engineer-to-order.

Make-to-stock are items made in bulk according to established standard operating procedures and production processes such as beer, soft drinks, pharmaceutical products and bulk printed maps. Make-to-order products also have standard operating procedures and production processes in place but will only do a production

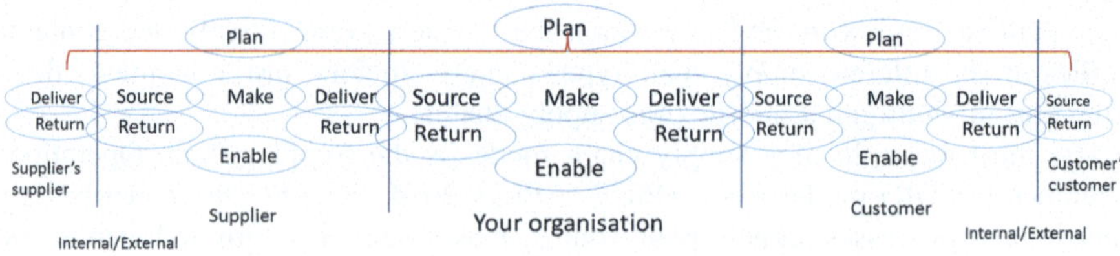

Fig. 19.6 SCOR model's management processes and scope (APICS 2014)

run once an order has been received. High-end luxury motor vehicles are examples of make-to-order items. In the realm of map production make-to-order maps are print-on-demand maps. Engineer-to-order products are products designed and produced for a specific reason based on very specific requirements such as bridges, concept vehicles and once-off specially designed maps. For the purpose of this chapter, bulk printed maps such as topographic maps, geological maps or nautical charts are make-to-stock products.

Level 3 process elements are used to configure Level 2 process categories, in this instance all the process elements required for producing maps in bulk. Process elements focuses on processes, inputs and outputs, performance, practices, technical and technology capabilities and staff skills (APICS 2014). The next section will briefly discuss some examples where SCOR was used to produce maps. The SCOR model concentrates on four major sections, namely supply chain performance, the processes as discussed above, each process have a set of best practices garnered from participating industries and people (APICS 2014). The latter consists of required skills to perform supply chain activities and it links with concept of lean supply chains where people are required to undergo continuous learning in order to improve the supply chain and their roles within the supply chain as well as respect for people (Jacobs and Chase 2011).

19.4.3 Examples

Three examples of the use of the SCOR model in Geographic Information Systems (GIS) and map production will be presented in this section. The first example looks at improving the GIS unit's efficiency to deliver GIS products (Schmitz et al. 2007), the second example is the use of SCOR to produce forensic maps for a fraud case (Schmitz 2016) and the last example is to map the production of nautical charts using the SCOR model (Schmitz et al. 2011).

19.4.3.1 ESI-GIS GIS Supply Chain

ESI-GIS is one of the GIS units in Eskom. Eskom is the state entity in South Africa that provides electricity to industry, rural areas, farming communities and municipalities. This was part of the author's Ph.D. research to improve the GIS unit's effectiveness and efficiency using the SCOR model. To do this study the SCOR model was slightly adjusted to accommodate spatial data maintenance, either as maintain-to-stock or as maintain-to-order. The SCOR model was used to determine ESI-GIS data creation and data maintenance processes. As shown in Fig. 19.7.

This was then mapped at Level 3 process elements and a disconnected analysis was run to identify areas of improvement within the supply chain. The savings in cost in 2007 was equivalent to junior GIS person's salary which enabled the unit to recruit the junior GIS person without having to budget for an extra person. The

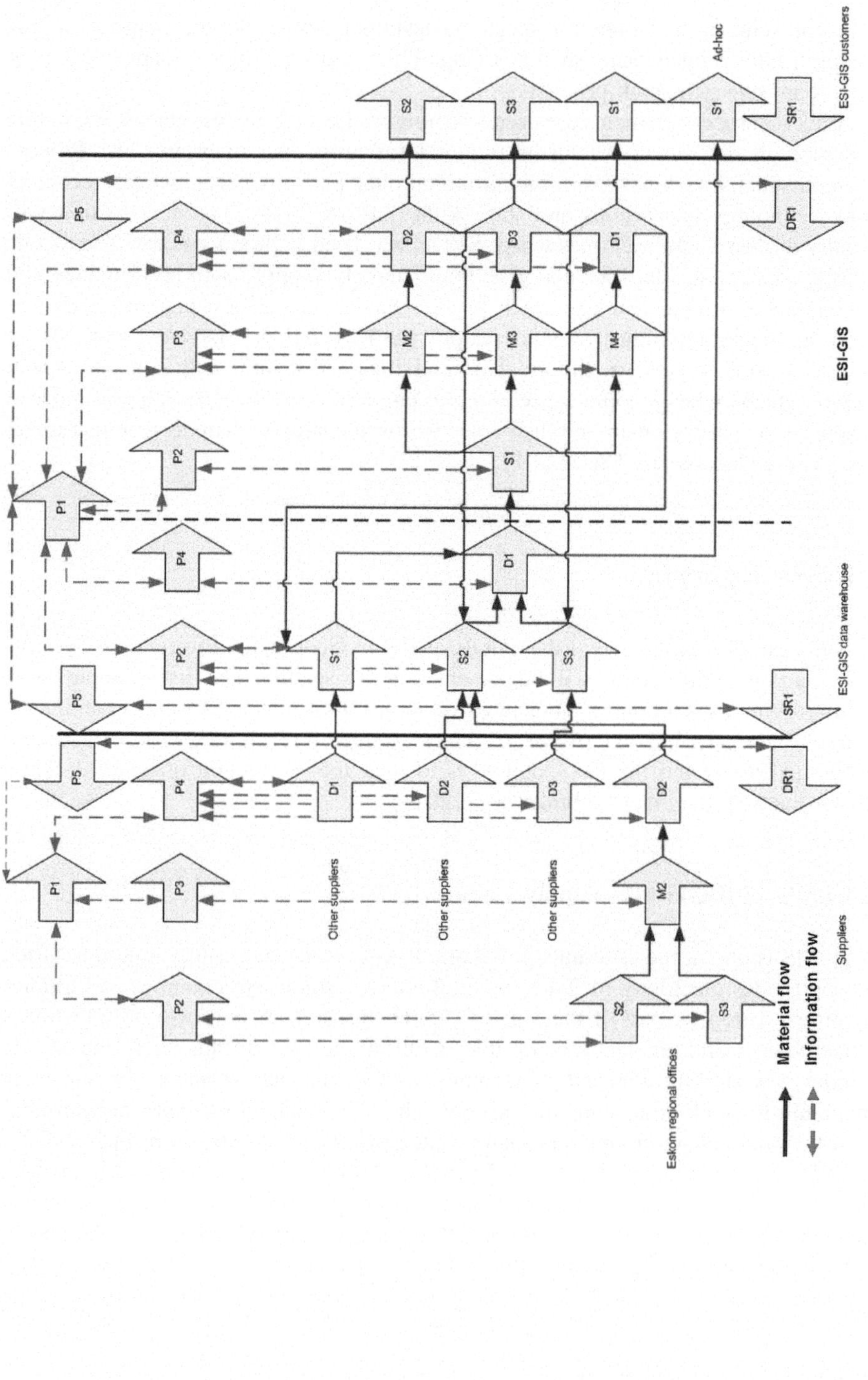

Fig. 19.7 ESI-GIS supply chain (Schmitz 2007)

main identified disconnect was the lack of skills and project management. With regards to skills, the personnel working with GIS had tertiary qualifications but lacked specific skills related to the use in GIS within the electricity distribution field and after attending several training courses the time spent on GIS projects were reduced (Schmitz 2007; Schmitz et al. 2007).

19.4.3.2 Producing Maps for Forensic Purposes

The second example is the production of maps that were used in a fraud case. During the last day of December 2011 and the first two days of January 2012 42 million Rand (USD 3,000,000 at R14.00 to the US dollar) were fraudulently transferred into faked accounts and then withdrawn from ATMs located in three of the nine provinces in South Africa. Figure 19.8 shows the supply chain using the SCOR model to model the production process from obtaining the raw data to the delivery of the final product as a report containing the maps.

Figure 19.9 shows a 3D map where the handover of the device to hack into the bank's mainframe was handed over to IT specialist that was co-opted into the crime syndicate (Schmitz 2016).

19.4.3.3 The Printing of Nautical Charts

This example is the closest to context of this chapter which is the use of the SCOR model to model the bulk printing of nautical charts on paper. The production and maintenance of nautical charts both digital and on paper falls under the auspices of the SA Navy Hydrographic Office located at Silvermine, Cape Town. Figure 19.10 shows the thread diagram at SCOR Level 2 for nautical paper charts from the various data suppliers through to verification, design, printing and selling the nautical charts. Printing on-demand maps are made when new information is added before the next cycle of chart production starts. Printing on-demand has been included in this supply chain example.

For the SCOR Level 3 a form of IDEF0 method had been used to describe inputs, outputs, enablers, controls and responsibilities for each process element. IDEF is the acronym for Integrated DEFinition method (IDEF 2017). The IDEF0 method uses box and arrow graphics to model the process and consists of five elements namely the function, inputs, outputs, mechanisms and controls (IDEF 2017). Figure 19.11 shows the sourcing (S2) of the survey data from the ship SAS Protea using the adapted IDEFo form at process element level.

Figure 19.12 gives an example of SCOR level 4, which is outside the scope of the SCOR model, to illustrate the process for verifying the survey data received from the survey vessel SAS Protea. For the purpose of this chapter Level 4 will be excluded since it will be a generic example to illustrate the use of supply chains and the SCOR model in decentralized map production which will be discussed in the next section.

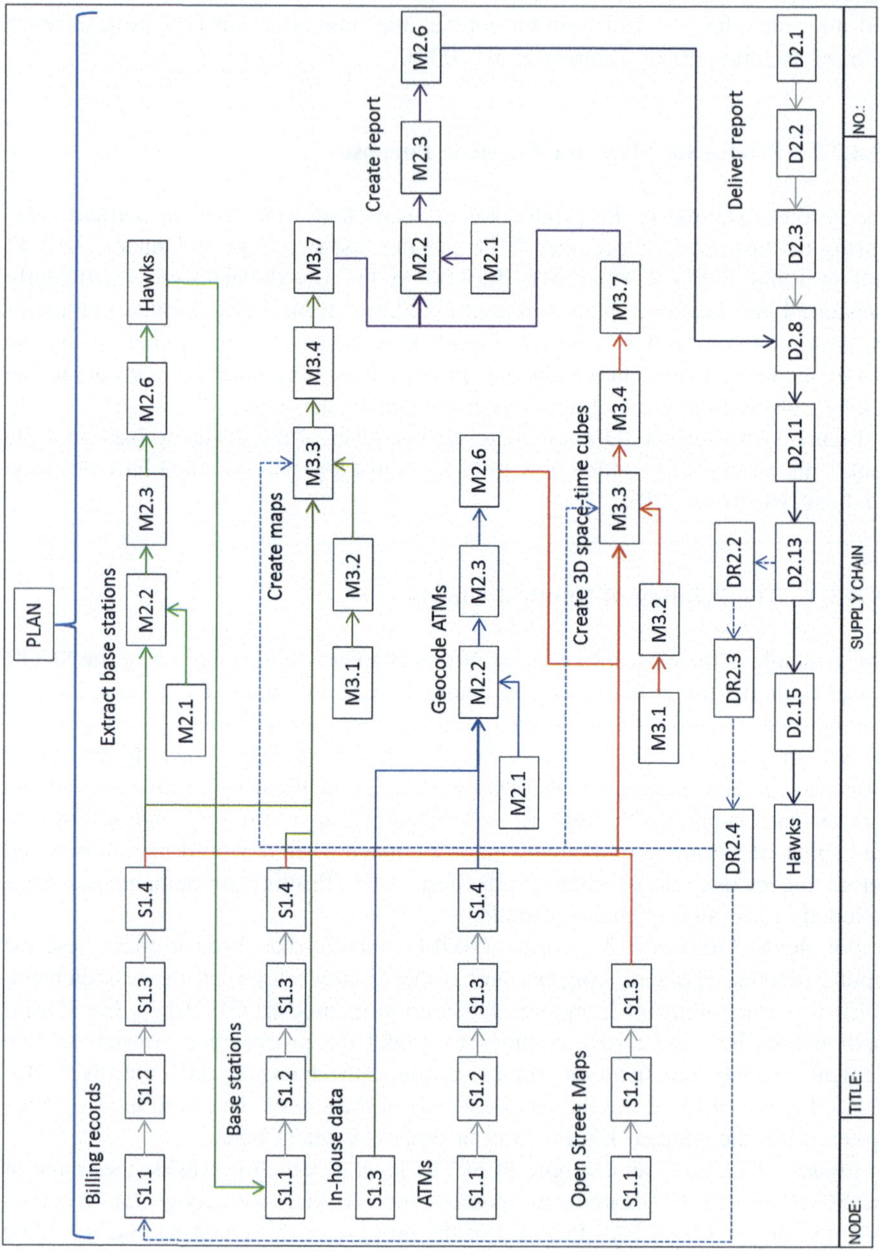

Fig. 19.8 The Forensic map production supply chain (Schmitz 2016)

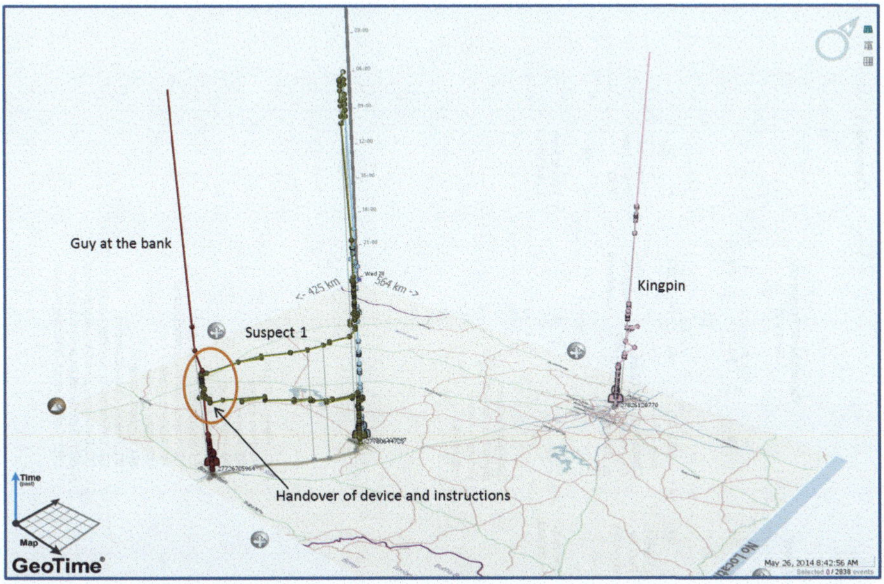

Fig. 19.9 The handover of the hacking device

19.5 SCOR Model and Decentralized Map Production

This section looks at the application of the SCOR model supply chain for decentralized map production. Figure 19.13 shows the various possible combinations for decentralized map production. It is based on alternative production locations within a global supply chain by Meixell and Gargeya (2005). The red thread in Fig. 19.13 is the decentralized map production supply chain used in this chapter.

There are three main groups of entities in this supply chain namely the cartographic unit, internal suppliers and service providers that are involved in producing a map that is given or sold to customer. The internal entities are within the same organization as the cartographic unit but not integrated within the cartographic unit, meaning these entities provide a service to other entities within the organization. External suppliers and service providers are completely outside the cartographic unit and/or the organization in which cartographic unit is located. These entities are engaged when needed to produce maps for the cartographic unit.

The example is a combination of these groupings that are engaged in decentralized map production. The data is sourced directly from the database within the cartographic unit, from various internal sources and external suppliers. The cartographic unit does the data compilation which includes verification and manipulation and gets the data ready for map design and final GIS layers which follows the data compilation phase. Once the final map has been designed it is sent to design enhancement section of the organization to which the cartographic unit belongs to. This section uses more sophisticated software such as Adobe or similar graphics

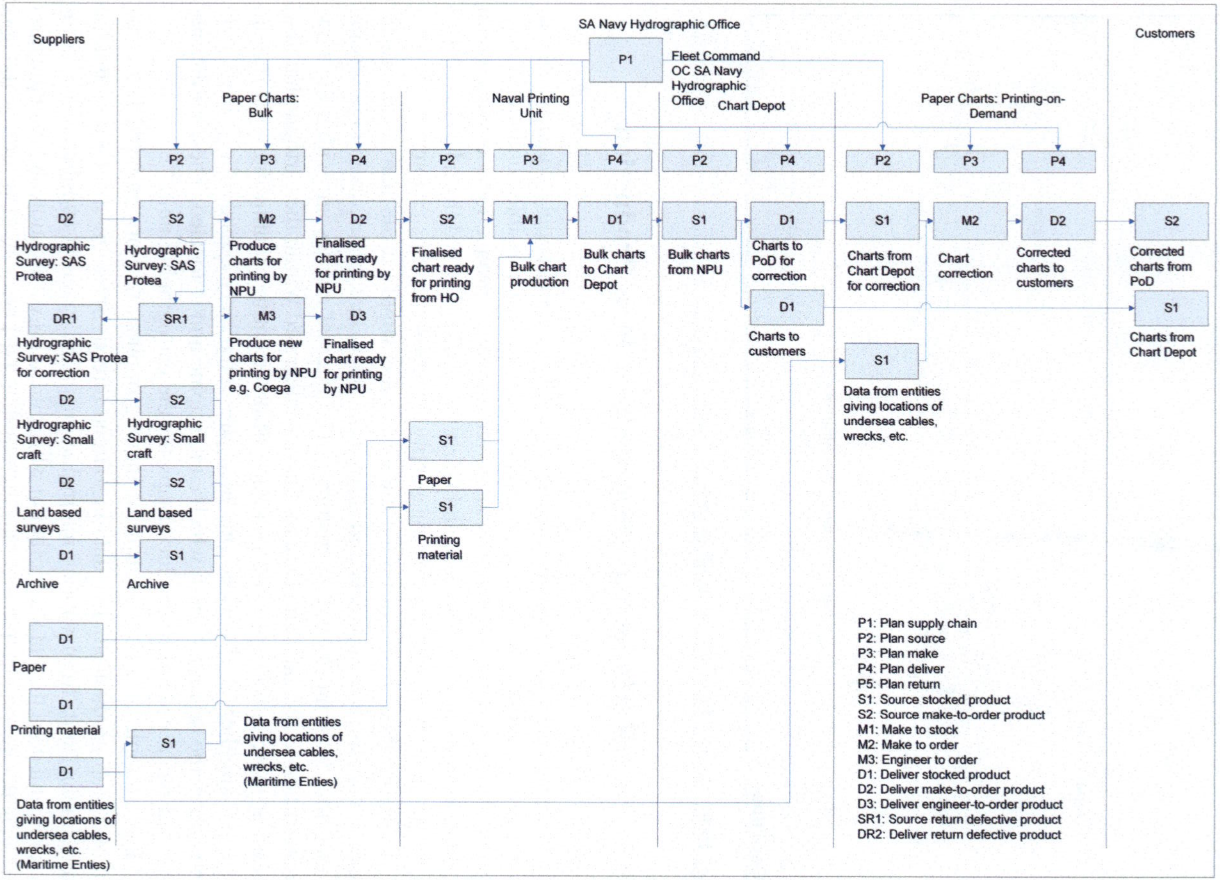

Fig. 19.10 Nautical paper chart production supply chain (reworked figure from Schmitz et al. 2011)

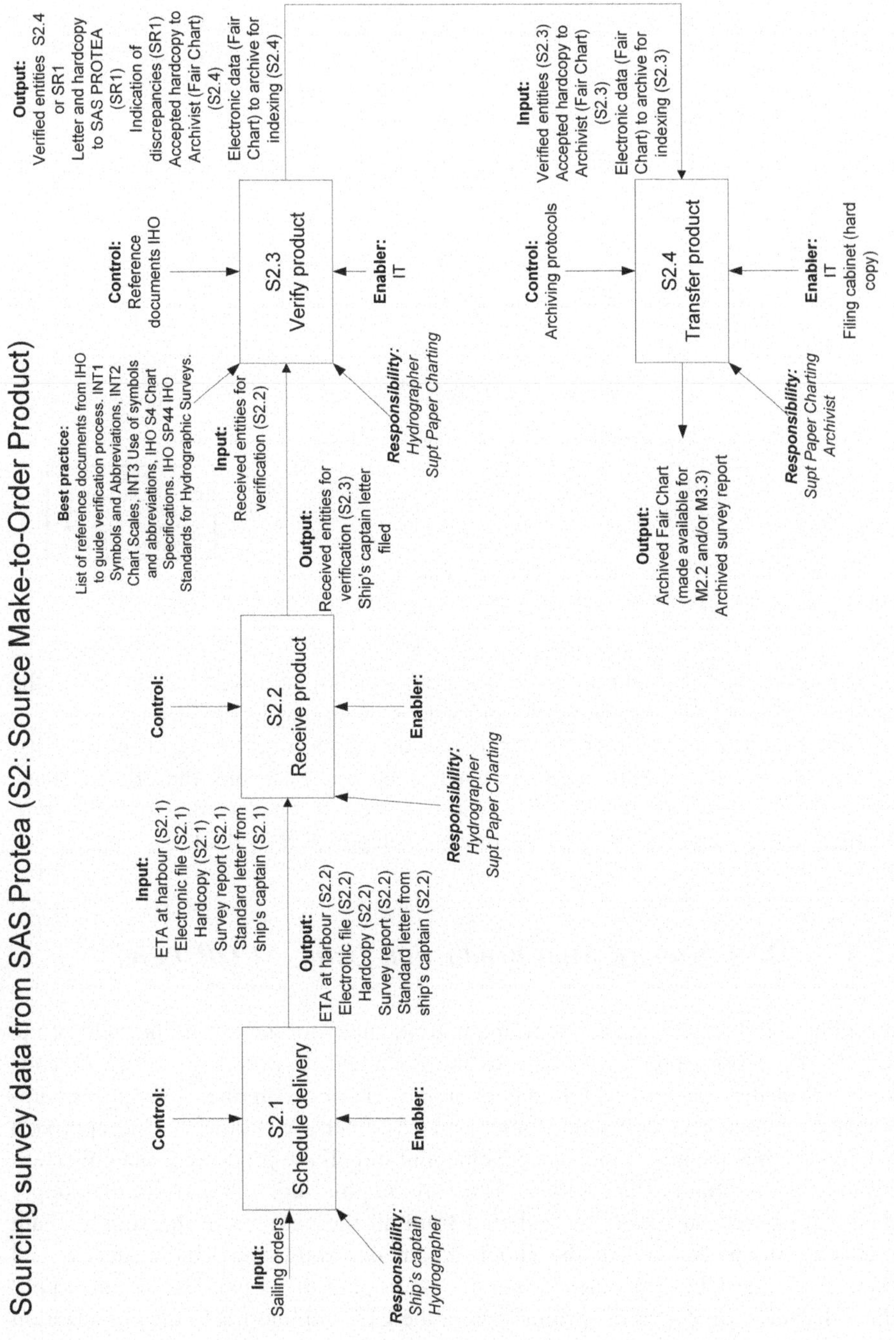

Fig. 19.11 Sourcing survey data from the survey vessel (Schmitz et al. 2011)

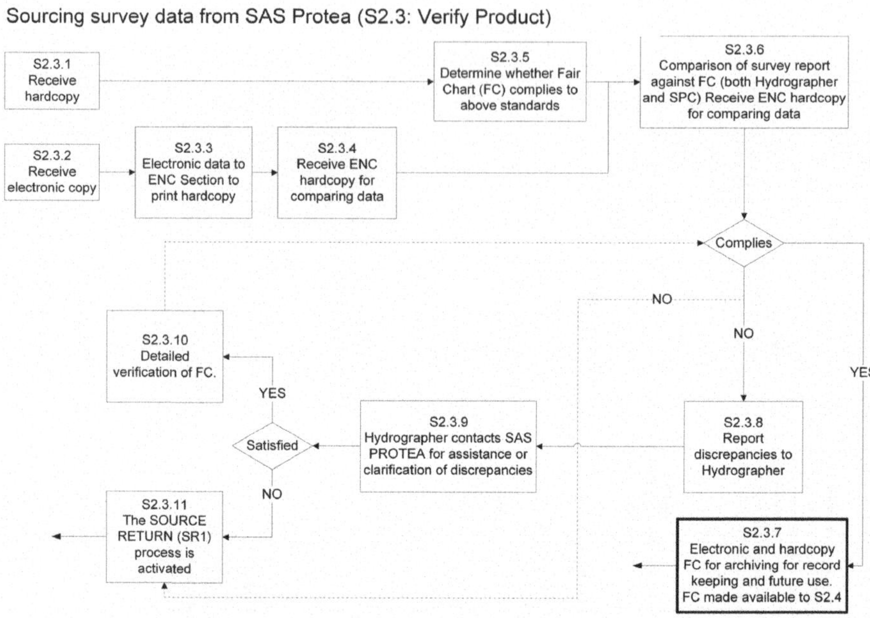

Fig. 19.12 SCOR Level 4 processes to verify survey data (Schmitz et al. 2011)

design software. The final product is an output which if approved will be sent to the printers for offset printing using the process as illustrated in Fig. 19.4.

The printed maps are then delivered to a distribution center which in turn distributes the maps to various retailers where map users can buy the printed maps. Figure 19.14 illustrates the SCOR version of the chosen decentralized map production supply chain using SCOR's process categories (Level 2).

19.5.1 Decentralized Map Production Using SCOR Level 2

Working Fig. 19.14 from the top to the bottom and from the left to the right of the figure: The cartographic unit manages the complete supply chain which extends further than the standard SCOR model supply chain of supplier's suppliers, suppliers, manufacturing entity, customer and customer's customers. This approach would have been suited if the cartographic unit did all the processes data collection through to distribution themselves. The cartographic unit (CU) plans the supply chain (P1) based on formation gathered from all the partners in the supply chain except the map user. Once the supply chain demands has been identified and planned for the CU distributes its supply chain plan to its partners in the supply chain. Partners are external, sections within the CU, and internal to the organization to which the CU belongs.

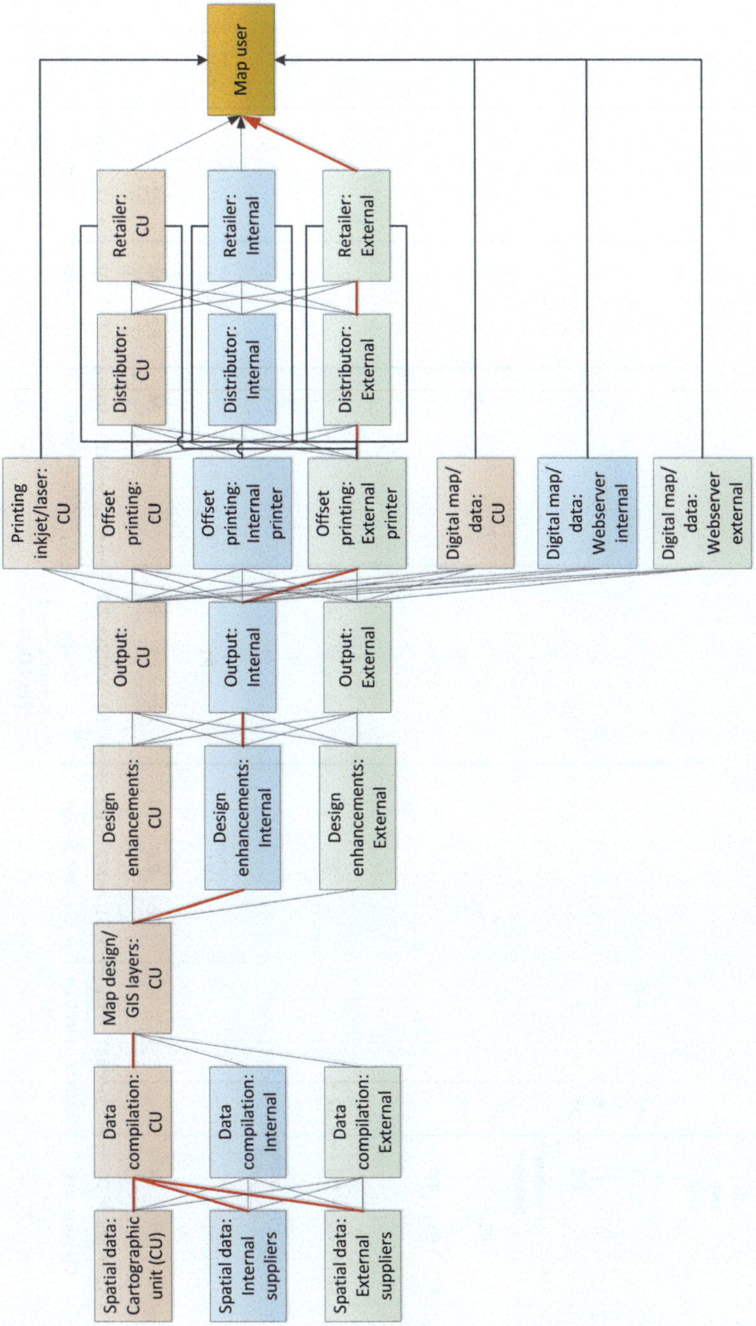

Fig. 19.13 Possible combinations of decentralized map production

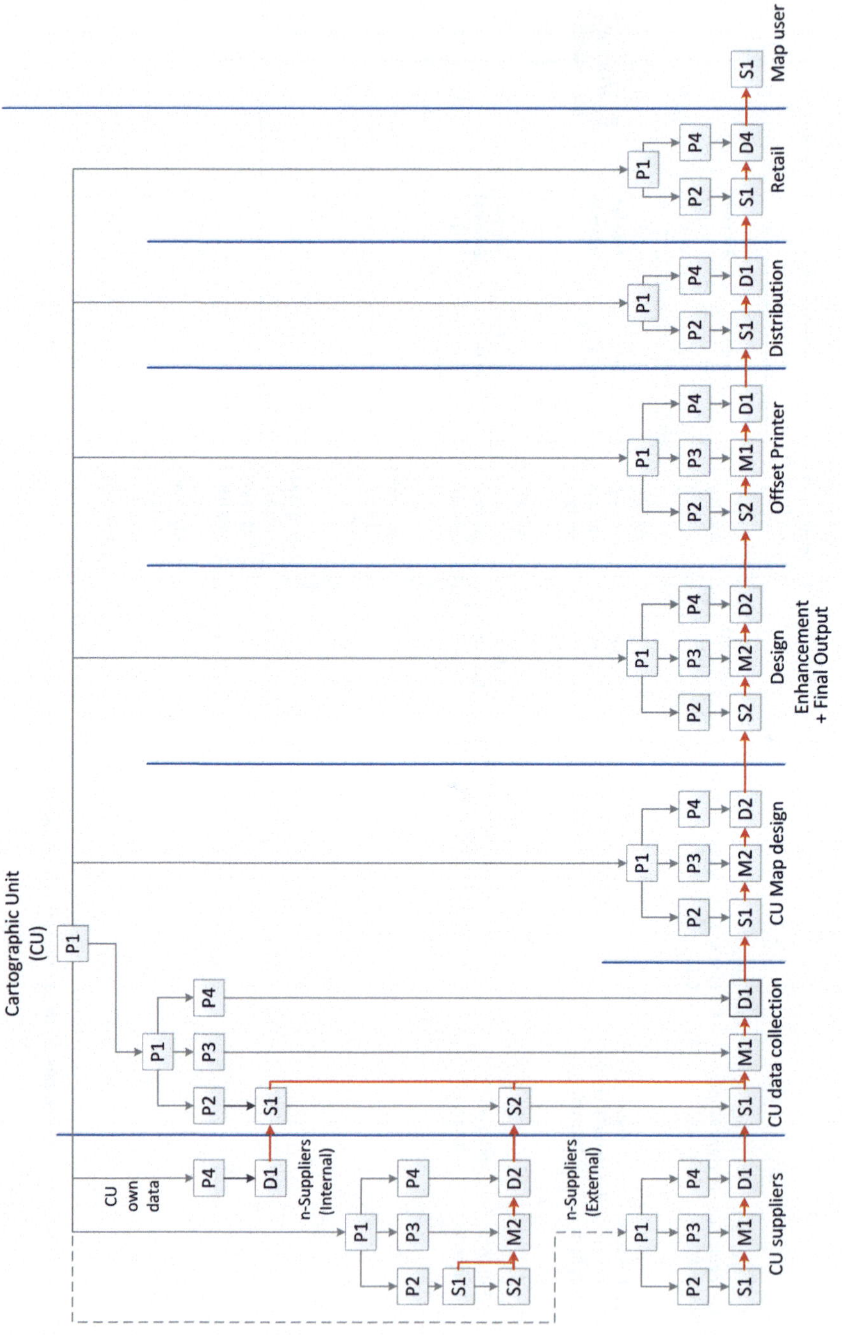

Fig. 19.14 The SCOR modelled decentralized map production supply chain

Each partner then incorporates the demands and plans from the CU into their own supply chain plans (P1) and communicates the plans back to the CU and into their respective Source, Make and Deliver Plans (P2, P3 and P4). The latter three impacts their sourcing, production and delivery of products.

The CU in this example has three types of suppliers, namely its own data sources such as map data from previously produced maps, internal suppliers from elsewhere in the organization the CU belongs to and various external suppliers. Most of the suppliers have stocked data collected for other purposes such as town planning units, transportation entities, etc. These stocked items process categories are S1 (Source stocked item), M1 (Make to stock) and D1 (Deliver stocked items). The Make part of the suppliers is to conform and check the data and make it ready for use by the CU. Some of the suppliers are delivering Make-to-order (M2) data which means that the spatial data will be created once an order has been received using standing operating procedures. The related process categories are Source Make-to-order product (S2) and Deliver Make-to-order product (D2).

The CU's data collection section collects the data from various suppliers as S1 and S2 products and prepares the data for use by the map designer. This data can also be used by other entities within the CU or organization hence the product is made to stock (M1) and is delivered as a stocked item (D1). The map designer will only create a map (M2) based on the demand on the supply chain using standard operating procedures such as which will be used, standard symbology, colour schemes, etc. to design the map for a specific purpose. The data are stocked items that are sourced (S1) and the final design is then delivered as a make-to-order product (D2) to the section that is outside the CU but within the CU's organization to enhance the map design for final output that will be delivered to the external printing unit for offset printing. The enhancement of the map is also a make-to-order product hence the S2, M2 and D2 process categories.

The printing unit receives the make-to-order final product (S2) for bulk printing of the maps which is then a make-to-stock process (M1) and delivers the maps to the distributors (D1). The distributors receive the printed maps (S1) and distribute them to the various retailers (D1). The retailers receives the maps from the distributors (S1), displays them and sells (D1) them to the map users (S1). Selected examples of the SCOR model level 3 process elements based on the mapped supply chain will be discussed next.

19.5.2 Selected SCOR Level 3 Process Elements

Figure 19.15 gives the adjusted IDEF0 box-and-arrow graphic that will be used to map the SCOR Level 3 process elements that are linked to the various process categories as shown in Fig. 19.14. The box-and-arrow graphics is an improved version of the graphic used in Fig. 19.11. It consists of six items that have impact on the process, namely inputs, outputs, controls, enablers, responsibilities and metrics.

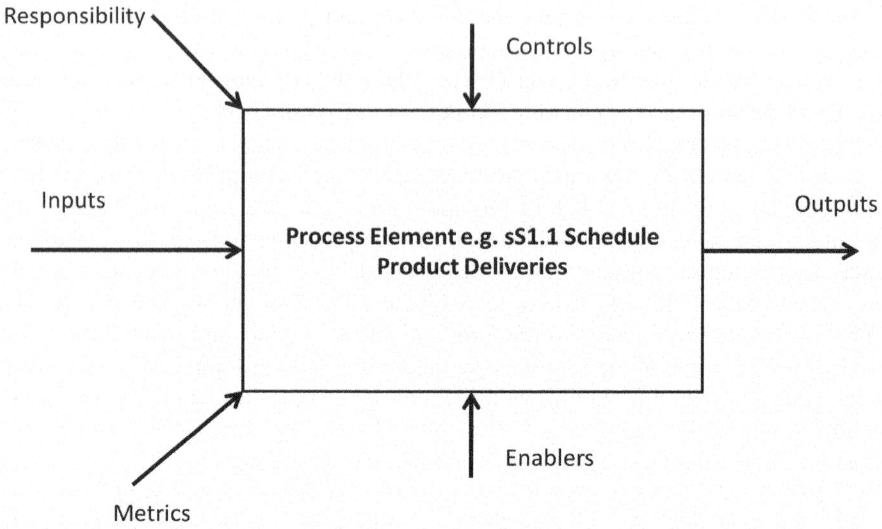

Fig. 19.15 The adapted IDEF0 box-and-arrow graphic to map SCOR Level 3 process elements

The example in Fig. 19.15 will be explained using SCOR v11 (APICS 2014) and is shown in Fig. 19.16 as adapted for map production. The example shown in Fig. 19.16 is sS1.1 Schedule Product Deliveries. The *responsible* person is the cartographer who is responsible for collecting the various spatial data sets required to produce the map. The *inputs* to this process element is the sourcing plans established in sP2 (Plan Source) and communicated in sP2.4 (Establish sourcing plans) and the production schedules as planned in sM1.1 and sM2.1 (Schedule production activities). The *outputs* are the various spatial data sets (products) that were ordered and when these will arrive at the cartographic unit for processing. This information is made available to sM1.1 and sM2.1 to alert the production when the spatial data will be available. The *controls* in this instance will be the cartographic unit's procurement processes. The *enablers* are a supplier database from which the cartographer gets all the details of the supplier to place and schedule an order and where applicable any supplier agreements such as service level agreements (SLAs) and memorandums of understanding (MoUs) that may impact on the procurement process. The *metrics* looks at the time spent to place the orders and scheduling the receipts, the cost of doing this and how long will it take the suppliers to deliver the data (suppliers lead times).

The "s" in front of the SCOR Level 2 process categories was introduced in SCOR version 10 to distinguish process categories from other models developed as part of the SCOR suite, namely Design Chain Operations Reference (DCOR) and Customer Chain Operations Reference (CCOR) models (SCC 2010). The "s" was excluded in the various process categories as shown in Fig. 19.14 owing to space issues. The "s" will be included in the subsequent figures and discussions. Figure 19.16 gave an example of mapping a SCOR Level 3 process element.

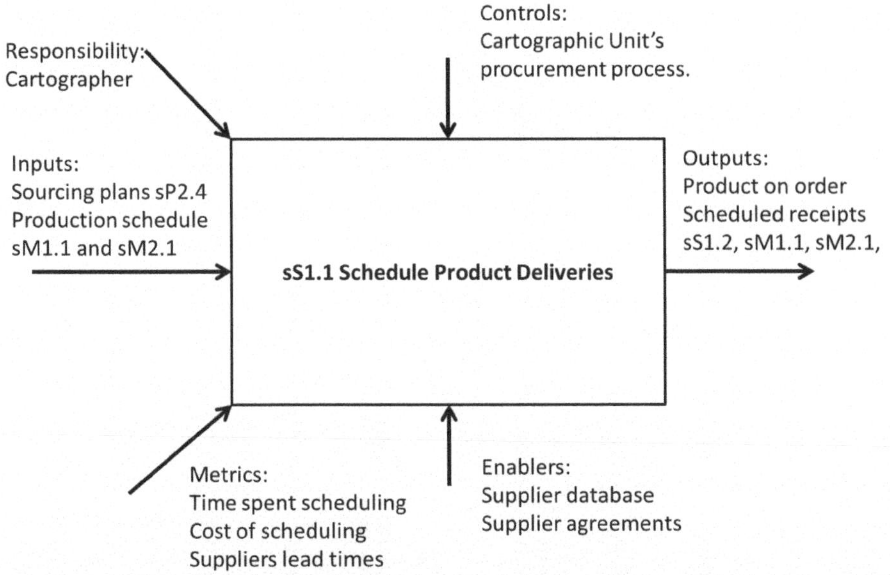

Fig. 19.16 Example SCOR Level 3 process element applied to map production

Figure 19.17 provides the sM2: Make-to-order Level 3 process elements in creating the map design by the cartographic unit as shown at Level 2 in Fig. 19.14.

When using the SCOR model, sM2 Make to order consists of seven process elements, namely sM2.1 Schedule production activities; sM2.2 Issue sourced or in-process product (issue product in Fig. 19.17); sM2.3 Produce and test; sM2.4 Package; sM2.5 Stage finished product; sM2.6 Release finished product to deliver and sM2.7 Waste disposal. One of the advantages of using the SCOR model is that process categories or process elements that are not applicable to a specific supply chain can be discarded, hence the absence of sM2.4; sM2.5 and sM2.7 in Fig. 19.17. The process elements shown in dashed lines are included to indicate the links between Source and Deliver with regards to producing a product (Make).

In the context of map design sM2.1 Schedule production activities, the inputs are the dates when the data will be received from the cartographic unit's data collection section that needs to get data ready for use in the map design. The second input is the production plans for doing the map design. The production plans are synchronized with the date when the data is made available the data collection section. There are currently no controls identified for this particular activity. The output is the production schedule that is synchronized with the receipts of the various data sets. The production schedule triggers sM2.2 to release the data sets required to design the map. The enablers are the availability of equipment, meaning software and computers and personnel to do the map design. If these are not available currently owing to other projects, then the project needs to be rescheduled to accommodate available resources. Certain equipment such as computers and

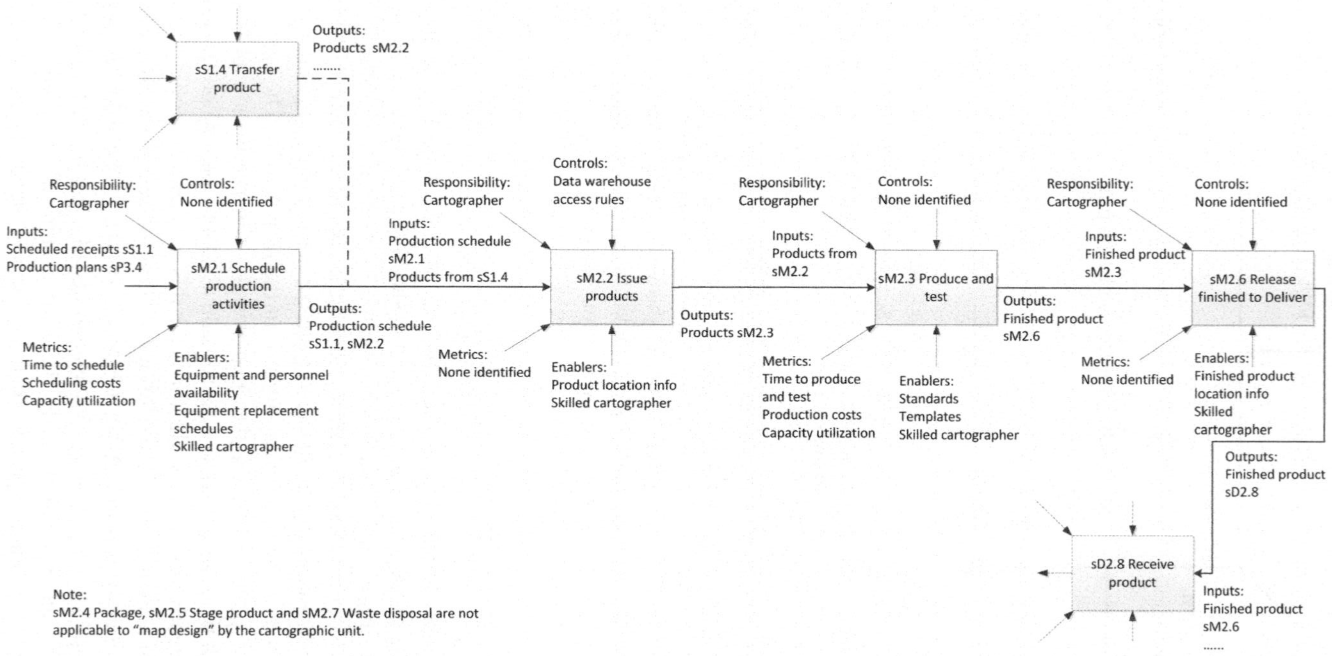

Fig. 19.17 "Map design" sM2 make to order production process at SCOR Level 3

printers have to be replaced after a specified time period and when scheduling a production run, its needs to take this into account as not to delay the completion of the map design. The metrics in this process element registers the time taken to setup the schedule and this can be translated into cost. Meaning if the scheduling of the production takes an hour and the hourly rate of the cartographer is 20 USD which includes direct and indirect costs, then the cost to schedule is 20USD × 1 h = 20USD. Capacity utilization is linked with availability, meaning if all personnel and equipment are utilized the schedule needs to be adjusted to the time when capacity is available to do the map design.

sM2.2 Issue products is the process where the cartographer downloads the required data to design the map. The data is made available via sS1.4 Transfer product which is showed as dashed lines to indicate where the link is between Source and Make. The controls for sM2.2 are the data warehouse access rules and the enabler is data location information within the data warehouse such directory and folder location information. The outputs are respective data layers that will be used in the design of the map. sM2.3 Produce and test is the process element where the actual production takes place. In this example it is the designing of the map using the data obtained from the data warehouse.

The responsibility is the cartographer's; the inputs are the data files; no controls were identified for the process; the enablers are the various standards and symbols required to create the map and the output is the designed map. The testing part of this process is the quality control applied by the cartographer to ensure the map is of a high standard. The metrics for sM2.3 are the time needed to design and test the map; the costs to design and test the map and capacity utilization. The latter looks at the percent usage of the equipment during the map design.

The last process element is sM2.6 Release finished product to deliver where the cartographer is going to deliver the designed map for map enhancement which is an internal "customer" of the cartographic unit. Map enhancement is done by a different unit with the same organization to which cartographic unit belongs to. The designed map is released to process element sD2.8 Receive product. sD2.8 is shown as dashed lines to indicate the link between Make and Deliver within the modelled supply chain. The input to sM2.6 is the designed map and the output is the map that will be delivered to the map enhancement unit. No controls and metrics were identified for this process element. The enabler is the location information from which the designed map will be downloaded from when it is delivered to the map enhancement unit. These SCOR Level 3 processes are mapped for each process category identified for each entity within the total supply chain as shown in Fig. 19.14. Once the whole supply chain has been mapped at SCOR Level 3 it can be translated into a spreadsheet to collect data and information for further analysis. Table 19.1 shows the metrics and costs for sM: Make-to-Order as indicated in Fig. 19.17

The mapping of each and every process category is outside the scope of this chapter. The aim of this chapter is to illustrate the use of the SCOR model for decentralized mapping. The conclusions and recommendations are discussed in the next section.

Table 19.1 Metrics and costs for sM2: Make-to-Order

sM2 Make-to-Order		Hours (planned)	Hours (actual)	Expenditure (planned)	Expenditure (actual)	Amount (planned)	Amount (actual)	Income (planned)	Income (actual)	Comment
sM2.1 Schedule production activities										
Responsibility	Cartographer: Jason Bourne									Staff rate: R475.00 per hour
Inputs	Scheduled receipts from sSl.1									
	Production plnas from sP3.4									
Outputs	Production schedule to sSl.1 and sM2.2									
Controls	None identified									
Enablers	Equipment, product and personnel availability									
	Equipment replacement schedules									
Metrics	Time to schedule	1	1							
	Scheduling costs			R475.00	R475.00					

(continued)

Table 19.1 (continued)

sM2 Make-to-Order		Hours (planned)	Hours (actual)	Expenditure (planned)	Expenditure (actual)	Amount (planned)	Amount (actual)	Income (planned)	Income (actual)	Comment
	Capacity utilisation					80%	75%			
sM2.2 Issue products										
Responsibility	Cartographer: Jason Bourne									
Inputs	Production schedule from sM2.1									
	Products from sSl.4									
Outputs	Issued products to SM2.3									
Controls	Data warehouse access rules									
Enablers	Product location information									
Metrics	None identified									In this context, it is part of sM2.3 produce and test with regards to time and cost

(continued)

Table 19.1 (continued)

sM2 Make-to-Order		Hours (planned)	Hours (actual)	Expenditure (planned)	Expenditure (actual)	Amount (planned)	Amount (actual)	Income (planned)	Income (actual)	Comment
sM2.3 Produce and test										Actual design of map
Responsibility	Cartographer: Jason Bourne									
Inputs	Issued products from sM2.2									
	Finished product: Designed map to									
Outputs	sM2.6									
Controls	None identified									
Enablers	Standards									
	Templates									
Metrics	Time to produce and test	64	60							
	Production costs			R30,400.00	R28,500.00					
	Capacity utilisation					90 K	90%			

(continued)

Table 19.1 (continued)

sM2 Make-to-Order		Hours (planned)	Hours (actual)	Expenditure (planned)	Expenditure (actual)	Amount (planned)	Amount (actual)	Income (planned)	Income (actual)	Comment
sM2.6 Release finished product to deliver										
Responsibility										Cartographer: Jason Bourne
nputs	Finished product from sM2.3									
Output	Finished product to sD2.B									
Controls	None identified									
Enablers	Finished product location information									
Metrics	None identified					1 n this context, it is part of sM2.3 produce and test with regards to time and cost				
sM2 Metrics and costs		65	61	R30,875.00	R28,975.00			R0.00	R0.00	

19.6 Conclusion

Maps have been part of humanity over the millennia as discussed in the introductory part of this chapter. The production of maps changed drastically over the last few centuries ranging from a single map to mass production of maps for various uses. In the beginning of mass production the maps were designed and printed in-house. Most government sponsored maps such as topographic map series, nautical charts and geological maps are examples of in-house designed and printed maps.

With the change in technology, the Internet and the Cloud it is now possible to do decentralized map production. This chapter described the use of the SCOR model to map decentralized map production's supply chain using a selected supply chain from various possibilities that were illustrated in Fig. 19.13. Figure 19.14 showed the selected supply chain using the SCOR model and Fig. 19.17 illustrated the modeling for a specific process category at SCOR Level 3 process elements. The advantage of using the SCOR model is that it is not software dependent and it is flexible in the sense that certain management processes (Level 1), process categories (Level 2) and process elements (Level 3) can be discarded when it is not applicable to a specific supply chain as illustrated in Figs. 19.14 and 19.17.

The SCOR model as illustrated in Figs. 19.14 and 19.17 and Table 19.1 can be used for gathering data to monitor and manage map production and based on the data collected to improve the supply chain for future decentralized map production.

This chapter introduced the concept of supply chains and decentralized map production and further research is required to investigate and analyze various map producers such as National Geo-spatial Information (NGI) in South Africa, Ordinance Survey in the United Kingdom, etc. to develop best practices, improved processes, skills required, etc. This information can then aid future map productions to be more efficient and cost effective without losing the quality of generated maps.

References

APICS (2014) Supply chain operations reference model version 11.0. APICS 8430 West Bryn Mawr Ave, Suite 1000, Chicago, IL 60631, USA

BBC (2014) Production methods. GCSE Bitesize. http://www.bbc.co.uk/schools/gcsebitesize/business/production/methodsofproductionrev1.shtml. Accessed 20 Apr 2017

Bolstorff P, Rosenbaum R (2003) Supply chain excellence: a handbook for dramatic improvement using the SCOR model. AMACOM (American Management Association), New York

Christopher M (2016) Logistics and supply chain management. Pearson Education Limited, United Kingdom. ISBN 9781292083827

Croxton KL, Garcia-Dastugue JG, Lambert DM, Rogers DS (2001) The supply chain management process. Int J Logistics Manage 12(2):13–36

Garrehy P (2015) Centralized versus decentralized manufacturing and where cloud ERP fits in. http://www.rootstock.com/erp-blog/centralized-vs-decentralized-manufacturing-where-cloud-erp-fits-in-part-2/#forward. Accessed 10 Apr 2017

Grough (2006) New maps issued for scottish explorers, 5 Sept 2006. https://www.grough.co.uk/magazine/2006/09/05/new-maps-issued-for-scottish-explorers?printview=1. Accessed 20 Apr 2017

Gruver A, Dutton JA (2017) e-Education Institute, College of Earth and Mineral Sciences, The Pennsylvania State University. https://www.e-education.psu.edu/geog486/l1_p6.html. Accessed 21 Apr 2017

Handfield RB, Nichols EL (1999) Introduction to supply chain management. Prentice Hall, Upper Saddle River

IDEF (2017) IDEF0 function modeling method. http://www.idef.com/idefo-function_modeling_method/. Accessed 10 May 2017

Jacobs FR, Chase RB (2011) Operations and supply chain management, 13th edn. McGraw-Hill Irwin, New York

Mentzer JT, DeWitt W, Keebler JS, Min S, Nix NW, Smith CD, Zacharia ZG (2001) What Is supply chain management? In: Mentzer JT (ed) Supply chain management. Sage Publications, Thousand Oaks

Meixell MJ, Gargeya VB (2005) Global supply chain design: a literature review and critique. Transp Res Part E 41(2005):531–550

Minhas S, Juzek C, Berger U (2012) Ontology based intelligent assistance system to support manufacturing activities in a distributed manufacturing environment. Procedia CIRP 3: 215–220. https://doi.org/10.1016/j.procir.2012.07.038

Mourtzis D, Doukas M (2014) Design and planning of manufacturing networks for mass customisation and personalisation: challenges and Outlook. Procedia CIRP Robust Manufacturing Conference (RoMaC 2014) 19(2014):1–13

Perez HD (2013) Supply chain strategies: which one hits the mark? Supply Chain Q, Quarter 1, 2013. http://www.supplychainquarterly.com/topics/Strategy/20130306-supply-chain-strategies-which-one-hits-the-mark/. Accessed 25 Apr 2017

Pusic M (2012) Cartography's history timeline: cartography=mapmaking how it evolved over time. https://prezi.com/cjaqnv8zrmnp/cartographys-history-timeline/. Accessed 21 April 2017

Schmitz PMU (2007) The use of supply chains and supply chain management to improve the efficiency and effectiveness of GIS units. Unpublished Ph.D. thesis, University of Johannesburg, Johannesburg

Schmitz PMU (2016) The use of supply chains and supply chain management in the production of forensic maps using data from a fraud case. S Afr J Geomatics 5(2):156–174. https://doi.org/10.4314/sajg.v5i2.5

Schmitz PMU, Kampfer A, Nelson M (2011) Using the supply-chain operations reference model (SCOR) to model the production of nautical charts by the SA Navy Hydrographic Office. In ICC2011_DIGITAL_PROCEEDINGS\Oral Presentations PDF of the 25th international cartographic conference in Paris, France, 3–8 July 2011. http://icaci.org/files/documents/ICC_proceedings/ICC2011/Oral%20Presentations%20PDF/E3-Production%20and%20management%20of%20thematic%20data%20base/CO-458.pdf

Schmitz PMU, Scheepers LGC, de Wit PWC, de la Rey A (2007) Understanding data supply chains by using the Supply-Chain Operations Refernece (SCOR) model. In: Global supply chains: developing skills, capabilities and networks. Proceedings of the logistics research network annual conference 2007, University of Hull, Hull, UK, 5–7 Sept 2007, ISBN 978-1-904564-19-5

Stein JC (2002) Information production and capital allocation: decentralized versus hierarchical firms. J Finan LVII(5):1891–1921

Stevens J, Smith JM, Bianchetti RA (2012) Mapping our changing world. In: MacEachren AM, Peuquet DJ. Department of Geography, The Pennsylvania State University, University Park

Supply-Chain Council (SCC) (2010) Supply chain operations reference model, version 10.0. The Supply Chain Council, Inc. www.supply-chain.org, ISBN 0-615-20259-4

The Economist (2012) A third industrial revolution. In: Special report: manufacturing and innovation. The Economist, 21 Apr 2012

Utrilla P, Mazo C, Sopena MC, Martínez-Bea M, Domingo R (2009) A palaeolithic map from 13,660 calBP: engraved stone blocks from the Late Magdalenian in Abauntz Cave (Navarra, Spain). J Hum Evol 57(2009):99–111

Whitford J, Zeitlin J (2004) Governing decentralized production: institutions, public policy, and the prospects for inter-firm collaboration in US manufacturing. Ind Innov 11(1–2):11–44. https://doi.org/10.1080/1366271042000200439

Wolodtschenko A, Forner T (2007) Prehistoric and Early Historic Maps in Europe: conception of Cd-Atlas. e-Perimetron 2(2):114–116

Chapter 20
Characterizing Potential User Groups for Versioned Geodata

Anita E. Locher

Abstract We explore the characteristics of different user groups for legacy geodata from the perspective of a long term archive. For the sake of this study, legacy geodata has been defined as all digital information useful for map creation including aerial photography, digital elevation models, LIDAR data, vector data bases etc., of which there exists at least one more recent version with the same characteristics. In the context of the ISO standard for open archival information systems (OAIS) potential user groups are called designated communities. The archive is supposed to adapt its service to their profiles and needs, which in the electronic environment includes taking into account their level of knowledge of technical aspects. A future technique, more precisely a Delphi study, has been used to predict the potential user groups and their need for geodata versions. In two rounds, two international Delphi groups have been questioned about user professions, frequency of access, amount of data needed, knowledge of GIS, age of the data they are interested in, preferred data set, scales, snapshot intervals and file formats. The answers allowed us to identify the following user types: geophysicists, commercial users, lawyers, policy makers, emergency response planning teams, architects and geo-related engineers, social scientists, the general public, archaeologists, historians, culture and arts professionals, conservation agents of the built and the natural environment, geodata creators and undergraduate teachers and students. We classified the user types by their characteristics into six clusters. The application of the user profiles showed that the method did not deliver sufficiently detailed answers for complying with all OAIS requirements, but that it was effective for gathering user characteristics which guide archives in strategic decisions about the designated communities they might serve.

Keywords Geodata · Spatial data · User study · User profile · Time series Designated community · Historic data · Legacy data

A. E. Locher (✉)
University of Barcelona, Barcelona, Spain
e-mail: anitalocher@gmail.com

J. Döllner et al. (eds.), *Service-Oriented Mapping*, Lecture Notes in Geoinformation and Cartography, https://doi.org/10.1007/978-3-319-72434-8_20

20.1 Introduction

In this study we will analyse geodata used or produced in the workflow of base map production and their users. Geodata include aerial photographs, digital elevation models, LIDAR data, vector data bases, raster maps etc. It is legacy data when at least one more recent version with the same characteristics exists. When legacy data are stored together with their temporal information time series emerge. A time series is defined as a data set where the compounds (versions) contain and present the same characteristics but represent the elements on earth at different moments in time. Versions exist not only of the map, but of all its by-products and many intermediate process stages.

20.1.1 Value of Historic Geospatial Data

The testimony provided by a map about the real objects on the ground is only true during a certain time. When major or continuous little changes occur, spatial data loses value for the primary user, as most GIS projects are focused on problems that require the use of the most current data (Morris et al. 2009; Morris 2006) and other user groups become interested in the legacy data. The longer the time span a series covers, the more possibilities for analysis it offers (Conway et al. 2013). The main argument for long-term preservation in the geosciences is longitudinal research (Moran et al. 2009; Sweetkind et al. 2006). Time series that are homogenous in the elaboration of their compounds ease analysis of evolution. Metadata can be responsible for a lack of homogeneity in semantics or technical elaboration standards. Nevertheless the value of time series could be reduced due to high requirements of time and human resources to restore sufficient homogeneity for automatic analysis (Comber et al. 2005). Climate change and other kinds of environmental change analysis are among the many subjects that benefit from long-term data preservation (Beruti et al. 2010; Erwin and Sweetkind-Singer 2009; Erwin et al. 2009; Harris 2001; Janée 2008; National Academy of Sciences 2002; Shaon et al. 2012). Understanding change on our planet can help predict natural events (Beruti et al. 2010) and help governments to better administer natural and human resources and manage planning. Some libraries (Erwin et al. 2009; Geospatial Multistate Archive and Preservation Partnership & Library of Congress 2010) have recognised the long-term value of geodata to their users and have engaged in preservation projects. One of the first challenges for these libraries was the selection of appropriate data. Part of the criteria given by the funding bodies for their preservation projects was that the information selected should be at risk and impossible or very difficult to recreate (Cruse and Sandore 2009; Lazorchak et al. 2008). Because change is unavoidable, geodata collected at a certain moment in time cannot be re-collected at a later time. In one case it was argued that almost all digital geospatial data are at risk due to the sheer amount of data produced, making it impossible for one institution to hold (Sweetkind et al. 2006; Erwin and Sweetkind-Singer 2009). It has also been mentioned that many geodata producers overwrite the old data with the new data (Morris 2010) because their primary users are not interested in legacy data or because they cannot handle the sheer amount that would be produced by a regular snapshot of a temporal-spatial data base (Gantner et al. 2013).

As geodata producers, ESA, NOAA and NASA are conscious of the value of legacy data. ESA perceives (Beruti et al. 2010) and Morris confirms (Morris 2013) a rising demand for time series which is why ESA recommends that its members archive geodata.

20.1.2 Problem

Knowing the value of legacy geodata for secondary users and assuming a constant curve of decreasing cost of digital storage space suggest that all data be kept. Nevertheless practical experience of archives shows that selection is needed.

The archive has the duty to make sure the information is safe and accessible by the future user. In contrast to information on paper, access to digital data presents a challenge due to its strong technological aspect. New guidelines have been appended to conventional archival practice to address this challenge. The most accepted standard for electronic archives is the ISO standard 14721, a reference model for an Open Archival Information System (OAIS) (Consultative Committee for Space Data Systems (CCSDS) 2002). In order to conform to OAIS, archives have to define a designated community for whom the contents are destined and adapt the archival system accordingly.

It lies in the nature of digital material that it requires mediation through a machine to be read and interpreted. If technological skills of its designated community are assumed to be weak, more mediation is needed. Mediation is assisted by the representation information; a concept of the OAIS that defines the necessary information to fill the gap between the technological data requirements and the tools and skills available to the designated community. Mediation can be served by the archive through technological assistance or automated rendering processes or by third parties specialized in rendering and serving geospatial data.

To face the new challenge of digital repositories, adapt the appraisal criteria and conform to OAIS, archives have taken added interest in user groups. This study is a proposal for a method to define designated communities and will test the method on legacy spatial data users.

20.2 Method

Digital preservation is a new field and the methods proposed in scientific literature to define the designated community lack tested and documented results. Nevertheless, it is important to analyse the appropriateness of proposals by Giaretta (2011) and Kärberg (2014) for our purpose of assisting appraisal and selection. Giaretta defines the designated community by its knowledge base which he proposes to divide into modules. A module can be software, an object or a skill. He says 'if u is a DC [Designated Community], its profile, denoted by T(u), is the set of modules assumed to be known from that community.' Giaretta's definition is intended to help determine the representation information and check its completeness periodically. Service levels can be built on the knowledge base of the designated community. Kärberg describes a way to define the designated community by web analytics. It includes measurement of three levels of properties: frequency of visit (to the digital archive), means used to find the archives webpage (link, search engine or organic) and length of stay. The combination of these properties results in 27 user profiles. Further analytics such as download statistics provide insight into what objects are most required, such that inferences about needs can be made. The intention of the Estonian project is to build a procedure that is repeatable and as automated as possible to monitor the designated community. Monitoring the designated community and checking understandability of holdings against it is a requirement of the OAIS reference model.

Giaretta's theoretical model does not propose a method and Kärberg's method is not complete enough to build basic knowledge of the user. If we were to rely only on web analytics, the question of 'what' the user requires would be biased towards digital data and we could not find out why and how customers use geodata. Therefore, we turned to market research for its long-standing practice focusing on user experience.

Market research can be performed at several stages in the implantation of a new product. At an early stage, it can help define the characteristics of the product or service. It can also evaluate user reaction to early design. In general, it explores the reaction of the market to the product and searches for a market segment—a user group that shares characteristics or needs.[1] A product or service is optimised for a market segment, as would be the archival service for the designated community. Nevertheless, market research has a short to medium time span, while digital archives must consider the long term. Several recent efforts bridge the gap between market research and the need for foresight over a longer term in an ever-changing society (Malanowski and Zweck 2007; Postma et al. 2007). Following these examples, we used the Delphi technique to get a better insight into prediction of technological changes and the profiles of presumed user groups.

[1]Source: Buisnessdictionnary.com (http://www.businessdictionary.com/definition/market-segment.html).

20.2.1 Delphi

The Delphi method is a predictive technique used when appropriate historic and technical data that could be used for a statistical prediction are missing. The goal of the Delphi method is to reach a consensus of opinion from a group of experts, who answer questions in subsequent rounds. After each round, they receive feedback and are asked to review or argue their opinions. This process is repeated until the differences of opinion become acceptably small for the contracting entity. In a classical Delphi study experts' names are not revealed to each other. This technique allows retrieving knowledge from a variety of sources, while avoiding negative effects of group interaction (Fildes and Allen 2011). The Delphi method was originally developed by the Rand Corporation to gain quantitative results. Nevertheless, it is usually combined with open questions that can be analysed qualitatively. In the first round, open questions are often asked to extract the most important subjects that subsequent rounds will quantify. The Delphi method is particularly well suited to new research areas and exploratory studies (Okoli and Pawlowski 2004). For this study, two international expert groups were formed: group one was for defining the user profiles and group two was for predicting the influences of technological changes on archiving and accessing geodata. Nevertheless, group two also answered some questions about user groups. In a Delphi study the quality of the experts is crucial. Every expert was given the opportunity to rate his or her own answers, according to the perceived understanding of the specific subject. Furthermore, as there were questions about the user groups defined in the first round, we asked how well each expert knew each user group. These two ratings were later used to weight the answers. A pre-question helped define the user groups of legacy geodata. In the first round the experts were asked whether the study should consider additional user types. The propositions resulted in three new user groups that were added in the second round. The first round took place in November 2013 and the second round from February to March 2014. The online survey contained an introduction that provided important definitions; a measure intended to avoid misunderstanding of the questions. Table 20.1 summarizes the questions asked to the experts and their profiles.

For the analysis of the Delphi study results, the answers were exported to spread sheets. Mathematical calculations were applied where possible. Where averages were calculated, preference was given to the mean instead of the median to avoid having outliers bias the result. Because it was not compulsory to answer every Delphi question some questions did not reach the minimum needed number of four answers. Many other Delphi studies recruit only four to five experts (Fildes and Allen 2011) which is why we considered four answers as sufficient.

Table 20.1 Structure of the Delphi study and questions asked to each expert group

	Expert group one	Expert group two
Expert profile	Information providers (Archives, libraries with map collections, and Open access and government data specialists)[a]	Technological experts (developer of GIS, developers of spatial and location services and geodata producers)[b]
Number of participants	12 recruited/10 participated Participation has fallen for the second round	13 recruited/11 participated Participation has fallen for the second round
Pre-question	Which user groups should we consider?	Which user groups should we consider?
Questions related to user profiles	Time frames for data that the user group is possibly interested in Percentage of the user group with GIS knowledge Percentage of access of each user group compared to all access activity on legacy geodata Desired frequency of archival snapshots of the database Ranking of user groups according to the amount of data they might request Type of data sets that are most needed Main activity the user groups conduct with the data Possible negative economic impact if they didn't have access to legacy geodata Which additional user groups did we not consider in the first round?	How the user groups would be affected by the systematic elimination of smaller scales legacy geodata Similarity between the different user profiles regarding the functionality they need in the data Which additional user groups did we not consider in the first round?

[a]Participating institutions (in expert group one): Biblioteca de Catalunya, Cartoteca de l'Institut Cartografic i Geologic de Catalunya, Technical University of Vienna, Federal Archives of Switzerland, Edina (UK), Landesarchiv Baden-Württemberg, University of Barcelona, desideData Data Company, Centro de Ciencias Humanas y Sociales - Consejo Superior de Investigaciones Científicas, International Cartographic Association

[b]List of participating institutions (in expert group two): Institut Cartogràfic i Geològic de Catalunya (E), gvSIG developer (E), Swisstopo (CH), ESRI, Deutsches Zentrum für Luft- und Raumfahrt (D), Data Archiving and Networked services (NL), Lokku ltd (UK), Ordnance Survey (UK), Google, Institut National de l'information Géografique et Forestière (F)

20.3 Results

The results derived from the answers of both Delphi groups are exposed as follows.

Eight experts named user profiles in response to a Delphi study pre-question. The answers provided the following suggestions for user groups, supposed to be interested in legacy geodata: General public, Archeologists, Historians, Geographers, Lawyers, Policy makers, Information officers in culture and arts (i.e. filmmakers and museum staff), Emergency response planning members and military, Conservation agents of the built environment, Environmentalists (public and private), Architects and geo-related engineers, Institutions that update and maintain geodata and Undergraduate teachers and students.

The following user groups accrued from the first round: Geophysicists (including seismologists, geologists), Social scientists (including economists, journalists, statisticians, people involved in development) and Commercial users (including banks, utility companies, retailers etc.).

For the following tables, if a question did not obtain four valid answers the respective user group is omitted or the table shows a question mark instead of an answer.

The first question explored the main tasks the user groups do with the legacy geodata they find. Do they consult or visualize it or do they query and amend it? The following table shows the opinion of the first Delphi group after the second round. It is indicated as 'both' when there was no majority for either option (Table 20.2).

Most user groups are said to consult the data. This would allow for offering them geographic information in the form of raster images that are independent of complex databases or software. As expected, geophysicists, geographers and institutions that update and maintain geodata need to query the data. Commercial users need to connect their data to the geographic information or offer new services by analysing and reprocessing it. It may be surprising that the experts include archaeologists in the category of users that query the data, because we expect archaeologists to use legacy photographs to detect archaeological sites and mostly amend recent data. This highlights one of the problems of the Delphi process: the answers sometimes seem to suggest that the experts forgot that the study concentrates on legacy data and seemed to consider geodata in general.

For the next question, we wanted to find out in which map-age range the user groups are most interested. The following table shows the minimum and maximum age in years the first Delphi group estimated for the user groups. Where '999' is indicated, it means that this user group is interested in data as old as possible (Table 20.3).

Some user groups, such as commercial users, lawyers, policy makers, emergency response planning members and geo-related architects and engineers are interested in rather young legacy data of up to 65 years. It is expected that they might never access this data through a long-term archive, but instead from data producers or services that also deliver the most current updates. It is important to add that the experts mainly agree that the availability of data influences the data age range in which a user group is interested. This means that some user groups would reach back even further, if data were available to them. These answers indicate that most users compare current with legacy data and do not only work with historic data, so that a service that provides both in the same place or through the same application would be appreciated.

Subsequently the experts were asked to select the user groups that might request the largest amounts of data (in bytes not in land coverage). The four most mentioned user groups were the following. Next to the name we state how many times that user group was selected.

Table 20.2 What do these user groups mainly do with geodata?

User group	Task
Social scientists	Both
Geophysicists	Query
Commercial users	Query
General public	Consult
Archaeologists	Query
Historians	Consult
Geographers	Query
Lawyers	Consult
Policy makers (government)	Consult
Information officers in culture and arts	Consult
Emergency response planning members	Consult
Conservation agents (non-natural environment)	Consult
Environmentalists of natural environment (public and private)	Consult
Architects and engineers	Both
Institutions that update and maintain geodata	Query
Undergraduate teachers and students	Consult

Table 20.3 Minimum and maximum age of data the user group is interested in

User group	Min. age	Max. age
Social scientists	?	999
Geophysicists	1	70
Commercial users	?	40
General public	1	509
Archaeologists	1	999
Historians	7,5	999
Geographers	1	150
Lawyers	1	50
Policy makers (government)	1	65
Information officers in culture and arts	1	750
Emergency response planning members	1	25
Conservation agents (non-natural environment)	?	170
Environmentalists of natural environment (public and private)	1	170
Architects and engineers	1	50
Institutions that update and maintain geodata	1	200
Undergraduate teachers and students	1	200

- Geographers (10)
- General public (9)
- Social scientists (8)
- Institutions that update and maintain geodata (7)

Table 20.4 User groups classified by frequency of use and amount of data they need

Class	User group
Frequent user, needs big amounts of data	Geophysicists Geographers Policy makers Institutions that update and maintain geodata
Frequent user, needs small amounts of data	Social Scientists Archaeologists Historians Environmentalists
Infrequent user, needs big amounts of data	No user groups in this category
Infrequent user, needs small amounts of data	General public Information agents in Culture and Arts Undergraduate teachers and students

We can contrast this information with the results of a question asked to the second Delphi group where experts classified the user groups into the following four profiles (Table 20.4).

For some user groups no classification dominated, but it could be determined that conservation agents of the non-natural environment probably were frequent users and emergency response planners probably were infrequent users. For commercial users, the expert opinions diverged and for lawyers we did not receive enough answers.

The next question was about the most required data set by the user groups. The following table shows the most and second-most mentioned data sets for the user groups (Table 20.5).

The options 'LIDAR data' and 'digital terrain model' did not reach the first two positions for any user group. We believe this is partly due to their relatively recent appearance compared to other types of geodata. In this table, on the one hand, maps dominate over photographs. On the other hand, counting vector maps and vector databases together, vector data slightly dominate over raster maps.

Table 20.5 Most used data sets by user group

User group	1st choice	2nd choice
Geophysicists	DB	VM
Commercial users	RM	VM
General public	RM	OR
Historians	RM	DB
Policy makers (government)	RM	?
Architects and engineers	DB	?
Institutions that update and maintain geodata	DB	RP
Undergraduate teachers and students	RM	OR

DB Vector data base; *VM* Vector map (shareable file); *RM* Raster map; *OR* Ortho-corrected photography; *RP* Raw aerial photography

Table 20.6 Percentage of users with GIS knowledge in each user group

User group	% with GIS
Social scientists	17.50
Commercial users	12.50
General public	5
Archaeologists	20
Historians	5
Geographers	90
Policy makers (government)	7.50
Information officers in culture and arts	10
Emergency response planning members	30
Environmentalists of natural environment (public and private)	30
Architects and engineers	30
Institutions that update and maintain geodata	95
Undergraduate teachers and students	7.50

We then asked about the percentage of users with knowledge of GIS within each user group. The estimates of the first Delphi group are shown in the following table and expressed as the mean of all estimates (Table 20.6).

We believe the results of Table 20.6 to be optimistic. It is true that nowadays during academic training geographers will learn to work with GIS, but not all geographers of previous generations employ GIS at their workplace. In the architects and engineers group, there are probably more members that know CAD than

Table 20.7 Estimated maximum acquisition interval of geodata useful for each user group

User group	Year interval
Social scientists	5
Geophysicists	5
Commercial users	0
General public	5
Archaeologists	25
Historians	10
Geographers	5
Lawyers	5
Policy makers (government)	10
Information officers in culture and arts	10
Emergency response planning members	0
Conservation agents (non-natural environment)	10
Environmentalists of natural environment (public and private)	10
Architects and engineers	5
Institutions that update and maintain geodata	1
Undergraduate teachers and students	10

GIS. Here the fact that we did not further define ‘knowledge’ limits our capacity to interpret results. Therefore, although the proportions between user groups are interesting, their explicit value is questionable.

Subsequently, participants established a ranking of the user groups by how the economy would be affected if no legacy geodata were available to them. The user groups that would have a bigger negative impact on the economy if they had no access to legacy data were ranked at the top.

1. Institutions that update and maintain geodata
2. Geographers
3. Environmentalists of natural environment (public and private)
4. Architects and engineers
5. Geophysicists
6. General public
7. Commercial users
8. Policy makers (government)
9. Emergency response planning members
10. Social Scientists
11. Historians and at the same level undergraduate teachers and students
12. Information officers in culture and arts
13. Conservation agents (non-natural environment)
14. Lawyers
15. Archaeologists

Geographers are a large user group that would have a big effect on the economy if they lacked legacy data. The impact of environmentalists probably lies more in the fact that catastrophes could not be predicted and prevented without time series. The same interpretation is made for emergency response planning members, but because this user group is significantly smaller and the consequences of not being able to predict are more local than in the case of environmentalists, the group has been ranked further down. The big impact of architects and engineers is related to the cost and scale of their work, because construction usually affects a wide community around the construction site. We think this is also true for geophysicists. Surprisingly, lawyers are ranked quite low considering that court decisions about property and construction can be quite expensive. This indicates a small user group with infrequent needs for small amounts of legacy data.

We also addressed the time interval of archived geodata the user groups might need (Table 20.7). In the first round, we asked about geodata in general, and in the second round we asked about remote sensed data and vector data separately. Unfortunately, not enough experts were aware of this separation and did not answer the question about vector data in the second round. The estimates were made in years. The experience the Delphi experts have with the various user groups has not been taken into account for weighting, because it would have made it impossible to obtain results for half of the user groups and the question about time interval preferences is crucial for archives. The results obtained in the second round are as shown in Table 20.7.

The figure zero was used to express 'This user group needs all updates'. The experts who estimated for both vector and raster data indicated in general bigger intervals for raster data or equal intervals compared to vector data. One expert thinks that social scientists, the general public, archaeologists and lawyers need more frequent updates of raster data than vector data. The other expert thinks this is true only for geophysicists.

We did not define 'raster data' in the beginning of the survey. It was intended to mean digital geographical information served in raster file formats. Analysis of the open questions revealed that some experts understood raster as data captured in raster formats such as LIDAR and other sensor data but excluding photography or maps rendered as raster. This ambiguity should be taken into account when interpreting the results of this question.

Finally, Delphi group two was asked to match user groups that might have similar necessities about functionalities in legacy geodata. Every user group could be matched to any other with the most similarities, so that an expert could set up a maximum of two associations for a matching pair. Each link between two user groups augmented the edge weight between them. Similarity is a non-directional relationship, because if A is similar to B, B is also similar to A. Because in our case all links are non-directional, integer and positive, the weighted graph can be mapped easily to a non-weighted multigraph (Newman 2004). Therefore it is possible to apply the Louvain method for community detection, which is based on modularity optimization (Blondel et al. 2008). Modularity is the degree of connectivity between the nodes of a cluster compared to the nodes of other clusters or communities. Modularity is high when nodes in a same cluster are highly connected while links to other clusters are sparse. We expected to find communities in the network and applied the Louvain clustering method. The method detected six clusters in the network diagram, which were then partitioned by modularity class and expressed in different colours (Fehler: Referenz nicht gefunden). For better visualization of the graph, edges with weights of one point are masked.

As shown in Fig. 20.1 there are several user groups with similar expectations towards legacy data sets. The clusters—also called communities—which were formed by the second Delphi group, are

Cluster one: General public, commercial users and undergraduate teachers and students
Cluster two: Policy makers and lawyers,
Cluster three: Archaeologists, historians and information officers in culture and arts
Cluster four: Geographers, social scientists and geophysicists
Cluster five: Environmentalists of the natural environment and conservation agents of the built environment

The similarity of three of the groups—institutions that create and update geodata, architects and emergency response planning teams and military—is low, so the edges between them are very thin. However, they have even lower modularity

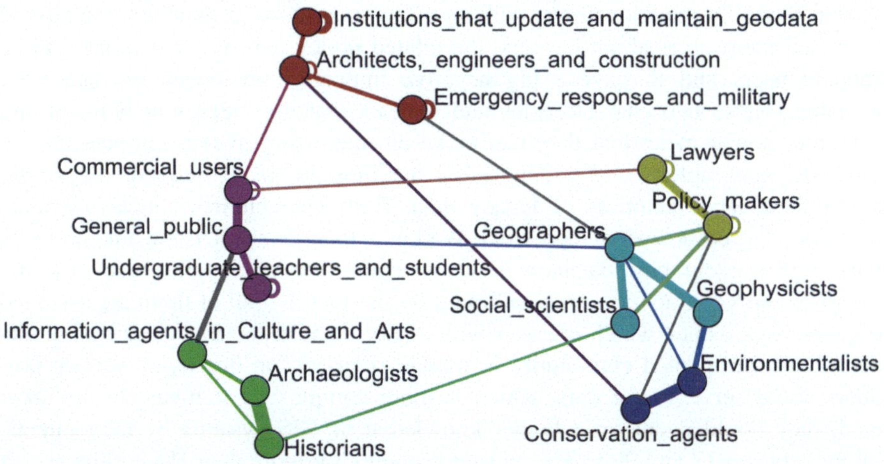

Fig. 20.1 Network where the nodes are the user groups and the edges represent similarity in needs for functionality in a data archive for legacy geodata. Thickness of the edges expresses the degree of similarity and colors, the affiliation to a cluster

to other user groups which is visualized by them being partitioned in a separate cluster (cluster six).

The case of the geographers, tied to both the social scientists and the geophysicists, shows the multidisciplinary nature of this subject. A geographer can specialize in human geography as well as physical geography. The manner in which the experts expressed the geographers' similarity with one or another related user group was affected depending on which specialists they had in mind.

20.4 Discussion

In this section we will discuss the characteristics of the user groups in the same cluster, to see if they should be considered for designated communities.

Commercial users and undergraduate teachers and students are considered similar to the general public. All members of cluster one would account for 41.5% of the access to legacy geodata, which makes this group an interesting target in terms of brand awareness of a potential service provider. Historians and archaeologists are very much seen as needing similar functionalities in legacy geodata. They are also considered similar to information officers acting in culture and arts. All three need ranges of recent to very old data and would be able to work with larger time intervals of 10–25 years. Another similarity is that they would all have a low impact on the economy if no legacy geodata were available, which might be a disincentive for a commercial service provider, due to related lack of economic power by the group. The policy makers and lawyers user groups are seen to have similar needs for functionalities. Although few experts have experience with and expressed themselves about lawyers, the related policy makers are estimated to be frequent users and to have a big negative impact if no legacy geodata were available. These two characteristics and the fact that they represent political and economic power mean that they can make an interesting market segment. Social scientists, geographers and geophysicists are thought to be frequent users who would need large amounts of legacy data. Both characteristics make them an interesting target for a service provider. We know that geophysicists would prefer to work with a geographic database or vector data. That this applies also to social scientists and geographers can be affirmed by the fact that all of them are assumed to query legacy data, which is easier with vector data. In the case of choosing this cluster as a designated community or market segment, on one hand, service providers could serve vector data, which is more complex to archive. On the other hand, they could count on a higher knowledge of GIS systems by their clients, falling between 17 and 90%. This cluster accounts for more than 18% of the access to legacy data.

Analysing the characteristics of environmentalists and conservation agents of the non-natural environment, we see that the needs of these user groups would be satisfied with a service designed for the general public. If institutions that create and maintain geodata do not hold the intellectual property of the legacy geodata they consume, or if they do not want to archive it by themselves for the period corresponding to their interest (200 years), they would be an interesting partner for a long-term service provider due to their economic impact and need for frequency and access. Finally, architects, engineers and construction workers, and emergency response planning teams and military are interested in geodata up to 25–50 years old and require rather small amounts of data. Therefore, we have to consider that they might never use the actual long-term data archive but get this data directly from the producer or through other short-term geographical service providers.

The results show that the user groups with similar needs towards legacy geodata can be grouped to form a user profile which could guide an archive in the elaboration of their designated community and services around the geodata archive. Some profiles present incentives to consider them as market segments for a commercial provider of access to long-term data series. In contrast to a commercial long-term access provider, a public archive has to define its designated communities in accordance with its social mission which might oblige it to deliver services to communities that do not present strong economic incentives such as the general public or historians.

The time intervals users are interested in can help the archive or access provider to limit the time frame in which data stays in a short-term service where it is not subject to selection. This time frame would ideally lie between 25 and 65 years. User groups interested in such recent data would probably never access a long-term archive. After that time span users who need very frequent updates are not interested in legacy geodata anymore. Commercial users, lawyers, policy makers, architects and emergency response planning teams are in this category. There are two exceptions to this: geographers and institutions that create and maintain geodata are considered to need frequent snapshots for a longer time span. These two user groups are also demanding in terms of complexity of the data. It should be envisioned to create a special service for this user profile.

20.5 Conclusion

The quality of a prediction is always difficult to assess unless the predicted events occur and can be measured. Nevertheless, we can speak to the limits of the method. A Delphi study relies on participants' understanding of the written concepts and questions. Even though the most important terms were defined in the beginning, some answers suggested that misunderstandings still occurred. Concepts such as what we meant by 'archiving' or 'raster data' should have been defined. Archiving has diverse meanings for the information science community and computer science people. The computer science term of a digital archive as a storage space or unit does not imply the standards and treatment an archivist would expect from a digital archive. Because people tend to be more familiar with computer science than with archival science, the long-term perspective gets lost when they think of a digital archive. Nevertheless, adding to the introduction of the questionnaire might have reduced the motivation to read it all, which would have entailed even lower agreement on the concepts.

The pre-question defined the first set of user groups. As the experts were already recruited we could not make sure enough experts were familiar with each user group. Therefore we were able to find out very little about some of the user groups, such as lawyers or conservation agents of the non-natural environment. Nevertheless, this can tell us something about the size of these user groups, which might be very small.

We had several options to receive more answers per question:

- recruit more experts
- make the answer to each question compulsory
- find an incentive to motivate the experts to stick with the project
- reduce the number of questions

The first three options were not prolific: The recruiting process was very long as appropriate experts are hard to find. As the recruiting process progressed, early recruits had to wait, putting at risk their availability and willingness to participate. Most of the experts were very busy, which is why we opted to leave them as much freedom as possible by leaving all the questions optional, with the exception of one. We think compulsory answers would also have increased the attrition rate. Where possible we met each participant personally to strengthen its motivation. This was not possible for participants residing outside of Europe or for late recruits. No financial incentive was given to the experts as the study took place within the framework of a non-sponsored PhD study. Finally we have realized that reducing the number of questions would have been a good way to increase the answer rate.

This study has shown that the Delphi method is appropriate for extracting characteristics of user groups that are known to sufficient experts if all concepts are clearly agreed upon between the experts and the conductor of the study. The Delphi technique has proved successful for determining characteristics such as access rates of the user group and the data age range required. To know the percentage of users with GIS knowledge seems useful for making strategic decisions, but is not precise enough to define the representation information. The Delphi study has failed to point out specific intermediate data sets of the production chain worthy of preservation. We were able to identify data sets that would be used primarily by certain user groups, but cannot exclude that the data sets not mentioned would not be used. The Delphi method could be more precise about acquisition intervals if it enquired on snapshots intervals for data bases or retention intervals of raster data for each data set individually as recommended by Bos et al. (2010).

Further research on specific user groups could be conducted when long term archives of spatial data are implemented. Established archival services would allow for user studies with participants in direct contact with the data; either analysis of user interaction logs, collection of opinions or usability studies on already existing services. A well-defined user universe could be addressed with a survey that would give quantifiable results that would allow comparing the Delphi method with survey techniques.[2]

As long as no sample population is available, qualitative approaches are more promising (Imms and Ereant 2009). As recommended and practiced in market research it would be useful to talk to members of the different user profiles. Interviews or focus groups could further develop or confirm the needs and expectations specific users have towards a long term archive.

References

Beruti V, Conway E, Forcada ME, Giaretta D, Albani M (2010) ESA plans—a pathfinder for long term data preservation. In iPres 2010. Retrieved from http://epubs.stfc.ac.uk/bitstream/6403/beruti-76.pdf

Blondel VD, Guillaume J, Lambiotte R, Lefebvre E (2008) Fast unfolding of communities in large networks. J Stat Mech Theory Exp v2(10):10008. https://doi.org/10.1088/1742-5468/2008/10/p10008

Bos M, Gollin H, Gerber U, Leuthold J, Meyer U (2010) Archiving of geodata: a joint preliminary study by swisstopo and the Swiss Federal Archives (p 71). Retrieved from http://www.swisstopo.admin.ch/internet/swisstopo/en/home/topics/geodata/geoarchive.parsysrelated1.59693.downloadList.93958.DownloadFile.tmp/preliminarystudyarchivingofgeodata.pdf

Comber A, Fisher P, Wadsworth R (2005) Comparing statistical and semantic approaches for identifying change from land cover datasets. J Environ Manage 77(1):47–55. https://doi.org/10.1016/j.jenvman.2005.02.009

[2]This might be currently possible in the UK where the project Digimap serves digital spatial legacy data of Ordnance Survey to an academic community.

Consultative Committee for Space Data Systems (CCSDS) (2002) Reference model for an open archival information system: CCSDS 650.0-B-1. NASA, Washington. Retrieved from http://public.ccsds.org/publications/archive/650x0b1.pdf
Conway E, Pepler S, Garland W, Hooper D, Marelli F, Liberti L, Badiali L (2013) Ensuring the long term impact of earth science data through data curation and preservation. Inf Stand Q 25(3):28. https://doi.org/10.3789/isqv25no3.2013.05
Cruse P, Sandore B (2009) Introduction: the library of congress national digital information infrastructure and preservation program. Libr Trends 57(3):301–314. https://doi.org/10.1353/lib.0.0055
Erwin T, Sweetkind-Singer J (2009) The national geospatial digital archive: a collaborative project to archive geospatial data. J Map Geogr Libr 6(1):6–25. https://doi.org/10.1080/15420350903432440
Erwin T, Sweetkind-Singer J, Larsgaard ML (2009) The national geospatial digital archives—collection development: lessons learned. Libr Trends 57(3):490–515. https://doi.org/10.1353/lib.0.0049
Fildes R, Allen PG (eds) (2011) Forecasting. In: Judgemental methods and forecasting competitions, vol III. SAGE, Los Angeles, p 390
Gantner F, Waldvogel B, Meile R, Laube P (2013) The basic formal ontology as a reference framework for modeling the evolution of administrative units. Trans GIS 17(2):206–226. https://doi.org/10.1111/j.1467-9671.2012.01356.x
Geospatial Multistate Archive and Preservation Partnership, & Library of Congress (2010) GeoMAPP interim report : 2007–2009 (p 71). Retrieved from http://www.geomapp.com/docs/GeoMAPP_InterimReport_Final.pdf
Giaretta D (2011) Advanced digital preservation. Berlin:Springer
Harris R (2001) Earth observation data archiving in the USA and Europe. Space Policy 17(1): 35–48. https://doi.org/10.1016/S0265-9646(00)00052-7
Imms M, Ereant G (2009) An introduction to qualitative market research (QMR), 1st edn. SAGE, Los Angeles
Janée G (2008) Preserving geospatial data: the national geospatial digital archive's approach. In: Archiving, pp 25–29
Kärberg T (2014) Digital preservation and knowledge in the public archives: for whom? Arch Rec 35(2):126–143
Lazorchak B, Nagy Z, Goreham D, Morris S, Bethune A, Peters M (2008) Creating a business plan for the archival preservation of geospatial data. In: Archiving, p 79
Malanowski N, Zweck A (2007) Bridging the gap between foresight and market research: integrating methods to assess the economic potential of nanotechnology. Technol Forecast Soc Chang 74(9):1805–1822. https://doi.org/10.1016/j.techfore.2007.05.010
Moran MS, Hutchinson BS, Marsh SE, McClaran MP, Olsson AD (2009) Archiving and distributing three long-term interconnected geospatial data sets. IEEE Trans Geosci Remote Sens 47(1):59–71. https://doi.org/10.1109/TGRS.2008.2002815
Morris SP (2006) Geospatial web services and geoarchiving: new opportunities and challenges in Geographic Information Service. Libr Trends 55(2):285–303. https://doi.org/10.1353/lib.2006.0059
Morris SP (2010) Appraisal and selection of geospatial data (p 23). Retrieved from http://www.digitalpreservation.gov/meetings/documents/othermeetings/AppraisalSelection_whitepaper_final.pdf
Morris SP (2013) Issues in the appraisal and selection of geospatial data: an NDSA report. (National Digital Stewardship Alliance. Geospatial Content Team, ed.). Retrieved from http://hdl.loc.gov/loc.gdc/lcpub.2013655112.1
Morris SP, Tuttle J, Essic J (2009) A partnership framework for geospatial data preservation in North Carolina. Libr Trends 57(3):516–540. https://doi.org/10.1353/lib.0.0050
National Academy of Sciences. National Research Council, & Committee on the Preservation of Geoscience Data and Collections (2002) Geoscience data and collections: national resources in

peril. The National Academy Press. Retrieved from http://www.nap.edu/catalog.php?record_id=10348

Newman MEJ (2004) Analysis of weighted networks. Phys Rev E 70(5):056131. https://doi.org/10.1103/PhysRevE.70.056131

Okoli C, Pawlowski SD (2004) The Delphi method as a research tool: an example, design considerations and applications. Inf Manage 42(1):15–29

Postma TJBM, Alers JC, Terpstra S, Zuurbier A (2007) Medical technology decisions in The Netherlands: how to solve the dilemma of technology foresight versus market research? Technol Forecast Soc Chang 74(9):1823–1833. https://doi.org/10.1016/j.techfore.2007.05.011

Shaon A, Rönsdorf C, Gerber U, Naumann K, Mason P, Woolf A, Bos M, Samuelsson G (2012) Long-term sustainability of spatial data infrastructures : a metadata framework and principles of geo-archiving. In iPres. Retrieved from https://ijpis.net/Global/Forskning/Center och institut/CEDIF/Geodata/GeoPres_IPRES_CR.pdf

Sweetkind J, Larsgaard ML, Erwin T (2006) Digital preservation of geospatial data. Libr Trends 55(2):304–314. https://doi.org/10.1353/lib.2006.0065

The manufacturer's authorised representative in the EU is Springer Nature Customer Service Centre GmbH, Europaplatz 3, 69115 Heidelberg, Germany. If you have any concerns regarding our products, please contact ProductSafety@springernature.com

Printed and bound by CPI Group (UK) Ltd, Croydon, CR0 4YY
30/09/2026
02225654-0016